CAMBRIDGE LIBRARY COLLECTION

Books of enduring scholarly value

Darwin

Two hundred years after his birth and 150 years after the publication of 'On the Origin of Species', Charles Darwin and his theories are still the focus of worldwide attention. This series offers not only works by Darwin, but also the writings of his mentors in Cambridge and elsewhere, and a survey of the impassioned scientific, philosophical and theological debates sparked by his 'dangerous idea'.

Zoonomia

Erasmus Darwin (1731–1802) is remembered not only as the grandfather of Charles but as a pioneering scientist in his own right. A friend and correspondent of Josiah Wedgwood, Joseph Priestley and Matthew Boulton, he practised medicine in Lichfield, but also wrote prolifically on scientific subjects. He organised the translation of Linnaeus from Latin into English prose, coining many plant names in the process, and also wrote a version in verse, The Loves of Plants. The aim of his Zoonomia, published in two volumes (1794–6), is to 'reduce the facts belonging to animal life into classes, orders, genera, and species; and by comparing them with each other, to unravel the theory of diseases'. The first volume describes human physiology, especially importance of motion, both voluntary and involuntary; the second is a detailed description of the symptoms of, the and the cures for, diseases, categorised according to his physiological classes. Many of his proposed treatments are not for the squeamish, but his attempt to classify symptoms and link them to human anatomy and physiology marked a new stage in the development of the science of medicine; and his theory that all living things may perhaps have evolved from 'a single living filament' prefigures in a remarkable way the subsequent work of his more famous grandson.

Cambridge University Press has long been a pioneer in the reissuing of out-of-print titles from its own backlist, producing digital reprints of books that are still sought after by scholars and students but could not be reprinted economically using traditional technology. The Cambridge Library Collection extends this activity to a wider range of books which are still of importance to researchers and professionals, either for the source material they contain, or as landmarks in the history of their academic discipline.

Drawing from the world-renowned collections in the Cambridge University Library, and guided by the advice of experts in each subject area, Cambridge University Press is using state-of-the-art scanning machines in its own Printing House to capture the content of each book selected for inclusion. The files are processed to give a consistently clear, crisp image, and the books finished to the high quality standard for which the Press is recognised around the world. The latest print-on-demand technology ensures that the books will remain available indefinitely, and that orders for single or multiple copies can quickly be supplied.

The Cambridge Library Collection will bring back to life books of enduring scholarly value (including out-of-copyright works originally issued by other publishers) across a wide range of disciplines in the humanities and social sciences and in science and technology.

Zoonomia

Or, the Laws of Organic Life

VOLUME 1

ERASMUS DARWIN

CAMBRIDGE
UNIVERSITY PRESS

CAMBRIDGE UNIVERSITY PRESS

Cambridge, New York, Melbourne, Madrid, Cape Town, Singapore,
São Paolo, Delhi, Dubai, Tokyo

Published in the United States of America by Cambridge University Press, New York

www.cambridge.org
Information on this title: www.cambridge.org/9781108005494

This edition first published 1794-6
This digitally printed version 2009

ISBN 978-1-108-00549-4 Paperback

ZOONOMIA;

OR,

THE LAWS

OF

ORGANIC LIFE.

VOL. I.

By ERASMUS DARWIN, M.D. F.R.S.
AUTHOR OF THE BOTANIC GARDEN.

Principiò cœlum, ac terras, campofque liquentes,
Lucentemque globum lunæ, titaniaque aftra,
Spiritus intùs alit, totamque infufa per artus
Mens agitat molem, et magno fe corpòre mifcet.

VIRG. Æn. vi.

Earth, on whofe lap a thoufand nations tread,
And Ocean, brooding his prolific bed,
Night's changeful orb, blue pole, and filvery zones,
Where other worlds encircle other funs,
One Mind inhabits, one diffufive Soul
Wields the large limbs, and mingles with the whole.

LONDON:
PRINTED FOR J. JOHNSON, IN ST. PAUL'S CHURCH-YARD.

1794.

DEDICATION.

To the candid and ingenious Members of the College of Phyſicians, of the Royal Philoſophical Society, of the Two Univerſities, and to all thoſe, who ſtudy the Operations of the Mind as a Science, or who practiſe Medicine as a Profeſſion, the ſubſequent Work is, with great reſpect, inſcribed by the Author.

DERBY, May 1, 1794.

CONTENTS.

SECT.

TO

ERASMUS DARWIN,

ON HIS WORK INTITLED

ZOONOMIA.

By DEWHURST BILSBORROW.

HAIL TO THE BARD! who fung, from Chaos hurl'd
How funs and planets form'd the whirling world;
How fphere on fphere Earth's hidden ftrata bend,
And caves of rock her central fires defend;
Where gems new-born their twinkling eyes unfold,   5
And young ores fhoot in arborefcent gold.

   How the fair Flower, by Zephyr woo'd, unfurls
Its panting leaves, and waves its azure curls;
Or fpreads in gay undrefs its lucid form
To meet the fun, and fhuts it to the ftorm;   10
While in green veins impaffion'd eddies move,
And Beauty kindles into life and love.

   How the firft embryon-fibre, fphere, or cube,
Lives in new forms,—a line,—a ring,—a tube;
Clofed in the womb with limbs unfinifh'd laves,   15
Sips with rude mouth the falutary waves;
Seeks round its cell the fanguine ftreams, that pafs,
And drinks with crimfon gills the vital gas;
Weaves with foft threads the blue meandering vein,
The heart's red concave, and the filver brain;   20
Leads the long nerve, expands the impatient fenfe,
And clothes in filken fkin the nafcent Ens.

   Erewhile, emerging from its liquid bed,
It lifts in gelid air its nodding head;
The light's firft dawn with trembling eyelid hails,   25
With lungs untaught arrefts the balmy gales;
Tries its new tongue in tones unknown, and hears
The ftrange vibrations with unpra&ctifed ears;

Seeks with fpread hands the bofom's velvet orbs,
With clofing lips the milky fount abforbs ;   30
And, as comprefs'd the dulcet ftreams diftil,
Drinks warmth and fragrance from the living rill ;—
Eyes with mute rapture every waving line,
Prints with adoring kifs the Paphian fhrine,
And learns erelong, the perfect form confefs'd,   35
Ideal Beauty from its mother's breaft.

    Now in ftrong lines, with bolder tints defign'd,
You fketch ideas, and portray the mind ;
Teach how fine atoms of impinging light
To ceafelefs change the vifual fenfe excite ;   40
While the bright lens collects the rays, that fwerve,
And bends their focus on the moving nerve.
How thoughts to thoughts are link'd with viewlefs chains,
Tribes leading tribes, and trains purfuing trains ;
With fhadowy trident how Volition guides,   45
Surge after furge, his intellectual tides ;
Or, Queen of Sleep, Imagination roves
With frantic Sorrows, or delirious Loves.

    Go on, O FRIEND ! explore with eagle-eye ;
Where wrapp'd in night retiring Caufes lie :   50
Trace their flight bands, their fecret haunts betray,
And give new wonders to the beam of day ;
Till, link by link with ftep afpiring trod,
You climb from NATURE to the throne of GOD,
——So faw the Patriarch with admiring eyes   55
From earth to heaven a golden ladder rife ;
Involved in clouds the myftic fcale afcends,
And brutes and angels crowd the diftant ends.

TRIN. COL. CAMBRIDGE, *Jan.* 1, 1794.

———

REFERENCES TO THE WORK.

Botanic Garden, Part I.		Line 18.	Sect. XVI. 2. and XXXVIII.
Line 1.	Canto I. l. 105.	—— 26.	—— XVI. 4.
—— 3.	—— IV. l. 402.	—— 30.	—— XVI. 4.
—— 4.	—— I. l. 140.	—— 36.	—— XVI. 6.
—— 5.	—— III. l. 401.	—— 38.	—— III. and VII.
—— 8.	—— IV. l. 452.	—— 43.	—— X.
—— 9.	—— I. l. 14.	—— 44.	—— XVIII. 17.
		—— 45.	—— XVII. 3. 7.
Zoonomia.		—— 47.	—— XVIII. 8.
		—— 50.	—— XXXIX. 4. 8.
—— 12.	Sect. XIII.	—— 51.	—— XXXIX. the Motto,
—— 13.	—— XXXIX. 4. 1.	—— 54.	—— XXXIX. 8.

PREFACE.

THE purport of the following pages is an endeavour to reduce the facts belonging to ANIMAL LIFE into claſſes, orders, genera, and ſpecies; and, by comparing them with each other, to unravel the theory of diſeaſes. It happened, perhaps unfortunately for the inquirers into the knowledge of diſeaſes, that other ſciences had received improvement previous to their own; whence, inſtead of comparing the properties belonging to animated nature with each other, they, idly ingenious, buſied themſelves in attempting to explain the laws of life by thoſe of mechaniſm and chemiſtry; they conſidered the body as an hydraulic machine, and the fluids as paſſing through a ſeries of chemical changes, forgetting that animation was its eſſential characteriſtic.

The great CREATOR of all things has infinitely diverſified the works of his hands, but has at the ſame time ſtamped a certain ſimilitude on the features of nature, that demonſtrates to us, that *the whole is one family of one parent.* On this ſimilitude is founded all rational analogy; which, ſo long as it is concerned in comparing the eſſential properties of bodies, leads us to many and important diſcoveries; but when with licentious activity it links together objects, otherwiſe diſcordant, by ſome fanciful ſimilitude; it may indeed collect ornaments for wit and poetry, but philoſophy and truth recoil from its combinations.

The want of a theory, deduced from ſuch ſtrict analogy, to conduct the practice of medicine is lamented by its profeſſors; for, as a great number of unconnected facts are difficult to be acquired, and to be reaſoned from, the art of medicine is in many inſtances leſs effica-

B   cious

cious under the direction of its wifeft practitioners; and by that bufy crowd, who either boldly wade in darknefs, or are led into endlefs error by the glare of falfe theory, it is daily practifed to the deftruction of thoufands; add to this the unceafing injury which accrues to the public by the perpetual advertifements of pretended noftrums; the minds of the indolent become fuperftitioufly fearful of difeafes, which they do not labour under; and thus become the daily prey of fome crafty empyric.

A theory founded upon nature, that fhould bind together the fcattered facts of medical knowledge, and converge into one point of view the laws of organic life, would thus on many accounts contribute to the intereft of fociety. It would capacitate men of moderate abilities to practife the art of healing with real advantage to the public; it would enable every one of literary acquirements to diftinguifh the genuine difciples of medicine from thofe of boaftful effrontery, or of wily addrefs; and would teach mankind in fome important fituations the *knowledge of themfelves.*

There are fome modern practitioners, who declaim againft medical theory in general, not confidering that to think is to theorize; and that no one can direct a method of cure to a perfon labouring under difeafe without thinking, that is, without theorizing; and happy therefore is the patient, whofe phyfician poffeffes the beft theory.

The words idea, perception, fenfation, recollection, fuggeftion, and affociation, are each of them ufed in this treatife in a more limited fenfe than in the writers of metaphyfic. The author was in doubt, whether he fhould rather have fubftituted new words inftead of them; but was at length of opinion, that new definitions of words already in ufe would be lefs burthenfome to the memory of the reader.

A great part of this work has lain by the writer above twenty years, as fome of his friends can teftify: he had hoped by frequent revifion to have made it more worthy the acceptance of the public;

this

this however his other perpetual occupations have in part prevented, and may continue to prevent, as long as he may be capable of revifing it; he therefore begs of the candid reader to accept of it in its prefent ftate, and to excufe any inaccuracies of expreffion, or of conclufion, into which the intricacy of his fubject, the general imperfection of language, or the frailty he has in common with other men, may have betrayed him; and from which he has not the vanity to believe this treatife to be exempt.

ZOONOMIA.

SECT. I.

OF MOTION.

THE WHOLE OF NATURE may be suppofed to confift of two ef-
fences or fubftances ; one of which may be termed fpirit, and the
other matter. The former of thefe poffeffes the power to commence
or produce motion, and the latter to receive and communicate it. So
that motion, confidered as a caufe, immediately precedes every effect ; and, confidered as an effect, it immediately fucceeds every
caufe.

The MOTIONS OF MATTER may be divided into two kinds, primary and fecondary. The fecondary motions are thofe, which are
given to or received from other matter in motion. Their laws have
been fuccefsfully inveftigated by philofophers in their treatifes on mechanic powers. Thefe motions are diftinguifhed by this circumftance, that the velocity multiplied into the quantity of matter of the
body acted upon is equal to the velocity multiplied into the quantity
of matter of the acting body.

The primary motions of matter may be divided into three claffes,
thofe belonging to gravitation, to chemiftry, and to life ; and each
clafs has its peculiar laws. Though thefe three claffes include the
motions of folid, liquid, and aerial bodies ; there is neverthelefs a
fourth divifion of motions ; I mean thofe of the fuppofed ethereal
fluids of magnetifm, electricity, heat, and light ; whofe properties
are not fo well inveftigated as to be claffed with fufficient accuracy.

1*st*. The gravitating motions include the annual and diurnal rotation of the earth and planets, the flux and reflux of the ocean, the descent of heavy bodies, and other phænomena of gravitation The unparalleled sagacity of the great NEWTON has deduced the laws of this class of motions from the simple principle of the general attraction of matter. These motions are distinguished by their tendency to or from the centers of the sun or planets.

2*d*. The chemical class of motions includes all the various appearances of chemistry. Many of the facts, which belong to these branches of science, are nicely ascertained, and elegantly classed ; but their laws have not yet been developed from such simple principles as those above-mentioned ; though it is probable, that they depend on the specific attractions belonging to the particles of bodies, or to the difference of the quantity of attraction belonging to the sides and angles of those particles. The chemical motions are distinguished by their being generally attended with an evident decomposition or new combination of the active materials.

3*d*. The third class includes all the motions of the animal and vegetable world ; as well those of the vessels, which circulate their juices, and of the muscles, which perform their locomotion, as those of the organs of sense, which constitute their ideas.

This last class of motion is the subject of the following pages ; which, though conscious of their many imperfections, I hope may give some pleasure to the patient reader, and contribute something to the knowledge and to the cure of diseases.

S E C T. II. 1.

EXPLANATIONS AND DEFINITIONS.

I. *Outline of the animal economy.*—II. 1. *Of the senforium.*   2. *Of the brain and nervous medulla.*   3. *A nerve.*   4. *A muscular fibre.*   5. *The immediate organs of sense.*   6. *The external organs of sense.*   7. *An idea or sensual motion.*   8. *Perception.*   9. *Sensation.*   10. *Recollection and suggestion.*   11. *Habit, causation, association, catenation.*   12. *Reflex ideas.*   13. *Stimulus defined.*

As some explanations and definitions will be necessary in the profecution of the work, the reader is troubled with them in this place, and is intreated to keep them in his mind as he proceeds, and to take them for granted, till an apt opportunity occurs to evince their truth; to which I shall premife a very short outline of the animal economy.

I.—1. THE nervous fyftem has its origin from the brain, and is diftributed to every part of the body. Thofe nerves, which ferve the fenfes, principally arife from that part of the brain, which is lodged in the head; and thofe, which ferve the purpofes of mufcular motion, principally arife from that part of the brain, which is lodged in the neck and back, and which is erroneoufly called the fpinal marrow. The ultimate fibrils of thefe nerves terminate in the immediate organs of fenfe and mufcular fibres, and if a ligature be put on any part of their paffage from the head or fpine, all motion and perception ceafe in the parts beneath the ligature.

2. The longitudinal mufcular fibres compofe the locomotive mufcles, whofe contractions move the bones of the limbs and trunk, to which their extremities are attached. The annular or fpiral mufcu-

lar

lar fibres compofe the vafcular mufcles, which conftitute the inteftinal canal, the arteries, veins, glands, and abforbent veffels.

3. The immediate organs of fenfe, as the retina of the eye, probably confift of moving fibrils, with a power of contraction fimilar to
that of the larger mufcles above defcribed.

4. The cellular membrane confifts of cells, which refemble thofe
of a fponge, communicating with each other, and connecting together all the other parts of the body.

5. The arterial fyftem confifts of the aortal and the pulmonary artery, which are attended through their whole courfe with their correfpondent veins. The pulmonary artery receives the blood from
the right chamber of the heart, and carries it to the minute extenfive
ramifications of the lungs, where it is expofed to the action of the air
on a furface equal to that of the whole external fkin, through the
thin moift coats of thofe veffels, which are fpread on the air-cells,
which conftitute the minute terminal ramifications of the wind-pipe.
Here the blood changes its colour from a dark red to a bright fcarlet.
It is then collected by the branches of the pulmonary vein, and conveyed to the left chamber of the heart.

6. The aorta is another large artery, which receives the blood
from the left chamber of the heart, after it has been thus aerated in
the lungs, and conveys it by afcending and defcending branches to
every other part of the fyftem; the extremities of this artery terminate either in glands, as the falivary glands, lacrymal glands, &c. or
in capillary veffels, which are probably lefs involuted glands; in
thefe fome fluid, as faliva, tears, perfpiration, are feparated from the
blood; and the remainder of the blood is abforbed or drank up by
branches of veins correfpondent to the branches of the artery; which
are furnifhed with valves to prevent its return; and is thus carried
back, after having again changed its colour to a dark red, to the
right chamber of the heart. The circulation of the blood in the liver
differs from this general fyftem; for the veins which drink up the re-

7   fluent

fluent blood from thofe arteries, which are fpread on the bowels and mefentery, unite into a trunk in the liver, and form a kind of artery, which is branched into the whole fubftance of the liver, and is called the vena portarum; and from which the bile is feparated by the numerous hepatic glands, which conftitute that vifcus.

7. The glands may be divided into three fyftems, the convoluted glands, fuch as thofe above defcribed, which feparate bile, tears, faliva, &c. Secondly, the glands without convolution, as the capillary veffels, which unite the terminations of the arteries and veins; and feparate both the mucus, which lubricates the cellular membrane, and the perfpirable matter, which preferves the fkin moift and flexible. And thirdly, the whole abforbent fyftem, confifting of the lacteals, which open their mouths into the ftomach and inteftines, and of the lymphatics, which open their mouths on the external furface of the body, and on the internal linings of all the cells of the cellular membrane, and other cavities of the body.

Thefe lacteal and lymphatic veffels are furnifhed with numerous valves to prevent the return of the fluids, which they abforb, and terminate in glands, called lymphatic glands, and may hence be confidered as long necks or mouths belonging to thefe glands. To thefe they convey the chyle and mucus, with a part of the perfpirable matter, and atmofpheric moifture; all which, after having paffed through thefe glands, and having fuffered fome change in them, are carried forward into the blood, and fupply perpetual nourifhment to the fyftem, or replace its hourly wafte. .

8. The ftomach and inteftinal canal have a conftant vermicular motion, which carries forwards their contents, after the lacteals have drank up the chyle from them; and which is excited into action by the ftimulus of the aliment we fwallow, -but which becomes occafionally inverted or retrograde, as in vomiting, and in the iliac paffion.

II. 1. The word *fenforium* in the following pages is defigned to exprefs not only the medullary part of the brain, fpinal marrow, nerves, organs of fenfe, and of the mufcles; but alfo at the fame time that living principle, or fpirit of animation, which refides throughout the body, without being cognizable to our fenfes, except by its effects. The changes which occafionally take place in the fenforium, as during the exertions of volition, or the fenfations of pleafure or pain, are termed *fenforial motions*.

2. The fimilarity of the texture of the brain to that of the pancreas, and fome other glands of the body, has induced the inquirers into this fubject to believe, that a fluid, perhaps much more fubtile than the electric aura, is feparated from the blood by that organ for the purpofes of motion and fenfation. When we recollect, that the electric fluid itfelf is actually accumulated and given out voluntarily by the torpedo and the gymnotus electricus, that an electric fhock will frequently ftimulate into motion a paralytic limb, and laftly that it needs no perceptible tubes to convey it, this opinion feems not without probability; and the fingular figure of the brain and nervous fyftem feems well adapted to diftribute it over every part of the body.

For the medullary fubftance of the brain not only occupies the cavities of the head and fpine, but paffes along the innumerable ramifications of the nerves to the various mufcles and organs of fenfe. In thefe it lays afide its coverings, and is intermixed with the flender fibres, which conftitute thofe mufcles and organs of fenfe. Thus all thefe diftant ramifications of the fenforium are united at one of their extremities, that is, in the head and fpine; and thus thefe central parts of the fenforium conftitute a communication between all the organs of fenfe and mufcles.

3. A *nerve* is a continuation of the medullary fubftance of the brain from the head or fpine towards the other parts of the body, wrapped in its proper membrane.

4. The *mufcular fibres* are moving organs intermixed with that
medullary

medullary fubftance which is continued along the nerves, as mentioned above. They are indued with the power of contraction, and are again elongated either by antagonift mufcles, by circulating fluids, or by elaftic ligaments. So the mufcles on one fide of the fore-arm bend the fingers by means of their tendons, and thofe on the other fide of the fore-arm extend them again. The arteries are diftended by the circulating blood; and in the necks' of quadrupeds there is a ftrong elaftic ligament, which affifts the mufcles, which elevate the head, to keep it in its horizontal pofition, and to raife it after it has been depreffed.

5. The *immediate organs of fenfe* confift in like manner of moving fibres enveloped in the medullary fubftance above mentioned; and are erroneoufly fuppofed to be fimply an expanfion of the nervous medulla, as the retina of the eye, and the rete mucofum of the fkin, which are the immediate organs of vifion, and of touch. Hence when we fpeak of the contractions of the fibrous parts of the body, we fhall mean both the contractions of the mufcles, and thofe of the immediate organs of fenfe. Thefe *fibrous motions* are thus diftinguifhed from the *fenforial motions* above mentioned.

6. The *external organs* of fenfe are the coverings of the immediate organs of fenfe, and are mechanically adapted for the reception or tranfmiffion of peculiar bodies, or of their qualities, as the cornea and humours of the eye, the tympanum of the ear, the cuticle of the fingers and tongue.

7. The word *idea* has various meanings in the writers of metaphyfic: it is here ufed fimply for thofe notions of external things, which our organs of fenfe bring us acquainted with originally; and is defined a contraction, or motion, or configuration, of the fibres, which conftitute the immediate organ of fenfe; which will be explained at large in another part of the work. Synonymous with the word idea, we fhall fometimes ufe the words *fenfual motion* in contradiftinction to *mufcular motion.*

8. The

8. The word *perception* includes both the action of the organ of sense in consequence of the impact of external objects, and our attention to that action; that is, it expresses both the motion of the organ of sense, or idea, and the pain or pleasure that succeeds or accompanies it.

9. The pleasure or pain which necessarily accompanies all those perceptions or ideas which we attend to, either gradually subsides, or is succeeded by other fibrous motions. In the latter case it is termed *sensation*, as explained in Sect. V. 2, and VI. 2.—The reader is intreated to keep this in his mind, that through all this treatise the word sensation is used to express pleasure or pain only in its active state, by whatever means it is introduced into the system, without any reference to the stimulation of external objects.

10. The vulgar use of the word *memory* is too unlimited for our purpose: those ideas which we voluntarily recall are here termed ideas of *recollection*, as when we will to repeat the alphabet backwards. And those ideas which are suggested to us by preceding ideas are here termed ideas of *suggestion*, as whilst we repeat the alphabet in the usual order; when by habits previously acquired B is suggested by A, and C by B, without any effort of deliberation.

11. The word *association* properly signifies a society or convention of things in some respects similar to each other. We never say in common language, that the effect is associated with the cause, though they necessarily accompany or succeed each other. Thus the contractions of our muscles and organs of sense may be said to be associated together, but cannot with propriety be said to be associated with irritations, or with volition, or with sensation; because they are caused by them, as mentioned in Sect. IV. When fibrous contractions succeed other fibrous contractions, the connection is termed *association*; when fibrous contractions succeed sensorial motions, the connection is termed *causation*; when fibrous and sensorial motions reciprocally introduce each other in progressive trains or tribes, it is termed *catena-*

tion

tion of animal motions. All thefe connections are faid to be produced by *habit*; that is, by frequent repetition.

12. It may be proper to obferve, that by the unavoidable idiom of our language the ideas of perception, of recollection, or of imagination, in the plural number fignify the ideas belonging to perception, to recollection, or to imagination; whilft the idea of perception, of recollection, or of imagination, in the fingular number is ufed for what is termed " a reflex idea of any of thofe operations of the fenforium."

13. By the word *ftimulus* is not only meant the application of external bodies to our organs of fenfe and mufcular fibres, which excites into action the fenforial power termed irritation; but alfo pleafure or pain, when they excite into action the fenforial power termed fenfation; and defire or averfion, when they excite into action the power of volition; and laftly, the fibrous contractions which precede affociation; as is further explained in Sect. XII. 2. 1.

SECT. III.

THE MOTIONS OF THE RETINA DEMONSTRATED BY EXPERIMENTS.

I. *Of animal motions and of ideas.* II. *The fibrous structure of the retina.* III. *The activity of the retina in vision.* 1. *Rays of light have no momentum.* 2. *Objects long viewed become fainter.* 3. *Spectra of black objects become luminous.* 4. *Varying spectra from gyration.* 5. *From long inspection of various colours.* IV. *Motions of the organs of sense constitute ideas.* 1. *Light from pressing the eye-ball, and sound from the pulsation of the caroted artery.* 2. *Ideas in sleep mistaken for perceptions.* 3. *Ideas of imagination produce pain and sickness like sensations.* 4. *When the organ of sense is destroyed, the ideas belonging to that sense perish.* V. *Analogy between muscular motions and sensual motions, or ideas.* 1. *They are both originally excited by irritations.* 2. *And associated together in the same manner.* 3. *Both act in nearly the same times.* 4. *Are alike strengthened or fatigued by exercise.* 5. *Are alike painful from inflammation.* 6. *Are alike benumbed by compression.* 7. *Are alike liable to paralysis.* 8. *To convulsion.* 9. *To the influence of old age.*—VI. *Objections answered.* 1. *Why we cannot invent new ideas.* 2. *If ideas resemble external objects.* 3. *Of the imagined sensation in an amputated limb.* 4. *Abstract ideas.*—VII. *What are ideas, if they are not animal motions?*

BEFORE the great variety of animal motions can be duly arranged into natural classes and orders, it is necessary to smooth the way to this yet unconquered field of science, by removing some obstacles which thwart our passage. I. To demonstrate that the retina and other immediate organs of sense possess a power of motion, and that these motions constitute our ideas, according to the fifth and seventh of the preceding assertions, claims our first attention.

Animal motions are distinguished from the communicated motions,

mentioned

mentioned in the firſt ſection, as they have no mechanical proportion to their cauſe ; for the goad of a ſpur on the ſkin of a horſe ſhall induce him to move a load of hay.   They differ from the gravitating motions there mentioned as they are exerted with equal facility in all directions, and they differ from the chemical claſs of motions, becauſe no apparent decompoſitions or new combinations are produced in the moving materials.

Hence, when we ſay animal motion is excited by irritation, we do not mean that the motion bears any proportion to the mechanical impulſe of the ſtimulus; nor that it is affected by the general gravitation of the two bodies ; nor by their chemical properties, but ſolely that certain animal fibres are excited into action by ſomething external to the moving organ.

In this ſenſe the ſtimulus of the blood produces the contractions of the heart ; and the ſubſtances we take into our ſtomach and bowels irritate them to perform their neceſſary functions.   The rays of light excite the retina into animal motion by their ſtimulus; at the ſame time that thoſe rays of light themſelves are phyſically converged to a focus by the inactive humours of the eye.   The vibrations of the air irritate the auditory nerve into animal action ; while it is probable that the tympanum of the ear at the ſame time undergoes a mechanical vibration.

To render this circumſtance more eaſy to be comprehended, *motion may be defined to be a variation of figure*; for the whole univerſe may be conſidered as one thing poſſeſſing a certain figure; the motions of any of its parts are a variation of this figure of the whole : this definition of motion will be further explained in Section XIV. 2. 2. on the production of ideas.

Now the motions of an organ of ſenſe are a ſucceſſion of configurations of that organ ; theſe configurations ſucceed each other quicker or flower; and whatever configuration of this organ of ſenſe, that is, whatever portion of the motion of it is, or has uſually been, attended

5

to,

to, conftitutes an idea.   Hence the configuration is not to be confidered as an effect of the motion of the organ, but rather as a part or
temporary termination of it; and that, whether a paufe fucceeds it,
or a new configuration immediately takes place.   Thus when a fucceffion of moving objects are prefented to our view, the ideas of
trumpets, horns, lords and ladies, trains and canopies, are configurations, that is, parts or links of the fucceffive motions of the organ of vifion.

Thefe motions or configurations of the organs of fenfe differ from
the fenforial motions to be defcribed hereafter, as they appear to be
fimply contractions of the fibrous extremities of thofe organs, and in
that refpect exactly refemble the motions or contractions of the larger
mufcles, as appears from the following experiment.   Place a circular
piece of red filk about an inch in diameter on a fheet of white paper
in a ftrong light, as in Plate I.—look for a minute on this area, or
till the eye becomes fomewhat fatigued, and then, gently clofing
your eyes, and fhading them with your hand, a circular green area
of the fame apparent diameter becomes vifible in the clofed eye.   This
green area is the colour reverfe to the red area, which had been previoufly infpected, as explained in the experiments on ocular fpectra
at the end of the work, and in Botanical Garden, P. I. additional
note, No. I.   Hence it appears, that a part of the retina, which had
been fatigued by contraction in one direction, relieves itfelf by exerting the antagonift fibres, and producing a contraction in an oppofite
direction, as is common in the exertions of our mufcles.   Thus when
we are tired with long action of our arms in one direction, as in holding a bridle on a journey, we occafionally throw them into an oppofite pofition to relieve the fatigued mufcles.

Mr. Locke has defined an idea to be " whatever is prefent to the
mind;" but this would include the exertions of volition, and the fenfations of pleafure and pain, as well as thofe operations of our fyftem, which acquaint us with external objects; and is therefore too
                                                              unlimited

unlimited for our purpofe. Mr. Lock feems to have fallen into a further error, by conceiving, that the mind could form a general or abftract idea by its own operation, which was the copy of no particular perception; as of a triangle in general, that was neither acute, obtufe, nor right angled. The ingenious Dr. Berkley and Mr. Hume have demonftrated, that fuch general ideas have no exiftence in nature, not even in the mind of their celebrated inventor. We fhall therefore take for granted at prefent, that our recollection or imagination of external objects confifts of a partial repetition of the perceptions, which were excited by thofe external objects, at the time we became acquainted with them; and that our reflex ideas of the operations of our minds are partial repetitions of thofe operations.

II. The following article evinces that the organ of vifion confifts of a fibrous part as well as of the nervous medulla, like other white mufcles; and hence, as it refembles the mufcular parts of the body in its ftructure, we may conclude, that it muft refemble them in poffeffing a power of being excited into animal motion.—The fubfequent experiments on the optic nerve, and on the colours remaining in the eye, are copied from a paper on ocular fpectra publifhed in the feventy-fixth volume of the Philof. Tranf. by Dr. R. Darwin of Shrewfbury; which, as I fhall have frequent occafion to refer to, is reprinted in this work, Sect. XL. The retina of an ox's eye was fufpended in a glafs of warm water, and forcibly torn in a few places; the edges of thefe parts appeared jagged and hairy, and did not contract and become fmooth like fimple mucus, when it is diftended till it breaks; which evinced that it confifted of fibres. This fibrous conftruction became ftill more diftinct to the fight by adding fome cauftic alcali to the water; as the adhering mucus was firft eroded, and the hair-like fibres remained floating in the veffel. Nor does the degree of tranfparency of the retina invalidate this evidence of its fi-

D   brous

brous ſtructure, ſince Leeuwenhoek has ſhewn, that the cryſtalline humour itſelf conſiſts of fibres. Arc. Nat. V. I. 70.

Hence it appears, that as the muſcles conſiſt of larger fibres inter-mixed with a ſmaller quantity of nervous medulla, the organ of vi-ſion conſiſts of a greater quantity of nervous medulla intermixed with ſmaller fibres. It is probable that the locomotive muſcles of micro-ſcopic animals may have greater tenuity than theſe of the retina; and there is reaſon to conclude from analogy, that the other immediate organs of ſenſe, as the portio mollis of the auditory nerve, and the rete mucoſum of the ſkin, poſſeſs a ſimilarity of ſtructure with the retina, and a ſimilar power of being excited into animal motion.

III. The ſubſequent articles ſhew, that neither mechanical im-preſſions, nor chemical combinations of light, but that the animal activity of the retina conſtitutes viſion.

1. Much has been conjectured by philoſophers about the momen-tum of the rays of light; to ſubject this to experiment a very light horizontal balance was conſtructed by Mr. Michel, with about an inch ſquare of thin leaf-copper ſuſpended at each end of it, as de-ſcribed in Dr. Prieſtley's Hiſtory of Light and Colours. The focus of a very large convex mirror was thrown by Dr. Powel, in his lec-tures on experimental philoſophy, in my preſence, on one wing of this delicate balance, and it receded from the light; thrown on the other wing, it approached towards the light, and this repeatedly; ſo that no ſenſible impulſe could be obſerved, but what might well be aſcribed to the aſcent of heated air.

Whence it is reaſonable to conclude, that the light of the day muſt be much too weak in its dilute ſtate to make any mechanical impreſ-ſion on ſo tenacious a ſubſtance as the retina of the eye.—Add to this, that as the retina is nearly tranſparent, it could therefore make leſs reſiſtance to the mechanical impulſe of light; which, according to the obſervations related by Mr. Melvil in the Edinburgh Literary
Eſſays,

Effays, only communicates heat, and fhould therefore only communicate momentum, where it is obftructed, reflected, or refracted.—From whence alfo may be collected the final caufe of this degree of tranfparency of the retina, viz. leaft by the focus of ftronger lights, heat and pain fhould have been produced in the retina, inftead of that ftimulus which excites it into animal motion.

2. On looking long on an area of fcarlet filk of about an inch in diameter laid on white paper, as in Plate I. the fcarlet colour becomes fainter, till at length it entirely vanifhes, though the eye is kept uniformly and fteadily upon it. Now if the change or motion of the retina was a mechanical impreffion, or a chemical tinge of coloured light, the perception would every minute become ftronger and ftronger,—whereas in this experiment it becomes every inftant weaker and weaker. The fame circumftance obtains in the continued application of found, or of fapid bodies, or of odorous ones, or of tangible ones, to their adapted organs of fenfe.

Thus when a circular coin, as a fhilling, is preffed on the palm of the hand, the fenfe of touch is mechanically compreffed; but it is the ftimulus of this preffure that excites the organ of touch into animal action, which conftitutes the perception of hardnefs and of figure: for in fome minutes the perception ceafes, though the mechanical preffure of the object remains.

3. Make with ink on white paper a very black fpot about half an inch in diameter, with a tail about an inch in length, fo as to refemble a tadpole, as in Plate II.; look fteadfaftly for a minute on the center of this fpot, and, on moving the eye a little, the figure of the tadpole will be feen on the white part of the paper; which figure of the tadpole will appear more luminous than the other part of the white paper; which can only be explained by fuppofing that a part of the retina, on which the tadpole was delineated, to have become more fenfible to light than the other parts of it, which were expofed to the

white

white paper; and not from any idea of mechanical impreſſion or chemical combination of light with the retina.

4. When any one turns round rapidly, till he becomes dizzy,
and falls upon the ground, the ſpectra of the ambient objects continue
to preſent themſelves in rotation, and he ſeems to behold the objects
ſtill in motion.   Now if theſe ſpectra were impreſſions on a paſſive
organ, they either muſt continue as they were received laſt, or not
continue at all.

5. Place a piece of red ſilk about an inch in diameter on a ſheet of
white paper in a ſtrong light, as in Plate I.; look ſteadily upon it
from the diſtance of about half a yard for a minute; then cloſing
your eye-lids, cover them with your hands and handkerchief, and a
green ſpectrum will be ſeen in your eyes reſembling in form the
piece of red ſilk.   After ſome ſeconds of time the ſpectrum will difappear, and in a few more ſeconds will reappear; and thus alternately three or four times, if the experiment be well made, till at
length it vaniſhes entirely.

6. Place a circular piece of white paper, about four inches in diameter, in the ſunſhine, cover the center of this with a circular
piece of black ſilk, about three inches in diameter; and the center
of the black ſilk with a circle of pink ſilk, about two inches in diameter; and the center of the pink ſilk with a circle of yellow ſilk,
about one inch in diameter; and the center of this with a circle of
blue ſilk, about half an inch in diameter; make a ſmall ſpot with ink
in the center of the blue ſilk, as in Plate III. look ſteadily for a minute on this central ſpot, and then cloſing your eyes, and applying
your hand at about an inch diſtance before them, ſo as to prevent too
much or too little light from paſſing through the eye-lids, and you
will ſee the moſt beautiful circles of colours that imagination can conceive; which are moſt reſembled by the colours occaſioned by pouring a drop or two of oil on a ſtill lake in a bright day.   But theſe
circular

circular irifes of colours are not only different from the colours of the filks above mentioned, but are at the fame time perpetually changing as long as they exift.

From all thefe experiments it appears, that thefe fpectra in the eye are not owing to the mechanical impulfe of light impreffed on the retina; nor to its chemical combination with that organ; nor to the abforption and emiffion of light, as is fuppofed, perhaps erroneoufly, to take place in calcined fhells and other phofphorefcent bodies, after having been expofed to the light : for in all thefe cafes the fpectra in the eye fhould either remain of the fame colour, or gradually decay, when the object is withdrawn; and neither their evanefcence during the prefence of their object, as in the fecond experiment, nor their change from dark to luminous, as in the third experiment, nor their rotation, as in the fourth experiment, nor the alternate prefence and evanefcence of them, as in the fifth experiment, nor the perpetual change of colours of them, as in the laft experiment, could exift.

IV. The fubfequent articles fhew, that thefe animal motions or configurations of our organs of fenfe conftitute our ideas.

1. If any one in the dark preffes the ball of his eye, by applying his finger to the external corner of it, a luminous appearance is obferved; and by a fmart ftroke on the eye great flafhes of fire are perceived. (Newton's Optics.) So that when the arteries, that are near the auditory nerve, make ftronger pulfations than ufual, as in fome fevers, an undulating found is excited in the ears. Hence it is not the prefence of the light and found, but the motions of the organ, that are immediately neceffary to conftitute the perception or idea of light and found.

2. During the time of fleep, or in delirium, the ideas of imagination are miftaken for the perceptions of external objects; whence it appears, that thefe ideas of imagination are no other than a reiteration of thofe motions of the organs of fenfe, which were originally excited by the ftimulus of external objects : and in our waking hours

the

the fimple ideas, that we call up by recollection or by imagination, as the colour of red, or the fmell of a rofe, are exact refemblances of the fame fimple ideas from perception; and in confequence muft be a repetition of thofe very motions.

3. The difagreeable fenfation called the tooth-edge is originally excited by the painful jarring of the teeth in biting the edge of the glafs, or porcelain cup, in which our food was given us in our infancy, as is further explained in the Section XVI. 10, on Inftinct.—This difagreeable fenfation is afterwards excitable not only by a repetition of the found, that was then produced, but by imagination alone, as I have myfelf frequently experienced; in this cafe the idea of biting a china cup, when I imagine it very diftinctly, or when I fee another perfon bite a cup or glafs, excites an actual pain in the nerves of my teeth. So that this idea and pain feem to be nothing more than the reiterated motions of thofe nerves, that were formerly fo difagreeably affected.

Other ideas that are excited by imagination or recollection in many inftances produce fimilar effects on the conftitution, as our perceptions had formerly produced, and are therefore undoubtedly a repetition of the fame motions. A ftory which the celebrated Baron Van Swieton relates of himfelf is to this purpofe. He was prefent when the putrid carcafe of a dead dog exploded with prodigious ftench; and fome years afterwards, accidentally riding along the fame road, he was thrown into the fame ficknefs and vomiting by the idea of the ftench, as he had before experienced from the perception of it.

4. Where the organ of fenfe is totally deftroyed, the ideas which were received by that organ feem to perifh along with it, as well as the power of perception. Of this a fatisfactory inftance has fallen under my obfervation. A gentleman about fixty years of age had been totally deaf for near thirty years: he appeared to be a man of good underftanding, and amufed himfelf with reading, and by converfing either by the ufe of the pen, or by figns made with his fin-

gers,

gers, to reprefent letters. I obferved that he had fo far forgot the pronunciation of the language, that when he attempted to fpeak, none of his words had diftinct articulation, though his relations could fometimes underftand his meaning. But, which is much to the point, he affured me, that in his dreams he always imagined that people converfed with him by figns or writing, and never that he heard any one fpeak to him. From hence it appears, that with the perceptions of founds he has alfo loft the ideas of them; though the organs of fpeech ftill retain fomewhat of their ufual habits of articulation.

This obfervation may throw fome light on the medical treatment of deaf people; as it may be learnt from their dreams whether the auditory nerve be paralytic, or their deafnefs be owing to fome defect of the external organ.

It rarely happens that the immediate organ of vifion is perfectly deftroyed. The moft frequent caufes of blindnefs are occafioned by defects of the external organ, as in cateracts and obfufcations of the cornea. But I have had the opportunity of converfing with two men, who had been fome years blind; one of them had a complete gutta ferena, and the other had loft the whole fubftance of his eyes. They both told me that they did not remember to have ever dreamt of vifible objects, fince the total lofs of their fight.

V. Another method of difcovering that our ideas are animal motions of the organs of fenfe, is from confidering the great analogy they bear to the motions of the larger mufcles of the body. In the following articles it will appear that they are originally excited into action by the irritation of external objects like our mufcles; are affociated together like our mufcular motions; act in fimilar time with them; are fatigued by continued exertion like them; and that the organs of fenfe are fubject to inflammation, numbnefs, palfy, convulfion, and the defects of old age, in the fame manner as the mufcular fibres.

1. All

1. All our perceptions or ideas of external objects are univerfally allowed to have been originally excited by the ftimulus of thofe external objects; and it will be fhewn in a fucceeding fection, that it is probable that all our mufcular motions, as well thofe that are become voluntary as thofe of the heart and glandular fyftem, were originally in like manner excited by the ftimulus of fomething external to the organ of motion.

2. Our ideas are alfo affociated together after their production precifely in the fame manner as our mufcular motions; which will likewife be fully explained in the fucceeding fection.

3. The time taken up in performing an idea is likewife much the fame as that taken up in performing a mufcular motion. A mufician can prefs the keys of an harpfichord with his fingers in the order of a tune he has been accuftomed to play, in as little time as he can run over thofe notes in his mind. So we many times in an hour cover our eye-balls with our eye-lids without perceiving that we are in the dark; hence the perception or idea of light is not changed for that of darknefs in fo fmall a time as the twinkling of an eye; fo that in this cafe the mufcular motion of the eye-lid is performed quicker than the perception of light can be changed for that of darknefs.—So if a fire-ftick be whirled round in the dark, a luminous circle appears to the obferver; if it be whirled fomewhat flower, this circle becomes interrupted in one part; and then the time taken up in fuch a revolution of the ftick is the fame that the obferver ufes in changing his ideas: thus the δολιχοςκοτον ευκος of Homer, the long fhadow of the flying javelin, is elegantly defigned to give us an idea of its velocity, and not of its length.

4. The fatigue that follows a continued attention of the mind to one object is relieved by changing the fubject of our thoughts; as the continued movement of one limb is relieved by moving another in its ftead. Whereas a due exercife of the faculties of the mind ftrengthens and improves thofe faculties, whether of imagination or

recollection

recollection ; as the exercife of our limbs in dancing or fencing in-creafes the ftrength and agility of the mufcles thus employed.

5. If the mufcles of any limb are inflamed, they do not move without pain ; fo when the retina is inflamed, its motions alfo are painful.   Hence light is as intolerable in this kind of ophthalmia, as preffure is to the finger in the paronychia.   In this difeafe the patients frequently dream of having their eyes painfully dazzled; hence the idea of ftrong light is painful as well as the reality.   The firft of thefe facts evinces that our perceptions are motions of the organs of fenfe ; and the latter, that our imaginations are alfo motions of the fame organs.

6. The organs of fenfe, like the moving mufcles, are liable to be-come benumbed, or lefs fenfible, from compreffion.   Thus, if any perfon on a light day looks on a white wall, he may perceive the ra-mifications of the optic artery, at every pulfation of it, reprefented by darker branches on the white wall ; which is evidently owing to its compreffing the retina during the diaftole of the artery.   Savage Nofolog.

7. The organs of fenfe and the moving mufcles are alike liable to be affected with palfy, as in the gutta ferena, and in fome cafes of deafnefs ; and one fide of the face has fometimes loft its power of fen-fation, but retained its power of motion ; other parts of the body have loft their motions but retained their fenfation, as in the common he-miplagia ; and in other inftances both thefe powers have perifhed together.

8. In fome convulfive difeafes a delirium or infanity fupervenes, and the convulfions ceafe ; and converfely the convulfions fhall fuper-vene, and the delirium ceafe.   Of this I have been a witnefs many times in a day in the paroxyfms of violent epilepfies ; which evinces that one kind of delirium is a convulfion of the organs of fenfe, and that our ideas are the motions of thefe organs : the fubfequent cafes will illuftrate this obfervation.

E

Mifs

Miſs G——, a fair young lady, with light eyes and hair, was ſeized with moſt violent convulſions of her limbs, with outrageous hic-cough, and moſt vehement efforts to vomit: after near an hour was elapſed this tragedy ceaſed, and a calm talkative delirium ſupervened for about another hour; and theſe relieved each other at intervals during the greateſt part of three or four days. After having carefully conſidered this diſeaſe, I thought the convulſions of her ideas leſs dan-gerous than thoſe of her muſcles; and having in vain attempted to make any opiate continue in her ſtomach, an ounce of laudanum was rubbed along the ſpine of her back, and a dram of it was uſed as an enema; by this medicine a kind of drunken delirium was continued many hours; and when it ceaſed the convulſions did not return; and the lady continued well many years, except ſome ſlighter relapſes, which were relieved in the ſame manner.

Miſs H——, an accompliſhed young lady, with light eyes and hair, was ſeized with convulſions of her limbs, with hiccough, and efforts to vomit, more violent than words can expreſs; theſe conti-nued near an hour, and were ſucceeded with a cataleptic ſpaſm of one arm, with the hand applied to her head; and after about twenty minutes theſe ſpaſms ceaſed, and a talkative reverie ſupervened for near another hour, from which no violence, which it was proper to uſe, could awaken her. Theſe periods of convulſions, firſt of the muſcles, and then of the ideas, returned twice a day for ſeveral weeks; and were at length removed by great doſes of opium, after a great va-riety of other medicines and applications had been in vain experienced. This lady was ſubject to frequent relapſes, once or twice a year for many years, and was as frequently relieved by the ſame method.

Miſs W——, an elegant young lady, with black eyes and hair, had ſometimes a violent pain of her ſide, at other times a moſt painful ſtrangury, which were every day ſucceeded by delirium; which gave a temporary relief to the painful ſpaſms. After the vain exhibition

7

of

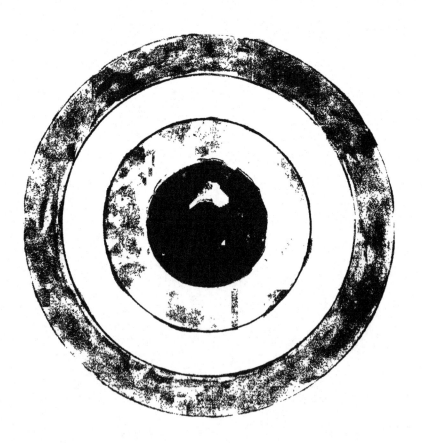

of variety of medicines and applications by different phyficians, for more than a twelvemonth, fhe was directed to take fome dofes of opium, which were gradually increafed, by which a drunken delirium was kept up for a day or two, and the pains prevented from returning. A flefh diet, with a little wine or beer, inftead of the low regimen fhe had previoufly ufed, in a few weeks completely eftablifhed her health ; which, except a few relapfes, has continued for many years.

9. Laftly, as we advance in life all the parts of the body become more rigid, and are rendered lefs fufceptible of new habits of motion, though they retain thofe that were before eftablifhed. This is fenfibly obferved by thofe who apply themfelves late in life to mufic, fencing, or any of the mechanic arts. In the fame manner many elderly people retain the ideas they had learned early in life, but find great difficulty in acquiring new trains of memory; infomuch that in extreme old age we frequently fee a forgetfulnefs of the bufinefs of yefterday, and at the fame time a circumftantial remembrance of the amufements of their youth ; till at length the ideas of recollection and activity of the body gradually ceafe together,—fuch is the condition of humanity !—and nothing remains but the vital motions and fenfations.

VI. 1. In oppofition to this doctrine of the production of our ideas, it may be afked, if fome of our ideas, like other animal motions, are voluntary, why can we not invent new ones, that have not been received by perception ? The anfwer will be better underftood after having perufed the fucceeding fection, where it will be explained, that the mufcular motions likewife are originally excited by the ftimulus of bodies external to the moving organ ; and that the will has only the power of repeating the motions thus excited.

2. Another objector may afk, Can the motion of an organ of fenfe refemble an odour or a colour ? To which I can only anfwer, that it has not been demonftrated that any of our ideas refemble the objects

E 2   that

that excite them; it has generally been believed that they do not; but this shall be discussed at large in Sect. XIV.

3. There is another objection that at first view would seem less easy to surmount. After the amputation of a foot or a finger, it has frequently happened, that an injury being offered to the stump of the amputated limb, whether from cold air, too great pressure, or other accidents, the patient has complained of a sensation of pain in the foot or finger, that was cut off. Does not this evince that all our ideas are excited in the brain, and not in the organs of sense? This objection is answered, by observing that our ideas of the shape, place, and solidity of our limbs, are acquired by our organs of touch and of sight, which are situated in our fingers and eyes, and not by any sensations in the limb itself.

In this case the pain or sensation, which formerly has arisen in the foot or toes, and been propagated along the nerves to the central part of the sensorium, was at the same time accompanied with a visible idea of the shape and place, and with a tangible idea of the solidity of the affected limb: now when these nerves are afterwards affected by any injury done to the remaining stump with a similar degree or kind of pain, the ideas of the shape, place, or solidity of the lost limb, return by association; as these ideas belong to the organs of sight and touch, on which they were first excited.

4. If you wonder what organs of sense can be excited into motion, when you call up the ideas of wisdom or benevolence, which Mr. Locke has termed abstracted ideas; I ask you by what organs of sense you first became acquainted with these ideas? And the answer will be reciprocal; for it is certain that all our ideas were originally acquired by our organs of sense; for whatever excites our perception must be external to the organ that perceives it, and we have no other inlets to knowledge but by our perceptions: as will be further explained in Section XIV. and XV. on the Productions and Classes of Ideas.

VII. If

VII. If our recollection or imagination be not a repetition of animal movements, I afk, in my turn, What is it? You tell me it confifts of images or pictures of things.   Where is this extenfive canvas hung up? or where are the numerous receptacles in which thofe are depofited? or to what elfe in the animal fyftem have they any fimilitude?

That pleafing picture of objects, reprefented in miniature on the retina of the eye, feems to have given rife to this illufive oratory! It was forgot that this reprefentation belongs rather to the laws of light, than to thofe of life; and may with equal elegance be feen in the camera obfcura as in the eye; and that the picture vanifhes for ever, when the object is withdrawn.

SECT. IV.

LAWS OF ANIMAL CAUSATION.

I. THE fibres, which conftitute the mufcles and organs of fenfe, poffefs a power of contraction.   The circumftances attending the exertion of this power of CONTRACTION conftitute the laws of animal motion, as the circumftances attending the exertion of the power of ATTRACTION conftitute the laws of motion of inanimate matter.

II. The fpirit of animation is the immediate caufe of the contraction of animal fibres, it refides in the brain and nerves, and is liable to general or partial diminution or accumulation.

III. The ftimulus of bodies external to the moving organ is the remote caufe of the original contractions of animal fibres.

IV. A certain quantity of ftimulus produces irritation, which is an exertion of the fpirit of animation exciting the fibres into contraction.

V. A certain quantity of contraction of animal fibres, if it be perceived at all, produces pleafure ; a greater or lefs quantity of contraction, if it be perceived at all, produces pain ; thefe conftitute fenfation.

VI. A certain quantity of fenfation produces defire or averfion ; thefe conftitute volition.

VII. All animal motions which have occurred at the fame time, or in immediate fucceffion, become fo connected, that when one of them is reproduced, the other has a tendency to accompany or fucceed it.   When fibrous contractions fucceed or accompany other fibrous contractions, the connection is termed affociation; when

fibrous

fibrous contractions fucceed fenforial motions, the connection is termed caufation ; when fibrous and fenforial motions reciprocally introduce each other, it is termed catenation of animal motions. All
thefe connections are faid to be produced by habit, that is, by frequent repetition. Thefe laws of animal caufation will be evinced by
numerous facts, which occur in our daily exertions ; and will afterwards be employed to explain the more recondite phænomena of the
production, growth, difeafes, and decay of the animal fyftem.

SECT.

SECT. V.

OF THE FOUR FACULTIES OR MOTIONS OF THE SENSORIUM.

*1. Four fenforial powers.   2. Irritation, fenfation, volition, affociation defined.
3. Senforial motions diftinguifhed from fibrous motions.*

1. THE fpirit of animation has four different modes of action, or in other words the animal fenforium poffeffes four different faculties, which are occafionally exerted, and caufe all the contractions of the fibrous parts of the body. Thefe are the faculty of caufing fibrous contractions in confequence of the irritations excited by external bodies, in confequence of the fenfations of pleafure or pain, in confequence of volition, and in confequence of the affociations of fibrous contractions with other fibrous contractions, which precede or accompany them.

Thefe four faculties of the fenforium during their inactive ftate are termed irritability, fenfibility, voluntarity, and affociability; in their active ftate they are termed as above, irritation, fenfation, volition, affociation.

2. IRRITATION is an exertion or change of fome extreme part of the fenforium refiding in the mufcles or organs of fenfe, in confequence of the appulfes of external bodies.

SENSATION is an exertion or change of the central parts of the fenforium, or of the whole of it, *beginning* at fome of thofe extreme parts of it, which refide in the mufcles or organs of fenfe.

VOLITION is an exertion or change of the central parts of the fenforium, or of the whole of it, *terminating* in fome of thofe extreme parts of it, which refide in the mufcles or organs of fenfe.

ASSOCIATION

ASSOCIATION is an exertion or change of fome extreme part of the fenforium refiding in the mufcles or organs of fenfe, in confequence of fome antecedent or attendant fibrous contractions.

3. Thefe four faculties of the animal fenforium may at the time of their exertions be termed motions without impropriety of language; for we cannot pafs from a ftate of infenfibility or inaction to a ftate of fenfibility or of exertion without fome change of the fenforium, and every change includes motion. We fhall therefore fometimes term the above defcribed faculties *fenforial motions* to diftinguifh them from *fibrous motions*; which latter expreffion includes the motions of the mufcles and organs of fenfe.

The active motions of the fibres, whether thofe of the mufcles or organs of fenfe, are probably fimple contractions; the fibres being again elongated by antagonift mufcles, by circulating fluids, or fometimes by elaftic ligaments, as in the necks of quadrupeds. The fenforial motions, which conftitute the fenfations of pleafure or pain, and which conftitute volition, and which caufe the fibrous contractions in confequence of irritation or of affociation, are not here fuppofed to be fluctuations or refluctuations of the fpirit of animation; nor are they fuppofed to be vibrations or revibrations, nor condenfations or equilibrations of it; but to be changes or motions of it peculiar to life.

F

SECT.

SECT. VI.

OF THE FOUR CLASSES OF FIBROUS MOTIONS.

I. Origin of fibrous contractions.   II. Distribution of them into four classes, irritative motions, senfitive motions, voluntary motions, and affociate motions, defined.

I. ALL the fibrous contractions of animal bodies originate from the fenforium, and refolve themfelves into four claffes, correfpondent with the four powers or motions of the fenforium above defcribed, and from which they have their caufation.

1. Thefe fibrous contractions were originally caufed by the irritations excited by objects, which are external to the moving organ. As the pulfations of the heart are owing to the irritations excited by the ftimulus of the blood; and the ideas of perception are owing to the irritations excited by external bodies.

2. But as painful or pleafurable fenfations frequently accompanied thofe irritations, by habit thefe fibrous contractions became caufeable by the fenfations, and the irritations ceafed to be neceffary to their production. As the fecretion of tears in grief is caufed by the fenfa-tion of pain; and the ideas of imagination, as in dreams or delirium, are excited by the pleafure or pain, with which they were formerly accompanied.

3. But as the efforts of the will frequently accompanied thefe pain-ful or pleafureable fenfations, by habit the fibrous contractions became caufable by volition; and both the irritations and fenfations ceafed to be neceffary to their production. As the deliberate locomotions of

the

the body, and the ideas of recollection, as when we will to repeat the alphabet backwards.

4. But as many of thefe fibrous contractions frequently accompanied other fibrous contractions, by habit they became caufable by their affociations with them; and the irritations, fenfations, and volition, ceafed to be neceffary to their production. As the actions of the mufcles of the lower limbs in fencing are affociated with thofe of the arms; and the ideas of fuggeftion are affociated with other ideas, which precede or accompany them; as in repeating carelefsly the alphabet in its ufual order after having began it.

II. We fhall give the following names to thefe four claffes of fibrous motions, and fubjoin their definitions.

1. Irritative motions. That exertion or change of the fenforium, which is caufed by the appulfes of external bodies, either fimply fubfides, or is fucceeded by fenfation, or it produces fibrous motions; it is termed irritation, and irritative motions are thofe contractions of the mufcular fibres, or of the organs of fenfe, that are immediately confequent to this exertion or change of the fenforium.

2. Senfitive motions. That exertion or change of the fenforium, which conftitutes pleafure or pain, either fimply fubfides, or is fucceeded by volition, or it produces fibrous motions; it is termed fenfation, and the fenfitive motions are thofe contractions of the mufcular fibres, or of the organs of fenfe, that are immediately confequent to this exertion or change of the fenforium.

3. Voluntary motions. That exertion or change of the fenforium, which conftitutes defire or averfion, either fimply fubfides, or is fucceeded by fibrous motions; it is then termed volition, and voluntary motions are thofe contractions of the mufcular fibres, or of the organs of fenfe, that are immediately confequent to this exertion or change of the fenforium.

4. Affociate

4. Affociate motions.   That exertion or change of the fenforium, which accompanies fibrous motions, either fimply fubfides, or is fucceeded by fenfation or volition, or it produces other fibrous motions; it is then termed affociation, and the affociate motions are thofe contractions of the mufcular fibres, or of the organs of fenfe, that are immediately confequent to this exertion or change of the fenforium.

SECT.

SECT. VII.

OF IRRITATIVE MOTIONS.

I. 1. *Some muscular motions are excited by perpetual irritations.* 2. *Others more frequently by sensations.* 3. *Others by volition. Case of involuntary stretchings in paralytic limbs.* 4. *Some sensual motions are excited by perpetual irritations.* 5. *Others more frequently by sensation or volition.*
II. 1. *Muscular motions excited by perpetual irritations occasionally become obedient to sensation and to volition.* 2. *And the sensual motions.*
III. 1. *Other muscular motions are associated with the irritative ones.* 2. *And other ideas with irritative ones. Of letters, language, hieroglyphics. Irritative ideas exist without our attention to them.*

I. 1. MANY of our muscular motions are excited by perpetual irritations, as those of the heart and arterial system by the circumfluent blood. Many other of them are excited by intermitted irritations, as those of the stomach and bowels by the aliment we swallow; of the bile-ducts by the bile; of the kidneys, pancreas, and many other glands, by the peculiar fluids they separate from the blood; and those of the lacteal and other absorbent vessels by the chyle, lymph, and moisture of the atmosphere. These motions are accelerated or retarded, as their correspondent irritations are increased or diminished, without our attention or consciousness, in the same manner as the various secretions of fruit, gum, resin, wax, and honey, are produced in the vegetable world, and as the juices of the earth and the moisture of the atmosphere are absorbed by their roots and foliage.

2. Other

2. Other mufcular motions, that are moſt frequently connected with our ſenſations, as thoſe of the ſphincters of the bladder and anus, and the muſculi erectores penis, were originally excited into motion by irritation, for young children make water, and have other evacuations without attention to theſe circumſtances; " et primis etiam ab incunabulis tenduntur ſæpius puerorum penes, amore nondum expergefacto." So the nipples of young women are liable to become turgid by irritation, long before they are in a ſituation to be excited by the pleaſure of giving milk to the lips of a child.

3. The contractions of the larger muſcles of our bodies, that are moſt frequently connected with volition, were originally excited into action by internal irritations: as appears from the ſtretching or yawning of all animals after long ſleep. In the beginning of ſome fevers this irritation of the muſcles produces perpetual ſtretching and yawning; in other periods of fever an univerſal reſtleſſneſs ariſes from the ſame cauſe, the patient changing the attitude of his body every minute. The repeated ſtruggles of the fœtus in the uterus muſt be owing to this internal irritation: for the fœtus can have no other inducement to move its limbs but the tœdium or irkſomeneſs of a continued poſture.

The following caſe evinces, that the motions of ſtretching the limbs after a continued attitude are not always owing to the power of the will. Mr. Dean, a maſon, of Auſtry in Leiceſterſhire had the ſpine of the third vertebra of the back enlarged; in ſome weeks his lower extremities became feeble, and at length quite paralytic: neither the pain of bliſters, the heat of fomentations, nor the utmoſt efforts of the will could produce the leaſt motion in theſe limbs; yet twice or thrice a day for many months his feet, legs, and thighs, were affected for many minutes with forceable ſtretchings, attended with the ſenſation of fatigue; and he at length recovered the uſe of his limbs, though the ſpine continued protuberant. The ſame circumſtance is frequently ſeen in a leſs degree in the common hemiplagia;

gia; and when this happens, I have believed repeated and ſtrong ſhocks of electricity to have been of great advantage.

4. In like manner the various organs of ſenſe are originally excited into motion by various external ſtimuli adapted to this purpoſe, which motions are termed perceptions or ideas; and many of theſe motions during our waking hours are excited by perpetual irritation, as thoſe of the organs of hearing and of touch. The former by the conſtant low indiſtinct noiſes that murmur around us, and the latter by the weight of our bodies on the parts which ſupport them; and by the unceaſing variations of the heat, moiſture, and preſſure of the atmoſphere; and theſe ſenſual motions, preciſely as the muſcular ones above mentioned, obey their correſpondent irritations without our attention or conſciouſneſs.

5. Other claſſes of our ideas are more frequently excited by our ſenſations of pleaſure or pain, and others by volition: but that theſe have all been originally excited by ſtimuli from external objects, and only vary in their combinations or ſeparations, has been fully evinced by Mr. Locke; and are by him termed the ideas of perception in contradiſtinction to thoſe, which he calls the ideas of reflection.

II. 1. Theſe muſcular motions, that are excited by perpetual irritation, are nevertheleſs occaſionally excitable by the ſenſations of pleaſure or pain, or by volition, as appears by the palpitation of the heart from fear, the increaſed ſecretion of ſaliva at the ſight of agreeable food, and the glow on the ſkin of thoſe who are aſhamed. There is an inſtance told in the Philoſophical Tranſactions of a man, who could for a time ſtop the motion of his heart when he pleaſed; and Mr. D. has often told me, he could ſo far increaſe the periſtaltic motion of his bowels by voluntary efforts, as to produce an evacuation by ſtool at any time in half an hour.

2. In like manner the ſenſual motions, or ideas, that are excited by perpetual irritation, are nevertheleſs occaſionally excitable by ſenſation or volition; as in the night, when we liſten under the influ-

ence

ence of fear, or from voluntary attention, the motions excited in the organ of hearing by the whifpering of the air in our room, the pulfation of our own arteries, or the faint beating of a diftant watch, become objects of perception.

III. Innumerable trains or tribes of other motions are affociated with thefe mufcular motions which are excited by irritation; as by the ftimulus of the blood in the right chamber of the heart, the lungs are induced to expand themfelves; and the pectoral and intercoftal mufcles, and the diaphragm, act at the fame time by their affociations with them. And when the pharinx is irritated by agreeable food, the mufcles of deglutition are brought into action by affociation. Thus when a greater light falls on the eye, the iris is brought into action without our attention; and the ciliary procefs, when the focus is formed before or behind the retina, by their affociations with the increafed irritative motions of the organ of vifion. Many common actions of life are produced in a fimilar manner. If a fly fettle on my forehead, whilft I am intent on my prefent occupation, I diflodge it with my finger, without exciting my attention or breaking the train of my ideas.

2. In like manner the irritative ideas fuggeft to us many other trains or tribes of ideas that are affociated with them. On this kind of connection, language, letters, hieroglyphics, and ever kind of fymbol, depend. The fymbols themfelves produce irritative ideas, or fenfual motions, which we do not attend to; and other ideas, that are fucceeded by fenfation, are excited by their affociation with them. And as thefe irritative ideas make up a part of the chain of our waking thoughts, introducing other ideas that engage our attention, though themfelves are unattended to, we find it very difficult to inveftigate by what fteps many of our hourly trains of ideas gain their admittance.

It may appear paradoxical, that ideas can exift, and not be attended to; but all our perceptions are ideas excited by irritation, and fuc-

7   ceeded

ceeded by fenfation.   Now when thefe ideas excited by irritation give us neither pleafure nor pain, we ceafe to attend to them.   Thus whilft I am walking through that grove before my window, I do not run againft the trees or the benches, though my thoughts are ftrenuoufly exerted on fome other object.   This leads us to a diftinct knowledge of irritative ideas, for the idea of the tree or bench, which I avoid, exifts on my retina, and induces by affociation the action of certain locomotive mufcles; though neither itfelf nor the actions of thofe mufcles engage my attention.

Thus whilft we are converfing on this fubject, the tone, note, and articulation of every individual word forms its correfpondent irritative idea on the organ of hearing; but we only attend to the affociated ideas, that are attached by habit to thefe irritative ones, and are fucceeded by fenfation; thus when we read the words " PRINTING-PRESS" we do not attend to the fhape, fize, or exiftence of the letters which compofe thefe words, though each of them excites a correfpondent irritative motion of our organ of vifion, but they introduce by affociation our idea of the moft ufeful of modern inventions; the capacious refervoir of human knowledge, whofe branching ftreams diffufe fciences, arts, and morality, through all nations and all ages.

SECT.

SECT. VIII.

OF SENSITIVE MOTIONS.

I. 1. *Senfitive mufcular motions were originally excited into action by irritation.* 2. *And fenfitive fenfual motions, ideas of imagination, dreams.* II. 1. *Senfitive mufcular motions are occafionally obedient to volition.* 2. *And fenfitive fenfual motions.* III. 1. *Other mufcular motions are affociated with the fenfitive ones.* 2. *And other fenfual motions.*

I. 1. MANY of the motions of our mufcles, that are excited into action by irritation, are at the fame time accompanied with painful or pleafurable fenfations; and at length become by habit caufable by the fenfations. Thus the motions of the fpincters of the bladder and anus were originally excited into action by irritation; for young children give no attention to thefe evacuations; but as foon as they become fenfible of the inconvenience of obeying thefe irritations, they fuffer the water or excrement to accumulate, till it difagreeably affects them; and the action of thofe fpincters is then in confequence of this difagreeable fenfation. So the fecretion of faliva, which in young children is copioufly produced by irritation, and drops from their mouths, is frequently attended with the agreeable fenfation produced by the maftication of tafteful food; till at length the fight of fuch food to a hungry perfon excites into action thefe falival glands; as is feen in the flavering of hungry dogs.

The motions of thofe mufcles, which are affected by lafcivious ideas, and thofe which are exerted in fmiling, weeping, ftarting from fear, and winking at the approach of danger to the eye, and at times the actions of every large mufcle of the body become caufable by our
fenfations.

fenfations. And all thefe motions are performed with ftrength and velocity in proportion to the energy of the fenfation that excites them, and the quantity of fenforial power.

2. Many of the motions of our organs of fenfe, or ideas, that were originally excited into action by irritation, become in like manner more frequently caufable by our fenfations of pleafure or pain. Thefe motions are then termed the ideas of imagination, and make up all the fcenery and tranfactions of our dreams. Thus when any painful or pleafurable fenfations poffefs us, as of love, anger, fear; whether in our fleep or waking hours, the ideas, that have been formerly excited by the objects of thefe fenfations, now vividly recur before us by their connection with thefe fenfations themfelves. So the fair fmiling virgin, that excited your love by her prefence, whenever that fenfation recurs, rifes before you in imagination; and that with all the pleafing circumftances, that had before engaged your attention. And in fleep, when you dream under the influence of fear, all the robbers, fires, and precipices, that you formerly have feen or heard of, arife before you with terrible vivacity. All thefe fenfual motions, like the mufcular ones above mentioned, are performed with ftrength and velocity in proportion to the energy of the fenfation of pleafure or pain, which excites them, and the quantity of fenforial power.

II. 1. Many of thefe mufcular motions above defcribed, that are moft frequently excited by our fenfations, are neverthelefs occafionally caufable by volition; for we can fmile or frown fpontaneoufly, can make water before the quantity or acrimony of the urine produces a difagreeable fenfation, and can voluntarily mafticate a naufeous drug, or fwallow a bitter draught, though our fenfation would ftrongly diffuade us.

2. In like manner the fenfual motions, or ideas, that are moft frequently excited by our fenfations, are neverthelefs occafionally caufeable by volition, as we can fpontaneoufly call up our laft night's dream before us, tracing it induftrioufly ftep by ftep through all its

G 2   variety

variety of fcenery and tranfaction; or can voluntarily examine or re-peat the ideas, that have been excited by our difguft or admiration.

III. 1. Innumerable trains or tribes of motions are affociated with thefe fenfitive mufcular motions above mentioned; as when a drop of water falling into the wind-pipe difagreeably affects the air-veffels of the lungs, they are excited into violent action; and with thefe fenfitive motions are affociated the actions of the pectoral and inter-coftal mufcles, and the diaphragm; till by their united and repeated fuccuffions the drop is returned through the larinx. The fame occurs when any thing difagreeably affects the noftrils, or the ftomach, or the uterus; variety of mufcles are excited by affociation into forcible action, not to be fuppreffed by the utmoft efforts of the will; as in fneezing, vomiting, and parturition.

2. In like manner with thefe fenfitive fenfual motions, or ideas of imagination, are affociated many other trains or tribes of ideas, which by fome writers of metaphyfics have been claffed under the terms of refemblance, caufation, and contiguity; and will be more fully treated of hereafter.

SECT.

SECT. IX.

OF VOLUNTARY MOTIONS.

I. 1. *Voluntary muscular motions are originally excited by irritations.*   2. *And voluntary ideas. Of reason.*   II. 1. *Voluntary muscular motions are occasionally causable by sensations.*   2. *And voluntary ideas.*   III. 1. *Voluntary muscular motions are occasionally obedient to irritations.*   2. *And voluntary ideas.* IV. 1. *Voluntary muscular motions are associated with other muscular motions.*   2. *And voluntary ideas.*

WHEN pleasure or pain affect the animal system, many of its motions both muscular and sensual are brought into action; as was shewn in the preceding section, and were called sensitive motions. The general tendency of these motions is to arrest and to possess the pleasure, or to dislodge or avoid the pain: but if this cannot immediately be accomplished, desire or aversion are produced, and the motions in consequence of this new faculty of the sensorium are called voluntary.

I. 1. Those muscles of the body that are attached to bones, have in general their principal connections with volition, as I move my pen or raise my body. These motions were originally excited by irritation, as was explained in the section on that subject, afterwards the sensations of pleasure or pain, that accompanied the motions thus excited, induced a repetition of them; and at length many of them

were

were voluntarily practised in succeffion or in combination for the common purpofes of life, as in learning to walk, or to fpeak; and are performed with ftrength and velocity in proportion to the energy of the volition, that excites them, and the quantity of fenforial power.

2. Another great clafs of voluntary motions confifts of the ideas of recollection. We will to repeat a certain train of ideas, as of the alphabet backwards; and if any ideas, that do not belong to this intended train, intrude themfelves by other connections, we will to reject them, and voluntarily perfift in the determined train. So at my approach to a houfe which I have but once vifited, and that at the diftance of many months, I will to recollect the names of the numerous family I expect to fee there, and I do recollect them.

On this voluntary recollection of ideas our faculty of reafon depends, as it enables us to acquire an idea of the diffimilitude of any two ideas. Thus if you voluntarily produce the idea of a right-angled triangle, and then of a fquare; and after having excited thefe ideas repeatedly, you excite the idea of their difference, which is that of another right-angled triangle inverted over the former; you are faid to reafon upon this fubject, or to compare your ideas.

Thefe ideas of recollection, like the mufcular motions above mentioned, were originally excited by the irritation of external bodies, and were termed ideas of perception: afterwards the pleafure or pain, that accompanied thefe motions, induced a repetition of them in the abfence of the external body, by which they were firft excited; and then they were termed ideas of imagination. At length they become voluntarily practised in succeffion or in combination for the common purpofes of life; as when we make ourfelves mafters of the hiftory of mankind, or of the fciences they have inveftigated; and are then called ideas of recollection; and

4

are performed with ftrength and velocity in proportion to the energy of the volition that excites them, and the quantity of fenforial power.

II. 1. The mufcular motions above defcribed, that are moft frequently obedient to the will, are neverthelefs occafionally caufable by painful or pleafurable fenfation, as in the ftarting from fear, and the contraction of the calf of the leg in the cramp.

2. In like manner the fenfual motions, or ideas, that are moft frequently connected with volition, are neverthelefs occafionally caufable by painful or pleafurable fenfation. As the hiftories of men, or the defcription of places, which we have voluntarily taken pains to remember, fometimes occur to us in our dreams.

III. 1. The mufcular motions that are generally fubfervient to volition, are alfo occafionally caufable by irritation, as in ftretching the limbs after fleep, and yawning. In this manner a contraction of the arm is produced by paffing the electric fluid from the Leyden phial along its mufcles; and that even though the limb is paralytic. The fudden motion of the arm produces a difagreeable fenfation in the joint, but the mufcles feem to be brought into action fimply by irritation.

2. The ideas, that are generally fubfervient to the will, are in like manner occafionally excited by irritation; as when we view again an object, we have before well ftudied, and often recollected.

IV. 1. Innumerable trains or tribes of motions are affociated with thefe voluntary mufcular motions above mentioned; as when I will to extend my arm to a diftant object, fome other mufcles are brought into action, and preferve the balance of my body. And when I wifh to perform any fteady exertion, as in threading a needle, or chopping with an ax, the pectoral mufcles are at the fame time brought

into

into action to preferve the trunk of the body motionlefs, and we ceafe to refpire for a time.

2. In like manner the voluntary fenfual motions, or ideas of recollection, are affociated with many other trains or tribes of ideas. As when I voluntarily recollect a gothic window, that I faw fome time ago, the whole front of the cathedral occurs to me at the fame time.

SECT.

SECT. X.

OF ASSOCIATE MOTIONS.

I. 1. *Many muscular motions excited by irritations in trains or tribes become associated.* 2. *And many ideas.* II. 1. *Many sensitive muscular motions become associated.* 2. *And many sensitive ideas.* III. 1. *Many voluntary muscular motions become associated.* 2. *And then become obedient to sensation or irritation.* 3. *And many voluntary ideas become associated.*

ALL the fibrous motions, whether muscular or sensual, which are frequently brought into action together, either in combined tribes, or in successive trains, become so connected by habit, that when one of them is reproduced the others have a tendency to succeed or accompany it.

I. 1. Many of our muscular motions were originally excited in successive trains, as the contractions of the auricles and of the ventricles of the heart; and others in combined tribes, as the various divisions of the muscles which compose the calf of the leg, which were originally irritated into synchronous action by the tædium or irksomeness of a continued posture. By frequent repetitions these motions acquire associations, which continue during our lives, and even after the destruction of the greatest part of the sensorium; for the heart of a viper or frog will continue to pulsate long after it is taken from the body; and when it has entirely ceased to move, if any part of it is goaded with a pin, the whole heart will again renew its pulsations. This kind of connection we shall term irri-

H   tative

tative affociation, to diftinguifh it from fenfitive and voluntary affociations.

2. In like manner many of our ideas are originally excited in tribes; as all the objects of fight, after we become fo well acquainted with the laws of vifion, as to diftinguifh figure and diftance as well as colour; or in trains, as while we pafs along the objects that furround us. The tribes thus received by irritation become affociated by habit, and have been termed complex ideas by the writers of metaphyfics, as this book, or that orange. The trains have received no particular name, but thefe are alike affociations of ideas, and frequently continue during our lives. So the tafte of a pine-apple, though we eat it blindfold, recalls the colour and fhape of it; and we can fcarcely think on folidity without figure.

II. 1. By the various efforts of our fenfations to acquire or avoid their objects, many mufcles are daily brought into fucceffive or fynchronous actions; thefe become affociated by habit, and are then excited together with great facility, and in many inftances gain indiffoluble connections. So the play of puppies and kittens is a reprefentation of their mode of fighting or of taking their prey; and the motions of the mufcles neceffary for thofe purpofes become affociated by habit, and gain a great adroitnefs of action by thefe early repetitions: fo the motions of the abdominal mufcles, which were originally brought into concurrent action with the protrufive motion of the rectum or bladder by fenfation, become fo conjoined with them by habit, that they not only eafily obey thefe fenfations occafioned by the ftimulus of the excrement and urine, but are brought into violent and unreftrainable action in the ftrangury and tenefmus. This kind of connection we fhall term fenfitive affociation.

2. So many of our ideas, that have been excited together or in fucceffion by our fenfations, gain fynchronous or fucceffive affoci-

ations,

ations, that are fometimes indiffoluble but with life. Hence the idea of an inhuman or difhonourable · action perpetually calls up before us the idea of the wretch that was guilty of it. And hence thofe unconquerable antipathies are formed, which fome people have to the fight of peculiar kinds of food, of which in their infancy they have eaten to excefs or by conftraint.

III. 1. In learning any mechanic art, as mufic, dancing, or the ufe of the fword, we teach many of our mufcles to act together or in fucceffion by repeated voluntary efforts ; which by habit become formed into tribes or trains of affociation, and ferve all our pur-pofes with great facility, and in fome inftances acquire an indif-foluble union.   Thefe motions are gradually formed into a habit of acting together by a multitude of repetitions, whilft they are yet feparately caufable by the will, as is evident from the long time that is taken up by children in learning to walk and to fpeak ; and is experienced by every one, when he firft attempts to fkate upon the ice or to fwim : thefe we fhall term voluntary affociations.

2. All thefe mufcular movements, when they are thus affociated into tribes or trains, become afterwards not only obedient to volition, but to the fenfations and irritations ; and the fame movement com-pofes a part of many different tribes or trains of motion.   Thus a fingle mufcle, when it acts in confort with its neighbours on one fide, affifts to move the limb in one direction ; and in an-other, when it acts with thofe in its neighbourhood on the other fide ; and in other directions, when it acts feparately or joint-ly with thofe that lie immediately under or above it ; and all thefe with equal facility after their affociations have been well eftablifhed.

The facility, with which each mufcle changes from one affociated tribe to another, and that either backwards or forwards, is well obfervable in the mufcles of the arm in moving the windlafs of an

air-

air pump; and the flowness of those muscular movements, that have not been associated by habit, may be experienced by any one, who shall attempt to saw the air quick perpendicularly with one hand, and horizontally with the other at the same time.

3. In learning every kind of science we voluntarily associate many tribes and trains of ideas, which afterwards are ready for all the purposes either of volition, sensation, or irritation; and in some instances acquire indissoluble habits of acting together, so as to affect our reasoning, and influence our actions. Hence the necessity of a good education.

These associate ideas are gradually formed into habits of acting together by frequent repetition, while they are yet separately obedient to the will; as is evident from the difficulty we experience in gaining so exact an idea of the front of St. Paul's church, as to be able to delineate it with accuracy, or in recollecting a poem of a few pages.

And these ideas, thus associated into tribes, not only make up the parts of the trains of volition, sensation, and irritation; but the same idea composes a part of many different tribes and trains of ideas. So the simple idea of whiteness composes a part of the complex idea of snow, milk, ivory; and the complex idea of the letter A composes a part of the several associated trains of ideas that make up the variety of words, in which this letter enters.

The numerous trains of these associated ideas are divided by Mr. Hume into three classes, which he has termed contiguity, causation, and resemblance. Nor should we wonder to find them thus connected together, since it is the business of our lives to dispose them into these three classes; and we become valuable to ourselves and our friends, as we succeed in it. Those who have combined an extensive class of ideas by the contiguity of time or place, are men learned in the history of mankind, and of the sciences they have cultivated. Those who have connected a great class of ideas

of

of refemblances, poffefs the fource of the ornaments of poetry and oratory, and of all rational analogy. While thofe who have connected great claffes of ideas of caufation, are furnifhed with the powers of producing effects. Thefe are the men of active wifdom, who lead armies to victory, and kingdoms to profperity; or difcover and improve the fciences, which meliorate and adorn the condition of humanity.

SECT.

SECT. XI.

ADDITIONAL OBSERVATIONS ON THE SENSORIAL POWERS.

I. *Stimulation is of various kinds adapted to the organs of sense, to the muscles, to hollow membranes, and glands. Some objects irritate our senses by repeated impulses.* II. 1. *Sensation and volition frequently affect the whole sensorium.* 2. *Emotions, passions, appetites.* 3. *Origin of desire and aversion. Criterion of voluntary actions, difference of brutes and men.* 4. *Sensibility and voluntarity.* III. *Associations formed before nativity, irritative motions mistaken for associated ones.*

Irritation.

I. THE various organs of sense require various kinds of stimulation to excite them into action; the particles of light penetrate the cornea and humours of the eye, and then irritate the naked retina; sapid particles, dissolved or diffused in water or saliva, and odorous ones, mixed or combined with the air, irritate the extremities of the nerves of taste and smell; which either penetrate, or are expanded on the membranes of the tongue and nostrils; the auditory nerves are stimulated by the vibrations of the atmosphere communicated by means of the tympanum and of the fluid, whether of air or of water, behind it; and the nerves of touch by the hardness of surrounding bodies, though the cuticle is interposed between these bodies and the medulla of the nerve.

As the nerves of the senses have each their appropriated objects, which stimulate them into activity; so the muscular fibres, which are the terminations of other sets of nerves, have their peculiar objects,

jects, which excite them into action; the longitudinal mufcles are ftimulated into contraction by extenfion, whence the ftretching or pandiculation after a long continued pofture, during which they have been kept in a ftate of extenfion; and the hollow mufcles are excited into action by diftention, as thofe of the rectum and bladder are induced to protrude their contents from their fenfe of the diftention rather than of the acrimony of thofe contents.

There are other objects adapted to ftimulate the nerves, which terminate in variety of membranes, and thofe efpecially which form the terminations of canals; thus the preparations of mercury particularly affect the falivary glands, ipecacuhana affects the fphincter of the anus, cantharides that of the bladder, and laftly every gland of the body appears to be indued with a kind of tafte, by which it felects or forms each its peculiar fluid from the blood; and by which it is irritated into activity.

Many of thefe external properties of bodies, which ftimulate our organs of fenfe, do not feem to effect this by a fingle impulfe, but by repeated impulfes; as the nerve of the ear is probably not excitable by a fingle vibration of air, nor the optic nerve by a fingle particle of light; which circumftance produces fome analogy between thofe two fenfes, at the fame time the folidity of bodies is perceived by a fingle application of a folid body to the nerves of touch, and that even through the cuticle; and we are probably poffeffed of a peculiar fenfe to diftinguifh the nice degrees of heat and cold.

The fenfes of touch and of hearing acquaint us with the mechanical impact and vibration of bodies, thofe of fmell and tafte feem to acquaint us with fome of their chemical properties, while the fenfe of vifion and of heat acquaint us with the exiftence of their peculiar fluids.

Senfation

Senfation and Volition.

II. Many motions are produced by pleafure or pain, and that even in contradiction to the power of volition, as in laughing, or in the ftrangury; but as no name has been given to pleafure or pain, at the time it is exerted fo as to caufe fibrous motions, we have ufed the term fenfation for this purpofe; and mean it to bear the fame analogy to pleafure and pain, that the word volition does to defire and averfion.

1. It was mentioned in the fifth Section, that what we have termed fenfation is a motion of the central parts, or of the whole fenforium, *beginning* at fome of the extremities of it. This appears firft, becaufe our pains and pleafures are always caufed by our ideas or mufcular motions, which are the motions of the extremities of the fenforium. And, fecondly, becaufe the fenfation of pleafure or pain frequently continues fome time after the ideas or mufcular motions which excited it have ceafed: for we often feel a glow of pleafure from an agreeable reverie, for many minutes after the ideas, that were the fubject of it, have efcaped our memory; and frequently experience a dejection of fpirits without being able to affign the caufe of it but by much recollection.

When the fenforial faculty of defire or averfion is exerted fo as to caufe fibrous motions, it is termed volition; which is faid in Sect. V. to be a motion of the central parts, or of the whole fenforium, *terminating* in fome of the extremities of it. This appears, firft, becaufe our defires and averfions always terminate in recollecting and comparing our ideas, or in exerting our mufcles; which are the motions of the extremities of the fenforium. And, fecondly, becaufe defire or averfion begins, and frequently continues for a time in the central parts of the fenforium, before it is peculiarly exerted at the

extremities

extremities of it: for we fometimes feel defire or averfion without immediately knowing their objects, and in confequence without immediately exerting any of our mufcular or fenfual motions to attain them : as in the beginning of the paffion of love, and perhaps of hunger, or in the ennui of indolent people.

Though fenfation and volition begin or terminate at the extremities or central parts of the fenforium, yet the whole of it is frequently influenced by the exertion of thefe faculties, as appears from their effects on the external habit : for the whole fkin is reddened by fhame, and an univerfal trembling is produced by fear : and every mufcle of the body is agitated in angry people by the defire of revenge.

There is another very curious circumftance, which fhews that fenfation and volition are movements of the fenforium in contrary directions; that is, that volition begins at the central parts of it, and proceeds to the extremities ; and that fenfation begins at the extremities, and proceeds to the central parts : I mean that thefe two fenforial faculties cannot be ftrongly exerted at the fame time ; for when we exert our volition ftrongly, we do not attend to pleafure or pain ; and converfely, when we are ftrongly affected with the fenfation of pleafure or pain, we ufe no volition. As will be further explained in Section XVIII. on fleep, and Section XXXIV. on volition.

2. All our emotions and paffions feem to arife out of the exertions of thefe two faculties of the animal fenforium. Pride, hope, joy, are the names of particular pleafures : fhame, defpair, forrow, are the names of peculiar pains : and love, ambition, avarice, of particular defires : hatred, difguft, fear, anxiety, of particular averfions. Whilft the paffion of anger includes the pain from a recent injury, and the averfion to the adverfary that occafioned it. And compaffion is the pain we experience at the fight of mifery, and the defire of relieving it.

There is another tribe of defires, which are commonly termed appetites, and are the immediate confequences of the abfence of fome

I   irritative

irritative motions. Thofe, which arife from defect of internal irritations, have proper names conferred upon them, as hunger, thirft,
luft, and the defire of air, when our refpiration is impaired by noxious vapours; and of warmth, when we are expofed to too great a
degree of cold. But thofe, whofe ftimuli are external to the body,
are named from the objects, which are by nature conftituted to excite them; thefe defires originate from our paft experience of the
pleafurable fenfations they occafion, as the fmell of an hyacinth, or the
tafte of a pine-apple.

Whence it appears, that our pleafures and pains are at leaft as various and as numerous as our irritations; and that our defires and averfions muft be as numerous as our pleafures and pains. And that as
fenfation is here ufed as a general term for our numerous pleafures and
pains, when they produce the contractions of our fibres; fo volition
is the general name for our defires and averfions, when they produce
fibrous contractions. Thus when a motion of the central parts, or of
the whole fenforium, terminates in the exertion of our mufcles, it is
generally called voluntary action; when it terminates in the exertion
of our ideas, it is termed recollection, reafoning, determining.

3. As the fenfations of pleafure and pain are originally introduced by
the irritations of external objects: fo our defires and averfions are originally introduced by thofe fenfations; for when the objects of our
pleafures or pains are at a diftance, and we cannot inftantaneoufly
poffefs the one, or avoid the other, then defire or averfion is produced, and a voluntary exertion of our ideas or mufcles fucceeds.

The pain of hunger excites you to look out for food, the tree, that
fhades you, prefents its odoriferous fruit before your eyes, you approach, pluck, and eat.

The various movements of walking to the tree, gathering the
fruit, and mafticating it, are affociated motions introduced by their
connection with fenfation; but if from the uncommon height of the
tree, the fruit be inacceffible, and you are prevented from quickly
                                                        poffeffing

poffeffing the intended pleafure, defire is produced. The confequence of this defire is, firft, a deliberation about the means to gain the object of pleafure in procefs of time, as it cannot be procured immediately; and, fecondly, the mufcular action neceffary for this purpofe.

You voluntarily call up all your ideas of caufation, that are related to the effect you defire, and voluntarily examine and compare them, and at length determine whether to afcend the tree, or to gather ftones from the neighbouring brook, is eafier to practife, or more promifing of fuccefs; and, finally, you gather the ftones, and repeatedly fling them to diflodge the fruit.

Hence then we gain a criterion to diftinguifh voluntary acts or thoughts from thofe caufed by fenfation. As the former are always employed about the *means* to acquire pleafurable objects, or the *means* to avoid painful ones; while the latter are employed in the poffeffion of thofe, which are already in our power.

Hence the activity of this power of volition produces the great difference between the human and the brute creation. The ideas and the actions of brutes are almoft perpetually employed about their prefent pleafures, or their prefent pains; and, except in the few inftances which are mentioned in Section XVI. on inftinct, they feldom bufy themfelves about the means of procuring future blifs, or of avoiding future mifery; fo that the acquiring of languages, the making of tools, and labouring for money, which are all only the means to procure pleafures; and the praying to the Deity, as another means to procure happinefs, are characteriftic of human nature.

4. As there are many difeafes produced by the quantity of the fenfation of pain or pleafure being too great or too little; fo are there difeafes produced by the fufceptibility of the conftitution to motions caufable by thefe fenfations being too dull or too vivid. This fufceptibility of the fyftem to fenfitive motions is termed fenfibility, to diftinguifh it from fenfation, which is the actual exiftence or exertion of pain or pleafure.

I 2

Other

Other claffes of difeafes are owing to the exceffive promptitude, or fluggifhnefs of the conftitution to voluntary exertions, as well as to the quantity of defire or of averfion. This fufceptibility of the fyftem to voluntary motions is termed voluntarity, to diftinguifh it from volition, which is the exertion of defire or averfion: thefe difeafes will be treated of at length in the progrefs of the work.

Affociation.

III. 1. It is not eafy to affign a caufe, why thofe animal movements, that have once occurred in fucceffion, or in combination, fhould afterwards have a tendency to fucceed or accompany each other. It is a property of animation, and diftinguifhes this order of being from the other productions of nature.

When a child firft wrote the word man, it was diftinguifhed in his mind into three letters, and thofe letters into many parts of letters; but by repeated ufe the word man becomes to his hand in writing it, as to his organs of fpeech in pronouncing it, but one movement without any deliberation, or fenfation, or irritation, interpofed between the parts of it. And as many feparate motions of our mufcles thus become united, and form, as it were, one motion; fo each feparate motion before fuch union may be conceived to confift of many parts or fpaces moved through; and perhaps even the individual fibres of our mufcles have thus gradually been brought to act in concert, which habits began to be acquired as early as the very formation of the moving organs, long before the nativity of the animal; as explained in the Section XVI. 2. on inftinct.

2. There are many motions of the body, belonging to the irritative clafs, which might by a hafty obferver be miftaken for affociated ones; as the periftaltic motion of the ftomach and inteftines, and the contractions of the heart and arteries; might be fuppofed to be affoci-

ated

ated with the irritative motions of their nerves of fenfe, rather than to be excited by the irritation of their mufcular fibres by the diften-tion, acrimony, or momentum of the blood. So the diftention or elongation of mufcles by objects external to them irritates them into contraction, though the cuticle or other parts may intervene between the ftimulating body and the contracting mufcle. Thus a horfe voids his excrement when its weight or bulk irritates the rectum or fphincter ani. The motion of thefe mufcles act from the irritation of diften-tion, when he excludes his excrement, but the mufcles of the abdo-men and diaphragm are brought into motion by affociation with thofe of the fphincter and rectum.

SECT.

SECT. XII.

OF STIMULUS, SENSORIAL EXERTION, AND FIBROUS CONTRACTION.

I. Of fibrous contraction.   1. *Two particles of a fibre cannot approach without the intervention of something, as in magnetism, electricity, elasticity.   Spirit of life is not electric ether.   Galvani's experiments.   2. Contraction of a fibre. 3. Relaxation succeeds.   4. Successive contractions, with intervals.   Quick pulse from debility, from paucity of blood.   Weak contractions performed in less time, and with shorter intervals.   5. Last situation of the fibres continues after contraction.   6. Contraction greater than usual induces pleasure or pain. 7. Mobility of the fibres uniform.   Quantity of sensorial power fluctuates. Constitutes excitability.* II. Of sensorial exertion.   1. *Animal motion includes stimulus, sensorial power, and contractile fibres.   The sensorial faculties act separately or conjointly.   Stimulus of four kinds.   Strength and weakness defined. Sensorial power perpetually exhausted and renewed.   Weakness from defect of stimulus.   From defect of sensorial power, the direct and indirect debility of Dr. Brown.   Why we become warm in Buxton bath after a time, and see well after a time in a darkish room.   Fibres may act violently, or with their whole force, and yet feebly.   Great exertion in inflammation explained.   Great muscular force of some insane people.   2. Occasional accumulation of sensorial power in muscles subject to constant stimulus.   In animals sleeping in winter. In eggs, seeds, schirrous tumours, tendons, bones.   3. Great exertion introduces pleasure or pain.   Inflammation.   Libration of the system between torpor and activity.   Fever-fits.   4. Desire and aversion introduced.   Excess of volition cures fevers.* III. Of repeated stimulus.   1. *A stimulus repeated too frequently loses effect.   As opium, wine, grief.   Hence old age.   Opium and aloes in small doses.   2. A stimulus not repeated too frequently does not lose effect.*

effect. Perpetual movement of the vital organs.   3. A stimulus repeated at uniform times produces greater effect.   Irritation combined with association.   4. A stimulus repeated frequently and uniformly may be withdrawn, and the action of the organ will continue.   Hence the bark cures agues, and strengthens weak constitutions.   5. Defect of stimulus repeated at certain intervals causes fever-fits.   6. Stimulus long applied ceases to act a second time.   7. If a stimulus excites sensation in an organ not usually excited into sensation, inflam- mation is produced.   IV. Of stimulus greater than natural.   1. *A stimulus greater than natural diminishes the quantity of sensorial power in general.   2. In particular organs.   3. Induces the organ into spasmodic actions.   4. Induces the antagonist fibres into action.   5. Induces the organ into convulsive or fixed spasms.   6. Produces paralysis of the organ.*   V. Of stimulus less than na-tural.   1. *Stimulus less than natural occasions accumulation of sensorial power in general.   2. In particular organs, flushing of the face in a frosty morning. In fibres subject to perpetual stimulus only.   Quantity of sensorial power in-versely as the stimulus.   3. Induces pain.   As of cold, hunger, head-ach.   4. Induces more feeble and frequent contraction.   As in low fevers.   Which are frequently owing to deficiency of sensorial power rather than to deficiency of stimulus.   5. Inverts successive trains of motion.   Inverts ideas.   6. In-duces paralysis and death.*   VI. Cure of increased exertion.   1. *Natural cure of exhaustion of sensorial power.   2. Decrease the irritations.   Venesection. Cold.   Abstinence.   3. Prevent the previous cold fit.   Opium.   Bark. Warmth.   Anger.   Surprise.   4. Excite some other part of the system. Opium and warm bath relieve pains both from defect and from excess of sti-* mulus.   5. *First increase the stimulus above, and then decrease it beneath the natural quantity.*   VII. Cure of decreased exertion.   1. *Natural cure by accumulation of sensorial power.   Ague-fits.   Syncope.   2. Increase the stimulation, by wine, opium, given so as not to intoxicate.   Cheerful ideas. 3. Change the kinds of stimulus.   4. Stimulate the associated organs.   Blisters of use in heart-burn, and cold extremities.   5. Decrease the stimulation for a time, cold bath.   6. Decrease the stimulation below natural, and then in-crease it above natural.   Bark after emetics.   Opium after venesection. Practice of Sydenham in chlorosis.   7. Prevent unnecessary expenditure of*

sensorial

senforial power. Decumbent posture, silence, darknefs. Pulse quickened by rising out of bed. 8. To the greatest degree of quiescence apply the least stimulus. Otherwise paralysis or inflammation of the organ ensues. Gin, wine, blisters, destroy by too great stimulation in fevers with debility. Intoxication in the slightest degree succeeded by debility. Golden rule for determining the best degree of stimulus in low fevers. Another golden rule for determining the quantity of spirit which those, who are debilitated by drinking it, may safely omit.

I. Of *fibrous contraction.*

1. IF two particles of iron lie near each other without motion, and afterwards approach each other; it is reafonable to conclude that fomething befides the iron particles is the caufe of their approximation; this invifible fomething is termed magnetifm. In the fame manner, if the particles, which compofe an animal mufcle, do not touch each other in the relaxed ftate of the mufcle, and are brought into contact during the contraction of the mufcle; it is reafonable to conclude, that fome other agent is the caufe of this new approximation. For nothing can act, where it does not exift; for to act includes to exift; and therefore the particles of the mufcular fibre (which in its ftate of relaxation are fuppofed not to touch) cannot affect each other without the influence of fome intermediate agent; this agent is here termed the fpirit of animation, or fenforial power, but may with equal propriety be termed the power, which caufes contraction; or may be called by any other name, which the reader may choofe to affix to it.

The contraction of a mufcular fibre may be compared to the following electric experiment, which is here mentioned not as a philofophical analogy, but as an illuftration or fimile to facilitate the conception of a difficult fubject. Let twenty very fmall Leyden phials properly

coated

coated be hung in a row by fine filk threads at a fmall diftance from
each other; let the internal charge of one phial be pofitive, and of the
other negative alternately, if a communication be made from the internal furface of the firft to the external furface of the laft in the row,
they will all of them inftantly approach each other, and thus fhorten
a line that might connect them like a mufcular fibre. See Botanic
Garden, p. 1. Canto I. l. 202, note on Gymnotus.

The attractions of electricity or of magnetifm do not apply philo-
fophically to the illuftration of the contraction of animal fibres, fince
the force of thofe attractions increafes in fome proportion inverfely as
the diftance, but in mufcular motion there appears no difference in
velocity or ftrength during the beginning or end of the contraction,
but what may be clearly afcribed to the varying mechanic advantage
in the approximation of one bone to another. Nor can mufcular motion be affimilated with greater plaufibility to the attraction of cohefion
or elafticity; for in bending a fteel fpring, as a fmall fword, a lefs
force is required to bend it the firft inch than the fecond; and the
fecond than the third; the particles of fteel on the convex fide of the
bent fpring endeavouring to reftore themfelves more powerfully the
further they are drawn from each other. See Botanic Garden, P. 1.
addit. Note XVIII.

I am aware that this may be explained another way, by fuppofing
the elafticity of the fpring to depend more on the compreffion of the
particles on the concave fide than on the extenfion of them on the convex fide; and by fuppofing the elafticity of the elaftic gum to depend
more on the refiftance to the lateral compreffion of its particles than
to the longitudinal extenfion of them. Neverthelefs in mufcular
contraction, as above obferved, there appears no difference in the velocity or force of it at its commencement or at its termination; from
whence we muft conclude that animal contraction is governed by
laws of its own, and not by thofe of mechanics, chemiftry, magnetifm, or electricity.

K

On

On thefe accounts I do not think the experiments conclufive, which were lately publifhed by Galvani, Volta, and others, to fhew a fimilitude between the fpirit of animation, which contracts the mufcular fibres, and the electric fluid. Since the electric fluid may act only as a more potent ftimulus exciting the mufcular fibres into action, and not by fupplying them with a new quantity of the fpirit of life. Thus in a recent hemiplegia I have frequently obferved, when the patient yawned and ftretched himfelf, that the paralytic limbs moved alfo, though they were totally difobedient to the will. And when he was electrified by paffing fhocks from the affected hand to the affected foot, a motion of the paralytic limbs was alfo produced. Now as in the act of yawning the mufcles of the paralytic limbs were excited into action by the ftimulus of the irkfomenefs of a continued pofture, and not by any additional quantity of the fpirit of life; fo we may conclude, that the paffage of the electric fluid, which produced a fimilar effect, acted only as a ftimulus, and not by fupplying any addition of fenforial power.

If neverthelefs this theory fhould ever become eftablifhed, a ftimulus muft be called an eductor of vital ether; which ftimulus may confift of fenfation or volition, as in the electric eel, as well as in the appulfes of external bodies; and by drawing off the charges of vital fluid may occafion the contraction or motions of the mufcular fibres, and organs of fenfe.

2. The immediate effect of the action of the fpirit of animation or fenforial power on the fibrous parts of the body, whether it acts in the mode of irritation, fenfation, volition, or affociation, is a contraction of the animal fibre, according to the fecond law of animal caufation. Sect. IV. Thus the ftimulus of the blood induces the contraction of the heart; the agreeable tafte of a ftrawberry produces the contraction of the mufcles of deglutition; the effort of the will contracts the mufcles, which move the limbs in walking; and by affociation other mufcles of the trunk are brought into contraction to

preferve

preferve the balance of the body.   The fibrous extremities of the organs of fenfe have been fhewn, by the ocular fpectra in Sect. III. to fuffer fimilar contraction by each of the above modes of excitation ; and by their configurations to conftitute our ideas.

3. After animal fibres have for fome time been excited into contraction, a relaxation fucceeds, even though the exciting caufe continues to act.   In refpect to the irritative motions this is exemplified in the periftaltic contractions of the bowels ; which ceafe and are renewed alternately, though the ftimulus of the aliment continues to be uniformly applied ; in the fenfitive motions, as in ftrangury, tenefmus, and parturition, the alternate contractions and relaxations of the mufcles exift, though the ftimulus is perpetual.   In our voluntary exertions it is experienced, as no one can hang long by the hands, however vehemently he wills fo to do ; and in the affociate motions the conftant change of our attitudes evinces the neceffity of relaxation to thofe mufcles, which have been long in action.

This relaxation of a mufcle after its contraction, even though the ftimulus continues to be applied, appears to arife from the expenditure or diminution of the fpirit of animation previoufly refident in the mufcle, according to the fecond law of animal caufation in Sect. IV. In thofe conftitutions, which are termed weak, the fpirit of animation becomes fooner exhaufted, and tremulous motions are produced, as in the hands of infirm people, when they lift a cup to their mouths. This quicker exhauftion of the fpirit of animation is probably owing to a lefs quantity of it refiding in the acting fibres, which therefore more frequently require a fupply from the nerves, which belong to them.

4. If the fenforial power continues to act, whether it acts in the mode of irritation, fenfation, volition, or affociation, a new contraction of the animal fibre fucceeds after a certain interval ; which interval is of fhorter continuance in weak people than in ftrong ones. This is exemplified in the fhaking of the hands of weak people, when

K 2   they

they attempt to write. In a manufcript epiftle of one of my cor-
refpondents, which is written in a fmall hand, I obferved from four
to fix zigzags in the perpendicular ftroke of every letter, which
fhews that both the contractions of the fingers, and intervals between them, muft have been performed in very fhort periods of
time.

The times of contraction of the mufcles of enfeebled people being
lefs, and the intervals between thofe contractions being lefs alfo, accounts for the quick pulfe in fevers with debility, and in dying animals. The fhortnefs of the intervals between one contraction and
another in weak conftitutions, is probably owing to the general deficiency of the quantity of the fpirit of animation, and that therefore
there is a lefs quantity of it to be received at each interval of the activity of the fibres. Hence in repeated motions, as of the fingers in
performing on the harpfichord, it would at firft fight appear, that
fwiftnefs and ftrength were incompatible; neverthelefs the fingle
contraction of a mufcle is performed with greater velocity as well as
with greater force by vigorous conftitutions, as in throwing a
javelin.

There is however another circumftance, which may often contribute to caufe the quicknefs of the pulfe in nervous fevers, as in animals bleeding to death in the flaughter-houfe; which is the deficient
quantity of blood; whence the heart is but half diftended, and in confequence fooner contracts. See Sect. XXXII. 2. 1.

For we muft not confound frequency of repetition with quicknefs
of motion, or the number of pulfations with the velocity, with which
the fibres, which conftitute the coats of the arteries, contract themfelves. For where the frequency of the pulfations is but feventy-five
in a minute, as in health; the contracting fibres, which conftitute the
fides of the arteries, may move through a greater fpace in a given
time, than where the frequency of pulfation is one hundred and fifty
in a minute, as in fome fevers with great debility. For if in thofe

fevers

fevers the arteries do not expand themfelves in their diaftole to more than half the ufual diameter of their diaftole in health, the fibres which conftitute their coats, will move through a lefs fpace in a minute than in health, though they make two pulfations for one.

Suppofe the diameter of the artery during its fyftole to be one line, and that the diameter of the fame artery during its diaftole is in health is four lines, and in a fever with great debility only two lines.   It follows, that the arterial fibres contract in health from a circle of twelve lines in circumference to a circle of three lines in circumference, that is they move through a fpace of nine lines in length.   While the arterial fibres in the fever with debility would twice contract from a circle of fix lines to a circle of three lines ; that is while they move through a fpace equal to fix lines.   Hence though the frequency of pulfation in fever be greater as two to one, yet the velocity of contraction in health is greater as nine to fix, or as three to two.

On the contrary in inflammatory difeafes with ftrength, as in the pleurify, the velocity of the contracting fides of the arteries is much greater than in health, for if we fuppofe the number of pulfations in a pleurify to be half as much more than in health, that is as one hundred and twenty to eighty, (which is about what generally happens in inflammatory difeafes) and if the diameter of the artery in diaftole be one third greater than in health, which I believe is near the truth, the refult will be, that the velocity of the contractile fides of the arteries will be in a pleurify as two and a half to one, compared to the velocity of their contraction in a ftate of health, for if the circumference of the fyftole of the artery be three lines, and the diaftole in health be twelve lines in circumference, and in a pleurify eighteen lines ; and fecondly, if the artery pulfates thrice in the difeafed ftate for twice in the healthy one, it follows, that the velocity of contraction in the difeafed ftate to that in the healthy ftate will be forty-five to eighteen, or as two and a half to one.

From

From hence it would appear, that if we had a criterion to determine the velocity of the arterial contractions, it would at the same time give us their ſtrength, and thus be of more ſervice in diſtinguiſhing diſeaſes, than the knowledge of their frequency. As ſuch a criterion cannot be had, the frequency of pulſation, the age of the patient being allowed for, will in ſome meaſure aſſiſt us to diſtinguiſh arterial ſtrength from arterial debility, ſince in inflammatory diſeaſes with ſtrength the frequency ſeldom exceeds one hundred and eighteen or one hundred and twenty pulſations in a minute; unleſs under peculiar circumſtance, as the great additional ſtimuli of wine or of external heat.

5. After a muſcle or organ of ſenſe has been excited into contraction, and the ſenſorial power ceaſes to act, the laſt ſituation or configuration of it continues; unleſs it be diſturbed by the action of ſome antagoniſt fibres, or other extraneous power. Thus in weak or languid people, wherever they throw their limbs on their bed or ſofa, there they lie, till another exertion changes their attitude; hence one kind of ocular ſpectra ſeems to be produced after looking at bright objects; thus when a fire-ſtick is whirled round in the night, there appears in the eye a complete circle of fire; the action or configuration of one part of the retina not ceaſing before the return of the whirling fire.

Thus if any one looks at the ſetting ſun for a ſhort time, and then covers his cloſed eyes with his hand, he will for many ſeconds of time perceive the image of the ſun on his retina. A ſimilar image of all other bodies would remain ſome time in the eye, but is effaced by the eternal change of the motions of the extremity of this nerve in our attention to other objects. See Sect. XVII. 1. 3. on Sleep. Hence the dark ſpots, and other ocular ſpectra, are more frequently attended to, and remain longer in the eyes of weak people, as after violent exerciſe, intoxication, or want of ſleep.

6. A contraction of the fibres somewhat greater than usual introduces pleasurable sensation into the system, according to the fourth law of animal causation. Hence the pleasure in the beginning of drunkenness is owing to the increased action of the system from the stimulus of vinous spirit or of opium. If the contractions be still greater in energy or duration, painful sensations are introduced, as in consequence of great heat, or caustic applications, or fatigue.

If any part of the system, which is used to perpetual activity, as the stomach, or heart, or the fine vessels of the skin, acts for a time with less energy, another kind of painful sensation ensues, which is called hunger, or faintness, or cold. This occurs in a less degree in the locomotive muscles, and is called wearysomeness. In the two former kinds of sensation there is an expenditure of sensorial power, in these latter there is an accumulation of it.

7. We have used the words exertion of sensorial power as a general term to express either irritation, sensation, volition, or association; that is, to express the activity or motion of the spirit of animation, at the time it produces the contractions of the fibrous parts of the system. It may be supposed that there may exist a greater or less mobility of the fibrous parts of our system, or a propensity to be stimulated into contraction by the greater or less quantity or energy of the spirit of animation; and that hence if the exertion of the sensorial power be in its natural state, and the mobility of the fibres be increased, the same quantity of fibrous contraction will be caused, as if the mobility of the fibres continues in its natural state, and the sensorial exertion be increased.

Thus it may be conceived, that in diseases accompanied with strength, as in inflammatory fevers with arterial strength, that the cause of greater fibrous contraction may exist in the increased mobility of the fibres, whose contractions are thence both more forceable and more frequent. And that in diseases attended with debility, as in nervous fevers, where the fibrous contractions are weaker, and

more

more frequent, it may be conceived that the caufe confifts in a decreafe of mobility of the fibres; and that thofe weak conftitutions, which are attended with cold extremities and large pupils of the eyes, may poffefs lefs mobility of the contractile fibres, as well as lefs quantity of exertion of the fpirit of animation.

In anfwer to this mode of reafoning it may be fufficient to obferve, that the contractile fibres confift of inert matter, and when the fenforial power is withdrawn, as in death, they poffefs no power of motion at all, but remain in their laft ftate, whether of contraction or relaxation, and muft thence derive the whole of this property from the fpirit of animation. At the fame time it is not improbable, that the moving fibres of ftrong people may poffefs a capability of receiving or containing a greater quantity of the fpirit of animation than thofe of weak people.

In every contraction of a fibre there is an expenditure of the fenforial power, or fpirit of animation; and where the exertion of this fenforial power has been for fome time increafed, and the mufcles or organs of fenfe have in confequence acted with greater energy, its propenfity to activity is proportionally leffened; which is to be afcribed to the exhauftion or diminution of its quantity. On the contrary, where there has been lefs fibrous contraction than ufual for a certain time, the fenforial power or fpirit of animation becomes accumulated in the inactive part of the fyftem. Hence vigour fucceeds reft, and hence the propenfity to action of all our organs of fenfe and mufcles is in a ftate of perpetual fluctuation. The irritability for inftance of the retina, that is, its quantity of fenforial power, varies every moment according to the brightnefs or obfcurity of the object laft beheld compared with the prefent one. The fame occurs to our fenfe of heat, and to every part of our fyftem, which is capable of being excited into action.

When this variation of the exertion of the fenforial power becomes much and permanently above or beneath the natural quantity, it be-

comes

comes a difeafe. If the irritative motions be too great or too little, it fhews that the ftimulus of external things affect this fenforial power too violently or too inertly. If the fenfitive motions be too great or too little, the caufe arifes from the deficient or exuberant quantity of fenfation produced in confequence of the motions of the mufcular fibres or organs of fenfe; if the voluntary actions are difeafed the caufe is to be looked for in the quantity of volition produced in confequence of the defire or averfion occafioned by the painful or pleafurable fenfations above mentioned. And the difeafes of affociations probably depend on the greater or lefs quantity of the other three fenforial powers by which they were formed.

From whence it appears that the propenfity to action, whether it be called irritability, fenfibility, voluntarity, or affociability, is only another mode of expreffion for the quantity of fenforial power refiding in the organ to be excited. And that on the contrary the words inirritability and infenfibility, together with inaptitude to voluntary and affociate motions, are fynonymous with deficiency of the quantity of fenforial power, or of the fpirit of animation, refiding in the organs to be excited.

II. Of fenforial Exertion.

1. There are three circumftances to be attended to in the production of animal motions. 1ft. The ftimulus. 2d. The fenforial power. 3d. The contractile fibre. 1ft. A ftimulus, external to the organ, originally induces into action the fenforial faculty termed irritation; this produces the contraction of the fibres, which, if it be perceived at all, introduces pleafure or pain; which in their active ftate are termed fenfation; which is another fenforial faculty, and occafionally produces contraction of the fibres; this pleafure or pain is therefore to be confidered as another ftimulus, which may either act alone or in conjunction with the former faculty of the fenforium termed irritation.

L   This

This new ſtimulus of pleaſure or pain either induces into action the ſenſorial faculty termed ſenſation, which then produces the contraction of the fibres; or it introduces deſire or averſion, which excite into action another ſenſorial faculty, termed volition, and may therefore be conſidered as another ſtimulus, which either alone or in conjunction with one or both of the two former faculties of the ſenſorium produces the contraction of animal fibres. There is another ſenſorial power, that of aſſociation, which perpetually, in conjunction with one or more of the above, and frequently ſingly, produces the contraction of animal fibres, and which is itſelf excited into action by the previous motions of contracting fibres.

Now as the ſenſorial power, termed irritation, reſiding in any particular fibres, is excited into exertion by the ſtimulus of external bodies acting on thoſe fibres; the ſenſorial power, termed ſenſation, reſiding in any particular fibres is excited into exertion by the ſtimulus of pleaſure or pain acting on thoſe fibres; the ſenſorial power, termed volition, reſiding in any particular fibres is excited into exertion by the ſtimulus of deſire or averſion; and the ſenſorial power, termed aſſociation, reſiding in any particular fibres, is excited into action by the ſtimulus of other fibrous motions, which had frequently preceded them. The word ſtimulus may therefore be uſed without impropriety of language, for any of theſe four cauſes, which excite the four ſenſorial powers into exertion. For though the immediate cauſe of volition has generally been termed *a motive*; and that of irritation only has generally obtained the name of *ſtimulus*; yet as the immediate cauſe, which excites the ſenſorial powers of ſenſation, or of aſſociation into exertion, have obtained no general name, we ſhall uſe the word ſtimulus for them all.

Hence the quantity of motion produced in any particular part of the animal ſyſtem will be as the quantity of ſtimulus and the quantity of ſenſorial power, or ſpirit of animation, reſiding in the contracting fibres. Where both theſe quantities are great, *ſtrength* is produced,

when

when that word is applied to the motions of animal bodies. Where either of them is deficient, *weakness* is produced, as applied to the motions of animal bodies.

Now as the senforial power, or spirit of animation, is perpetually exhausted by the expenditure of it in fibrous contractions, and is perpetually renewed by the secretion or production of it in the brain and spinal marrow, the quantity of animal strength must be in a perpetual state of fluctuation on this account; and if to this be added the unceasing variation of all the four kinds of stimulus above described, which produce the exertions of the senforial powers, the ceaseless viciffitude of animal strength becomes easily comprehended.

If the quantity of senforial power remains the same, and the quantity of stimulus be leffened, a weakness of the fibrous contractions enfues, which may be denominated *debility from defect of stimulus*. If the quantity of stimulus remains the same, and the quantity of senforial power be leffened, another kind of weakness enfues, which may be termed *debility from defect of senforial power*; the former of these is called by Dr. Brown, in his Elements of Medicine, direct debility, and the latter indirect debility. The coincidence of some parts of this work with correspondent deductions in the Brunonian Elementa Medicina, a work (with some exceptions) of great genius, must be considered as confirmations of the truth of the theory, as they were probably arrived at by different trains of reasoning.

Thus in those who have been exposed to cold and hunger there is a deficiency of stimulus. While in nervous fever there is a deficiency of senforial power. And in habitual drunkards, in a morning before their usual potation, there is a deficiency both of stimulus and of senforial power. While, on the other hand, in the beginning of intoxication there is an excess of stimulus; in the hot-ach, after the hands have been immersed in snow, there is a redundancy of senforial power; and in inflammatory difeafes with arterial strength, there is an excess of both.

L 2

Hence

Hence if the fenforial power be leffened, while the quantity of ftimulus remains the fame as in nervous fever, the frequency of repetition of the arterial contractions may continue, but their force in refpect to removing obftacles, as in promoting the circulation of the blood, or the velocity of each contraction, will be diminifhed, that is, the animal ftrength will be leffened. And fecondly, if the quantity of fenforial power be leffened, and the ftimulus be increafed to a certain degree, as in giving opium in nervous fevers, the arterial contractions may be performed more frequently than natural, yet with lefs ftrength.

And thirdly, if the fenforial power continues the fame in refpect to quantity, and the ftimulus be fomewhat diminifhed, as in going into a darkifh room, or into a coldifh bath, fuppofe of about eighty degrees of heat, as Buxton-bath, a temporary weaknefs of the affected fibres is induced, till an accumulation of fenforial power gradually fucceeds, and counterbalances the deficiency of ftimulus, and then the bath ceafes to feel cold, and the room ceafes to appear dark; becaufe the fibres of the fubcutaneous veffels, or of the organs of fenfe, act with their ufual energy.

A fet of mufcular fibres may thus be ftimulated into violent exertion, that is, they may act frequently, and with their whole fenforial power, but may neverthelefs not act ftrongly; becaufe the quantity of their fenforial power was originally fmall, or was previoufly exhaufted. Hence a ftimulus may be great, and the irritation in confequence act with its full force, as in the hot paroxifms of nervous fever; but if the fenforial power, termed irritation, be fmall in quantity, the force of the fibrous contractions, and the times of their continuance in their contracted ftate, will be proportionally fmall.

In the fame manner in the hot paroxifm of putrid fevers, which are fhewn in Sect. XXXIII. to be inflammatory fevers with arterial debility, the fenforial power termed fenfation is exerted with great activity, yet the fibrous contractions, which produce the circulation of the blood, are performed without ftrength, becaufe the quantity of fenforial power then refiding in that part of the fyftem is fmall.

Thus in irritative fever with arterial ftrength, that is, with excefs of fpirit of animation, the quantity of exertion during the hot part of the paroxifm is to be eftimated from the quantity of ftimulus, and the quantity of fenforial power. While in fenfitive (or inflammatory) fever with arterial ftrength, that is, with excefs of fpirit of animation, the violent and forcible actions of the vafcular fyftem during the hot part of the paroxifm are induced by the exertions of two fenforial powers, which are excited by two kinds of ftimulus. Thefe are the fenforial power of irritation excited by the ftimulus of bodies external to the moving fibres, and the fenforial power of fenfation excited by the pain in confequence of the increafed contractions of thofe moving fibres.

And in infane people in fome cafes the force of their mufcular actions will be in proportion to the quantity of fenforial power, which they poffefs, and the quantity of the ftimulus of defire or averfion, which excites their volition into action. At the fame time in other cafes the ftimulus of pain or pleafure, and the ftimulus of external bodies, may excite into action the fenforial powers of fenfation and irritation, and thus add greater force to their mufcular actions.

2. The application of the ftimulus, whether that ftimulus be fome quality of external bodies, or pleafure or pain, or defire or averfion, or a link of affociation, excites the correfpondent fenforial power into action, and this caufes the contraction of the fibre. On the contraction of the fibre a part of the fpirit of animation becomes expended, and the fibre ceafes to contract, though the ftimulus continues to be applied; till in a certain time the fibre having received a fupply of fenforial power is ready to contract again, if the ftimulus continues to be applied. If the ftimulus on the contrary be withdrawn, the fame quantity of quiefcent fenforial power becomes refident in the fibre as before its contraction; as appears from the readinefs for action

tion of the large locomotive muscles of the body in a short time after common exertion.

But in those muscular fibres, which are subject to constant stimulus, as the arteries, glands, and capillary vessels, another phenomenon occurs, if their accustomed stimulus be withdrawn; which is, that the sensorial power becomes accumulated in the contractile fibres, owing to the want of its being perpetually expended, or carried away, by their usual unremitted contractions. And on this account those muscular fibres become afterwards excitable into their natural actions by a much weaker stimulus; or into unnatural violence of action by their accustomed stimulus, as is seen in the hot fits of intermittent fevers, which are in consequence of the previous cold ones. Thus the minute vessels of the skin are constantly stimulated by the fluid matter of heat; if the quantity of this stimulus of heat be a while diminished, as in covering the hands with snow, the vessels cease to act, as appears from the paleness of the skin; if this cold application of snow be continued but a short time, the sensorial power, which had habitually been supplied to the fibres, becomes now accumulated in them, owing to the want of its being expended by their accustomed contractions. And thence a less stimulus of heat will now excite them into violent contractions.

If the quiescence of fibres, which had previously been subject to perpetual stimulus, continues a longer time; or their accustomed stimulus be more completely withdrawn; the accumulation of sensorial power becomes still greater, as in those exposed to cold and hunger; pain is produced, and the organ gradually dies from the chemical changes, which take place in it; or it is at a great distance of time restored to action by stimulus applied with great caution in small quantity, as happens to some larger animals and to many insects, which during the winter months lie benumbed with cold, and are said to sleep, and to persons apparently drowned, or apparently frozen to death. Snails have been said to revive by throwing them

into

into water after having been many years fhut up in the cabinets of the curious ; and eggs and feeds in general are reftored to life after many months of torpor by the ftimulus of warmth and moifture.

The inflammation of fchirrous tumours, which have long exifted in a ftate of inaction, is a procefs of this kind ; as well as the fenfibility acquired by inflamed tendons and bones, which had at their formation a fimilar fenfibility, which had fo long lain dormant in their uninflamed ftate.

3. If after long quiefcence from defect of ftimulus the fibres, which had previoufly been habituated to perpetual ftimulus, are again expofed to but their ufual quantity of it ; as in thofe who have fuffered the extremes of cold or hunger ; a violent exertion of the affected organ commences, owing, as above explained, to the great accumulation of fenforial power. This violent exertion not only diminifhes the accumulated fpirit of animation, but at the fame time induces pleafure or pain into the fyftem, which, whether it be fucceeded by inflammation or not, becomes an additional ftimulus, and acting along with the former one, produces ftill greater exertions ; and thus reduces the fenforial power in the contracting fibres beneath its natural quantity.

When the fpirit of animation is thus exhaufted by ufelefs exertions, the organ becomes torpid or unexcitable into action, and a fecond fit of quiefcence fucceeds that of abundant activity. During this fecond fit of quiefcence the fenforial power becomes again accumulated, and another fit of exertion follows in train. Thefe viciffitudes of exertion and inertion of the arterial fyftem conftitute the paroxifms of remittent fevers ; or intermittent ones, when there is an interval of the natural action of the arteries between the exacerbations.

In thefe paroxifms of fevers, which confift of the libration of the arterial fyftem between the extremes of exertion and quiefcence, either the fits become lefs and lefs violent from the contractile fibres becoming

coming lefs excitable to the ftimulus by habit, that is, by becoming
accuftomed to it, as explained below XII. 3. 1. or the whole fenforial power becomes exhaufted, and the arteries ceafe to beat, and the
patient dies in the cold part of the paroxifm. Or fecondly, fo much
pain is introduced into the fyftem by the violent contractions of the
fibres, that inflammation arifes, which prevents future cold fits by
expending a part of the fenforial power in the extenfion of old veffels
or the production of new ones; and thus preventing the too great accumulation or exertion of it in other parts of the fyftem ; or which
by the great increafe of ftimulus excites into great action the whole
glandular fyftem as well as the arterial, and thence a greater quantity
of fenforial power is produced in the brain, and thus its exhauftion in
any peculiar part of the fyftem ceafes to be affected.

4. Or thirdly, in confequence of the painful or pleafurable fenfation above mentioned, defire and averfion are introduced, and inordinate
volition fucceeds; which by its own exertions expends fo much of the
fpirit of animation, that the two other fenforial faculties, or irritation
and fenfation, act fo much feebler ; that the paroxifms of fever, or
that libration between the extremes of exertion and inactivity of the
arterial fyftem, gradually fubfides. On this account a temporary infanity is a favourable fign in fevers, as I have had fome opportunities
of obferving.

III. *Of repeated Stimulus.*

1. When a ftimulus is repeated more frequently than the expenditure of fenforial power can be renewed in the acting organ, the effect
of the ftimulus becomes gradually diminifhed. Thus if two grains of
opium be fwallowed by a perfon unufed to fo ftrong a ftimulus, all
the vafcular fyftems in the body act with greater energy, all the fecretions and the abforption from thofe fecreted fluids are increafed in
quantity

quantity; and pleafure or pain are introduced into the fyftem, which adds an additional ftimulus to that already too great. After fome hours the fenforial power becomes diminifhed in quantity, expended by the great activity of the fyftem; and thence, when the ftimulus of the opium is withdrawn, the fibres will not obey their ufual degree of natural ftimulus, and a confequent torpor or quiefcence fucceeds, as is experienced by drunkards, who on the day after a great excefs of fpirituous potation feel indigeftion, head-ach, and general debility.

In this fit of torpor or quiefcence of a part or of the whole of the fyftem, an accumulation of the fenforial power in the affected fibres is formed, and occafions a fecond paroxyfm of exertion by the application only of the natural ftimulus, and thus a libration of the fenforial exertion between one excefs and the other continues for two or three days, where the ftimulus was violent in degree; and for weeks in fome fevers, from the ftimulus of contagious matter.

But if a fecond dofe of opium be exhibited before the fibres have regained their natural quantity of fenforial power, its effect will be much lefs than the former, becaufe the fpirit of animation or fenforial power is in part exhaufted by the previous excefs of exertion. Hence all medicines repeated too frequently gradually lofe their effect, as opium and wine. Many things of difagreeable tafte at firft ceafe to be difagreeable by frequent repetition, as tobacco; grief and pain gradually diminifh, and at length ceafe altogether, and hence life itfelf becomes tolerable.

Befides the temporary diminution of the fpirit of animation or fenforial power, which is naturally ftationary or refident in every living fibre, by a fingle exhibition of a powerful ftimulus, the contractile fibres themfelves, by the perpetual application of a new quantity of ftimulus, before they have regained their natural quantity of fenforial power, appear to fuffer in their capability of receiving fo much as the natural quantity of fenforial power; and hence a permanent defici-

ency

ency of fpirit of animation takes place, however long the ftimulus may have been withdrawn. On this caufe depends the permanent debility of thofe, who have been addicted to intoxication, the general weaknefs of old age, and the natural debility or inirritability of thofe, who have pale fkins and large pupils of their eyes.

There is a curious phenomenon belongs to this place, which has always appeared difficult of folution ; and that is, that opium or aloes may be exhibited in fmall dofes at firft, and gradually increafed to very large ones without producing ftupor or diarrhœa. In this cafe, though the opium and aloes are given in fuch fmall dofes as not to produce intoxication or catharfis, yet they are exhibited in quantities fufficient in fome degree to exhauft the fenforial power, and hence a ftronger and a ftronger dofe is required; otherwife the medicine would foon ceafe to act at all.

On the contrary, if the opium or aloes be exhibited in a large dofe at firft, fo as to produce intoxication or diarrhœa ; after a few repetitions the quantity of either of them may be diminifhed, and they will ftill produce this effect. For the more powerful ftimulus diffevers the progreffive catenations of animal motions, defcribed in Sect. XVII. and introduces a new link between them ; whence every repetition ftrengthens this new affociation or catenation, and the ftimulus may be gradually decreafed, or be nearly withdrawn, and yet the effect fhall continue ; becaufe the fenforial power of affociation or catenation being united with the ftimulus, increafes in energy with every repetition of the catenated circle ; and it is by thefe means that all the irritative affociations of motions are originally produced.

2. When a ftimulus is repeated at fuch diftant intervals of time, that the natural quantity of fenforial power becomes completely reftored in the acting fibres, it will act with the fame energy as when firft applied. Hence thofe who have lately accuftomed themfelves to large dofes of opium by beginning with fmall ones, and gradually increafing them, and repeating them frequently, as mentioned in the

preceding

preceding paragraph; if they intermit the ufe of it for a few days only, muft begin again with as fmall dofes as they took at firft, other-wife they will experience the inconveniences of intoxication.

On this circumftance depend the conftant unfailing effects of the various kinds of ftimulus, which excite into action all the vafcular fyftems in the body; the arterial, venous, abforbent, and glandular veffels, are brought into perpetual unwearied action by the fluids, which are adapted to ftimulate them; but thefe have the fenforial power of affociation added to that of irritation, and even in fome degree that of fenfation, and even of volition, as will be fpoken of in their places; and life itfelf is thus carried on by the production of fenforial power being equal to its wafte or expenditure in the perpetual movement of the vafcular organization.

3. When a ftimulus is repeated at uniform intervals of time with fuch diftances between them, that the expenditure of fenforial power in the acting fibres becomes completely renewed, the effect is produced with greater facility or energy. For the fenforial power of affociation is combined with the fenforial power of irritation, or, in common language, the acquired habit affifts the power of the ftimulus.

This circumftance not only obtains in the annual and diurnal cate-nations of animal motions explained in Sect. XXXVI. but in every lefs circle of actions or ideas, as in the burthen of a fong, or the ite-rations of a dance; and conftitutes the pleafure we receive from re-petition and imitation; as treated of in Sect. XXII. 2.

4. When a ftimulus has been many times repeated at uniform in-tervals, fo as to produce the complete action of the organ, it may then be gradually diminifhed, or totally withdrawn, and the action of the organ will continue. For the fenforial power of affociation becomes united with that of irritation, and by frequent repetition be-comes at length of fufficient energy to carry on the new link in

the

the circle of actions, without the irritation which at firft introduced it.

Hence, when the bark is given at ftated intervals for the cure of
intermittent fevers, if fixty grains of it be given every three hours for
the twenty-four hours preceding the expected paroxyfm, fo as to ftimulate the defective part of the fyftem into action, and by that means
to prevent the torpor or quiefcence of the fibres, which conftitutes the
cold fit; much lefs than half the quantity, given before the time at
which another paroxyfm of quiefcence would have taken place, will
be fufficient to prevent it; becaufe now the fenforial power, termed
affociation, acts in a twofold manner. Firft, in refpect to the period
of the catenation in which the cold fit was produced, which is now
diffevered by the ftronger ftimulus of the firft dofes of the bark; and,
fecondly, becaufe each dofe of bark being repeated at periodical times,
has its effect increafed by the fenforial faculty of affociation being combined with that of irritation.

Now, when fixty grains of Peruvian bark are taken twice a day,
fuppofe at ten o'clock and at fix, for a fortnight, the irritation excited by this additional ftimulus becomes a part of the diurnal circle
of actions, and will at length carry on the increafed action of the
fyftem without the affiftance of the ftimulus of the bark. On this
theory the bitter medicines, chalybeates, and opiates in appropriated dofes, exhibited for a fortnight, give permanent ftrength to pale
feeble children, and other weak conftitutions.

5. When a defect of ftimulus, as of heat, recurs at certain diurnal
intervals, which induces fome torpor or quiefcence of a part of the
fyftem, the diurnal catenation of actions becomes difordered, and a
new affociation with this link of torpid action is formed; on the next
period the quantity of quiefcence will be increafed, fuppofe the fame
defect of ftimulus to recur, becaufe now the new affociation con-

fpires

fpires with the defective irritation in introducing the torpid action of this part of the diurnal catenation. In this manner many fever-fits commence, where the patient is for fome days indifpofed at certain hours, before the cold paroxifm of fever is completely formed. See Sect. XVII. 3. 3. on Catenation of Animal Motions.

6. If a ftimulus, which' at firft excited the affected organ into fo great exertion as to produce fenfation, be continued for a certain time, it will ceafe to produce fenfation both then and when repeated, though the irritative motions in confequence of it may continue or be re-excited.

Many catenations of irritative motions were at firft fucceeded by fenfation, as the apparent motions of objects when we walk paft them, and probably the vital motions themfelves in the early ftate of our exiftence. But as thofe fenfations were followed by no movements of the fyftem in confequence of them, they gradually ceafed to be produced, not being joined to any fucceeding link of catenation. Hence contagious matter, which has for fome weeks ftimulated the fyftem into great and permanent fenfation, ceafes afterwards to produce general fenfation, or inflammation, though it may ftill induce topical irritations. See Sect. XXXIII. 2. 8. XIX. 10.

Our abforbent fyftem then feems to receive thofe contagious matters, which it has before experienced, in the fame manner as it imbibes common moifture or other fluids; that is, without being thrown into fo violent action as to produce fenfation; the confequence of which is an increafe of daily energy or activity, till inflammation and its confequences fucceed.

7. If a ftimulus excites an organ into fuch violent contractions as to produce fenfation, the motions of which organ had not ufually produced fenfation, this new fenforial power, added to the irritation occafioned by the ftimulus, increafes the activity of the organ. And if this activity be catenated with the diurnal circle of actions, an increafing inflammation is produced; as in the evening paroxyfms of

fmall-ſmall-pox, and other fevers with inflammation. And hence ſchirrous tumours, tendons and membranes, and probably the arteries themſelves become inflamed, when they are ſtrongly ſtimulated.

IV. *Of Stimulus greater than natural.*

1. A quantity of ſtimulus greater than natural, producing an increaſed exertion of ſenſorial power, whether that exertion be in the mode of irritation, ſenſation, volition, or aſſociation, diminiſhes the general quantity of it. This fact is obſervable in the progreſs of intoxication, as the increaſed quantity or energy of the irritative motions, owing to the ſtimulus of vinous ſpirit, introduces much pleaſurable ſenſation into the ſyſtem, and much exertion of muſcular or ſenſual motions in conſequence of this increaſed ſenſation; the voluntary motions, and even the aſſociate ones, become much impaired or diminiſhed; and delirium and ſtaggering ſucceed. See Sect. XXI. on Drunkenneſs. And hence the great proſtration of the ſtrength of the locomotive muſcles in ſome fevers, is owing to the exhauſtion of ſenſorial power by the increaſed action of the arterial ſyſtem.

In like manner a ſtimulus greater than natural, applied to a part of the ſyſtem, increaſes the exertion of ſenſorial power in that part, and diminiſhes it in ſome other part. As in the commencement of ſcarlet fever, it is uſual to ſee great redneſs and heat on the faces and breaſts of children, while at the ſame time their feet are colder than natural; partial heats are obſervable in other fevers with debility, and are generally attended with torpor or quieſcence of ſome other part of the ſyſtem. But theſe partial exertions of ſenſorial power are ſometimes attended with increaſed partial exertions in other parts of the ſyſtem, which ſympathize with them, as the fluſhing of the face after a full meal. Both theſe therefore are to be aſcribed to ſympathetic aſſociations,

ations, explained in Sect. XXXV. and not to general exhauftion or accumulation of fenforial power.

2. A quantity of ftimulus greater than natural, producing an increafed exertion of fenforial power in any particular organ, diminifhes the quantity of it in that organ. This appears from the contractions of animal fibres being not fo eafily excited by a lefs ftimulus after the organ has been fubjected to a greater. Thus after looking at any luminous object of a fmall fize, as at the fetting fun, for a fhort time, fo as not much to fatigue the eye, this part of the retina becomes lefs fenfible to fmaller quantities of light; hence when the eyes are turned on other lefs luminous parts of the fky, a dark fpot is feen refembling the fhape of the fun, or other luminous object which we laft behold. See Sect. XL. No. 2.

Thus we are fome time before we can diftinguifh objects in an obfcure room after coming from bright day-light, though the iris prefently contracts itfelf. We are not able to hear weak founds after loud ones. And the ftomachs of thofe who have been much habituated to the ftronger ftimulus of fermented or fpirituous liquors, are not excited into due action by weaker ones.

3. A quantity of ftimulus fomething greater than the laft mentioned, or longer continued, induces the organ into fpafmodic action, which ceafes and recurs alternately. Thus on looking for a time on the fetting fun, fo as not greatly to fatigue the fight, a yellow fpectrum is feen when the eyes are clofed and covered, which continues for a time, and then difappears and recurs repeatedly before it entirely vanifhes. See Sect. XL. No. 5. Thus the action of vomiting ceafes and is renewed by intervals, although the emetic drug is thrown up with the firft effort. A tenefmus continues by intervals fome time after the exclufion of acrid excrement; and the pulfations of the heart of a viper are faid to continue fome time after it is cleared from its blood.

In thefe cafes the violent contractions of the fibres produce .pain according to law 4; and this pain conftitutes an additional kind or quantity of excitement, which again induces the fibres into contraction, and which painful excitement is again renewed, and again induces contractions of the fibres with gradually diminifhing effect.

4. A quantity of ftimulus greater than that laft mentioned, or longer continued, induces the antagonift mufcles into fpafmodic action. This is beautifully illuftrated by the ocular fpectra defcribed in Sect. XL. No. 6. to which the reader is referred. From thofe experiments there is reafon to conclude that the fatigued part of the retina throws itfelf into a contrary mode of action like ofcitation or pandiculation, as foon as the ftimulus, which has fatigued it, is withdrawn; but that it ftill remains liable to be excited into action by any other colours except the colour with which it has been fatigued. Thus the yawning and ftretching the limbs after a continued action or attitude feems occafioned by the antagonift mufcles being ftimulated by their extenfion during the contractions of thofe in action, or in the fituation in which that action laft left them.

5. A quantity of ftimulus greater than the laft, or longer continued, induces variety of convulfions or fixed fpafms either of the affected organ or of the moving fibres in the other parts of the body. In refpect to the fpectra in the eye, this is well illuftrated in No. 7 and 8, of Sect. XL. EpileCtic convulfions, as the emprofthotonos and opifthotonos, with the cramp of the calf of the leg, locked jaw, and other cataleptic fits, appear to originate from pain, as fome of thefe patients fcream aloud before the convulfion takes place; which feems at firft to be an effort to relieve painful fenfation, and afterwards an effort to prevent it.

In thefe cafes the violent contractions of the fibres produce fo much pain, as to conftitute a perpetual excitement; and that in fo great a degree as to allow but fmall intervals of relaxation of the con-

tracting

tracting fibres as in convulfions, or no intervals at all as in fixed fpafms.

6. A quantity of ftimulus greater than the laft, or longer con-tinued, produces a paralyfis of the organ. In many cafes this para-lyfis is only a temporary effect, as on looking long on a fmall area of bright red filk placed on a fheet of white paper on the floor in a ftrong light, the red filk gradually becomes paler, and at length difappears; which evinces that a part of the retina, by being violently excited, becomes for a time unaffected by the ftimulus of that colour. Thus cathartic medicines, opiates, poifons, contageous matter, ceafe to in-fluence our fyftem after it has been habituated to the ufe of them, except by the exhibition of increafed quantities of them; our fibres not only become unaffected by ftimuli, by which they have previ-oufly been violently irritated, as by the matter of the fmall-pox or meafles; but they alfo become unaffected by fenfation, where the violent exertions, which difabled them, were in confequence of too great quantity of fenfation. And laftly the fibres, which become difobedient to volition, are probably difabled by their too violent exertions in confequence of too great a quantity of vo-lition.

After every exertion of our fibres a temporary paralyfis fucceeds, whence the intervals of all mufcular contractions, as mentioned in No. 3 and 4 of this Section; the immediate caufe of thefe more per-manent kinds of paralyfis is probably owing in the fame manner to the too great exhauftion of the fpirit of animation in the affected part; fo that a ftronger ftimulus is required, or one of a different kind from that, which occafioned thofe too violent contractions, to again excite the affected organ into activity; and if a ftronger ftimulus could be applied, it muft again induce paralyfis.

For thefe powerful ftimuli excite pain at the fame time, that they produce irritation; and this pain not only excites fibrous motions by its ftimulus, but it alfo produces volition; and thus all thefe ftimuli

N   acting

acting at the fame time, and fometimes with the addition of their af-fociations, produce fo great exertion as to expend the whole of the fenforial power in the affected fibres.

V. *Of Stimulus lefs than natural.*

1. A quantity of ftimulus lefs than natural, producing a decreafed exertion of fenforial power, occafions an accumulation of the general quantity of it.   This circumftance is obfervable in the hæmiplagia, in which the patients are perpetually moving the mufcles, which are unaffected.   On this account we awake with greater vigour after fleep, becaufe during fo many hours, the great ufual expenditure of fenforial power in the performance of voluntary actions, and in the exertions of our organs of fenfe, in confequence of the irritations oc-cafioned by external objects had been fufpended, and a confequent ac-cumulation had taken place.

In like manner the exertion of the fenforial power lefs than natural in one part of the fyftem, is liable to produce an increafe of the ex-ertion of it in fome other part.   Thus by the action of vomiting, in which the natural exertion of the motions of the ftomach are de-ftroyed or diminifhed, an increafed abforption of the pulmonary and cellular lymphatics is produced, as is known by the increafed abforp-tion of the fluid depofited in them in dropfical cafes.   But thefe par-tial quiefcences of fenforial power are alfo fometimes attended with other partial quiefcences, which fympathize with them, as cold and pale extremities from hunger.   Thefe therefore are to be afcribed to the affociations of fympathy explained in Sect. XXXV. and not to the general accumulation of fenforial power.

2. A quantity of ftimulus lefs than natural, applied to fibres pre-vioufly accuftomed to perpetual ftimulus, is fucceeded by accumula-

tion

tion of fenforial power in the affected organ.   The truth of this pro-
pofition is evinced, becaufe a ftimulus lefs than natural, if it be
fomewhat greater than that above mentioned, will excite the organ
fo circumftanced into violent activity.   Thus on a frofty day with
wind, the face of a perfon expofed to the wind is at firft pale and
fhrunk; but on turning the face from the wind, it becomes foon of
a glow with warmth and flufhing.   The glow of the fkin in emerging from the cold-bath is owing to the fame caufe.

It does not appear, that an accumulation of fenforial power above
the natural quantity is acquired by thofe mufcles, which are not fubject to perpetual ftimulus, as the locomotive mufcles: thefe, after the
greateft fatigue, only acquire by reft their ufual aptitude to motion;
whereas the vafcular fyftem, as the heart and arteries, after a fhort
quiefcence, are thrown into violent action by their natural quantity of
ftimulus.

Neverthelefs by this accumulation of fenforial power during the
application of decreafed ftimulus, and by the exhauftion of it during
the action of increafed ftimulus, it is wifely provided, that the actions of the vafcular mufcles and organs of fenfe are not much de-
ranged by fmall variations of ftimulus; as the quantity of fenforial
power becomes in fome meafure inverfely as the quantity of ftimulus.

3. A quantity of ftimulus lefs than that mentioned above, and
continued for fome time, induces pain in the affected organ, as the
pain of cold in the hands, when they are immerfed in fnow, is owing
to a deficiency of the ftimulation of heat.   Hunger is a pain from the
deficiency of the ftimulation of food.   Pain in the back at the commencement of ague-fits, and the head-achs which attend feeble people, are pains from defect of ftimulus, and are hence relieved by opium, effential oils, fpirit of wine.

As the pains, which originate from defect of ftimulus, only occur
in thofe parts of the fyftem, which have been previoufly fubjected to

N 2   perpetual

perpetual ſtimulus; and as an accumulation of ſenſorial power is produced in the quieſcent organ along with the pain, as in cold or
hunger, there is reaſon to believe, that the pain is owing to the accumulation of ſenſorial power. For, in the locomotive muſcles, in
the retina of the eye, and other organs of ſenſes, no pain occurs from
the abſence of ſtimulus, nor any great accumulation of ſenſorial
power beyond their natural quantity, ſince theſe organs have not
been uſed to a perpetual ſupply of it. There is indeed a greater accumulation occurs in the organ of viſion after its quieſcence, becauſe
it is ſubject to more conſtant ſtimulus.

4. A certain quantity of ſtimulus leſs than natural induces the
moving organ into feebler and more frequent contractions, as mentioned in No. I. 4. of this Section. For each contraction moving
through a leſs ſpace, or with leſs force, that is, with leſs expenditure of the ſpirit of animation, is ſooner relaxed, and the ſpirit of animation derived at each interval into the acting fibres being leſs, theſe
intervals likewiſe become ſhorter. Hence the tremours of the hands
of people accuſtomed to vinous ſpirit, till they take their uſual ſtimulus; hence the quick pulſe in fevers attended with debility, which
is greater than in fevers attended with ſtrength; in the latter the
pulſe ſeldom beats above 120 times in a minute, in the former it frequently exceeds 140.

It muſt be obſerved, that in this and the two following articles the
decreaſed action of the ſyſtem is probably more frequently occaſioned
by deficiency in the quantity of ſenſorial power, than in the quantity
of ſtimulus. Thus thoſe feeble conſtitutions which have large pupils
of their eyes, and all who labour under nervous fevers, ſeem to owe
their want of natural quantity of activity in the ſyſtem to the deficiency of ſenſorial power; ſince, as far as can be ſeen, they frequently
poſſeſs the natural quantity of ſtimulus.

5. A certain quantity of ſtimulus, leſs than that above mentioned,
inverts the order of ſucceſſive fibrous contractions; as in vomiting
the

the vermicular motions of the ftomach and duodenum are inverted, and their contents ejected, which is probably owing to the exhauftion of the fpirit of animation in the acting mufcles by a previous exceffive ftimulus, as by the root of ipecacuanha, and the confequent defect of fenforial power. The fame retrograde motions affect the whole inteftinal canal in ileus; and the œfophagus in globus hyftericus. See this further explained in Sect. XXIX. No. 11. on Retrograde Motions.

I muft obferve, alfo, that fomething fimilar happens in the production of our ideas, or fenfual motions, when they are too weakly excited; when any one is thinking intenfely about one thing, and carelefsly converfing about another, he is liable to ufe the word of a contrary meaning to that which he defigned, as cold weather for hot weather, fummer for winter.

6. A certain quantity of ftimulus, lefs than that above mentioned, is fucceeded by paralyfis, firft of the voluntary and fenfitive motions, and afterwards of thofe of irritation and of affociation, which conftitutes death.

VI. *Cure of increafed Exertion.*

1. The cure, which nature has provided for the increafed exertion of any part of the fyftem, confifts in the confequent expenditure of the fenforial power. But as a greater torpor follows this exhauftion of fenforial power, as explained in the next paragraph, and a greater exertion fucceeds this torpor, the conftitution frequently finks under thefe increafing librations between exertion and quiefcence; till at length complete quiefcence, that is, death, clofes the fcene.

For, during the great exertion of the fyftem in the hot fit of fever, an increafe of ftimulus is produced from the greater momentum of

the

the blood, the greater diftention of the heart and arteries, and the in-creafed production of heat, by the violent actions of the fyftem oc-cafioned by this augmentation of ftimulus, the fenforial power be-comes diminifhed in a few hours much beneath its natural quantity, the veffels at length ceafe to obey even thefe great degrees of ftimulus, as fhewn in Sect. XL. 9. 1. and a torpor of the whole or of a part of the fyftem enfues.

Now as this fecond cold fit commences with a greater deficiency of fenforial power, it is alfo attended with a greater deficiency of ftimu-lus than in the preceding cold fit, that is, with lefs momentum of blood, lefs diftention of the heart. On this account the fecond cold fit becomes more violent and of longer duration than the firft; and as a greater accumulation of fenforial power muft be produced before the fyftem of veffels will again obey the diminifhed ftimulus, it follows, that the fecond hot fit of fever will be more violent than the former one. And that unlefs fome other caufes counteract either the violent exertions in the hot fit, or the great torpor in the cold fit, life will at length be extinguifhed by the expenditure of the whole of the fenforial power. And from hence it appears, that the true means of curing fevers muft be fuch as decreafe the action of the fyftem in the hot fit, and increafe it in the cold fit; that is, fuch as prevent the too great diminution of fenforial power in the hot fit, and the too great accumulation of it in the cold one.

2. Where the exertion of the fenforial powers is much increafed, as in the hot-fits of fever or inflammation, the following are the ufual means of relieving it. Decreafe the irritations by blood-letting, and other evacuations; by cold water taken into the ftomach, or injected as an enema, or ufed externally; by cold air breathed into the lungs, and diffufed over the fkin; with food of lefs ftimulus than the patient has been accuftomed to.

3. As a cold fit, or paroxyfm of inactivity of fome parts of the fyftem, generally precedes the hot fit, or paroxyfm of exertion, by

which

which the fenforial power becomes accumulated, this cold paroxyfm fhould be prevented by ftimulant medicines and diet, as wine, opium, bark, warmth, cheerfulnefs, anger, furprife.

4. Excite into greater action fome other part of the fyftem, by which means the fpirit of animation may be in part expended, and thence the inordinate actions of the difeafed part may be leffened. Hence when a part of the fkin acts violently, as of the face in the eruption of the fmall-pox, if the feet be cold they fhould be covered. Hence the ufe of a blifter applied near a topical inflammation. Hence opium and warm bath relieve pains both from excefs and defect of ftimulus.

5. Firft increafe the general ftimulation above its natural quantity, which may in fome degree exhauft the fpirit of animation, and then decreafe the ftimulation beneath its natural quantity. Hence after fudorific medicines and warm air, the application of refrigerants may have greater effect, if they could be adminiftered without danger of producing too great torpor of fome part of the fyftem; as frequently happens to people in health from coming out of a warm room into the cold air, by which a topical inflammation in confequence of torpor of the mucous membrane of the noftril is produced, and is termed a cold in the head.

VII. Cure of decreafed Exertion.

1. Wʜᴇʀᴇ the exertion of the fenforial powers is much decreafed, as in the cold fits of fever, a gradual accumulation of the fpirit of animation takes place; as occurs in all cafes where inactivity or torpor of a part of the fyftem exifts; this accumulation of fenforial power increafes, till ftimuli lefs than natural are fufficient to throw it

into

into action, then the cold fit ceafes; and from the action of the natural ftimuli a hot one fucceeds with increafed activity of the whole fyftem.

So in fainting fits, or fyncope, there is a temporary deficiency of fenforial exertion, and a confequent quiefcence of a great part of the fyftem. This quiefcence continues, till the fenforial power becomes again accumulated in the torpid organs; and then the ufual diurnal ftimuli excite the revivefcent parts again into action.; but as this kind of quiefcence continues but a fhort time compared to the cold paroxyfm of an ague, and lefs affects the circulatory fyftem, a lefs fuperabundancy of exertion fucceeds in the organs previoufly torpid, and a lefs excefs of arterial activity. See Sect. XXXIV. 1. 6.

2. In the difeafes occafioned by a defect of fenforial exertion, as in cold fits of ague, hyfteric complaint, and nervous fever, the following means are thofe commonly ufed. 1. Increafe the ftimulation above its natural quantity for fome weeks, till a new habit of more energetic contraction of the fibres is eftablifhed. This is to be done by wine, opium, bark, fteel, given at exact periods, and in appropriate quantities; for if thefe medicines be given in fuch quantity, as to induce the leaft degree of intoxication, a debility fucceeds from the ufelefs exhauftion of fpirit of animation in confequence of too great exertion of the mufcles or organs of fenfe. To thefe irritative ftimuli fhould be added the fenfitive ones of cheerful ideas, hope, affection.

3. Change the kinds of ftimulus. The habits acquired by the conftitution depend on fuch nice circumftances, that when one kind of ftimulus ceafes to excite the fenforial power into the quantity of exertion neceffary to health, it is often fufficient to change the ftimulus for another apparently fimilar in quantity and quality. Thus when wine ceafes to ftimulate the conftitution, opium in appropriate dofes fupplies the defect; and the contrary. This is alfo obferved in the

the effects of cathartic medicines, when one lofes its power, another, apparently lefs efficacious, will fucceed. Hence a change of diet, drink, and ftimulating medicines, is often advantageous in difeafes of debility.

4. Stimulate the organs, whofe motions are affociated with the torpid parts of the fyftem. The actions of the minute veffels of the various parts of the external fkin are not only affociated with each other, but are ftrongly affociated with thofe of fome of the internal membranes, and particularly of the ftomach. Hence when the exertion of the ftomach is lefs than natural, and indigeftion and heartburn fucceed, nothing fo certainly removes thefe fymptoms as the ftimulus of a blifter on the back. The coldnefs of the extremities, as of the nofe, ears, or fingers, are hence the beft indication for the fuccefsful application of blifters.

5. Decreafe the ftimulus for a time. By leffening the quantity of heat for a minute or two by going into the cold bath, a great accumulation of fenforial power is produced; for not only the minute veffels of the whole external fkin for a time become inactive, as appears by their palenefs; but the minute veffels of the lungs lofe much of their activity alfo by concert with thofe of the fkin, as appears from the difficulty of breathing at firft going into cold water. On emerging from the bath the fenforial power is thrown into great exertion by the ftimulus of the common degree of the warmth of the atmofphere, and a great production of animal heat is the confequence. The longer a perfon continues in the cold bath the greater muft be the prefent inertion of a great part of the fyftem, and in confequence a greater accumulation of fenforial power. Whence M. Pomè recommends fome melancholy patients to be kept from two to fix hours in fpring-water, and in baths ftill colder.

6 Decreafe the ftimulus for a time below the natural, and then increafe it above natural. The effect of this procefs, improperly ufed, is feen in giving much food, or applying much warmth, to thofe

who

who have been previoufly expofed to great hunger, or to great cold. The accumulated fenforial power is thrown into fo violent exertion, that inflammations and mortifications fupervene, and death clofes the cataftrophe. In many difeafes this method is the moft fuccefsful; hence the bark in agues produces more certain effect after the previous exhibition of emetics. In difeafes attended with violent pain, opium has double the effect, if venefection and a cathartic have been previoufly ufed. On this feems to have been founded the fuccefsful practice of Sydenham, who ufed venefection and a cathartic in chlorofis before the exhibition of the bark, fteel, and opiates.

7. Prevent any unneceffary expenditure of fenforial power. Hence in fevers with debility, a decumbent pofture is preferred, with filence, little light, and fuch a quantity of heat as may prevent any chill fenfation, or any coldnefs of the extremities. The pulfe of patients in fevers with debility increafes in frequency above ten pulfations in a minute on their rifing out of bed. For the expenditure of fenforial power to preferve an erect pofture of the body adds to the general deficiency of it, and thus affects the circulation.

8. The longer in time and the greater in degree the quiefcence or inertion of an organ has been, fo that it ftill retains life or excitability, the lefs ftimulus fhould at firft be applied to it. The quantity of ftimulation is a matter of great nicety to determine, where the torpor or quiefcence of the fibres has been experienced in a great degree, or for a confiderable time, as in cold-fits of the ague, in continued fevers with great debility, or in people famifhed at fea, or perifhing with cold. In the two laft cafes, very minute quantities of food fhould be firft fupplied, and very few additional degrees of heat. In the two former cafes, but little ftimulus of wine or medicine, above what they had been lately accuftomed to, fhould be exhibited, and this at frequent and ftated intervals, fo that the effect of one quantity may be obferved before the exhibition of another.

If thefe circumftances are not attended to, as the fenforial power

5

becomes

becomes accumulated in the quiefcent fibres, an inordinate exertion takes place by the increafe of ftimulus acting on the accumulated quantity of fenforial power, and either the paralyfis, or death of the contractile fibres enfues, from the total expenditure of the fenforial power in the affected organ, owing to this increafe of exertion, like the debility after intoxication.   Or, fecondly, the violent exertions above mentioned produce painful fenfation, which becomes a new ftimulus, and by thus producing inflammation, and increafing the activity of the fibres already too great, fooner exhaufts the whole of the fenforial power in the acting organ, and mortification, that is, the death of the part, fupervenes.

Hence there have been many inftances of people, whofe limbs have been long benumbed by expofure to cold, who have loft them by mortification on their being too haftily brought to the fire; and of others, who were nearly famifhed at fea, who have died foon after having taken not more than an ufual meal of food.   I have heard of two well-attefted inftances of patients in the cold fit of ague, who have died from the exhibition of gin and vinegar, by the inflammation which enfued.   And in many fevers attended with debility, the un-limited ufe of wine, and the wanton application of blifters, I believe, has deftroyed numbers by the debility confequent to too great ftimulation, that is, by the exhauftion of the fenforial power by its in-ordinate exertion.

Wherever the leaft degree of intoxication exifts, a proportional de-bility is the confequence; but there is a golden rule by which the neceffary and ufeful quantity of ftimulus in fevers with debility may be afcertained.   When wine or beer are exhibited either alone or di-luted with water, if the pulfe becomes flower the ftimulus is of a proper quantity; and fhould be repeated every two or three hours, or when the pulfe again becomes quicker.

In the chronical debility brought on by drinking fpirituous or fer-mented liquors, there is another golden rule by which I have fuccefs-

fully

fully directed the quantity of spirit which they may safely leffen, for there is no other means by which they can recover their health. It should be premifed, that where the power of digeftion in thefe patients is totally deftroyed, there is not much reafon to expect a return to healthful vigour.

I have directed feveral of thefe patients to omit one fourth part of the quantity of vinous fpirit they have been lately accuftomed to, and if in a fortnight their appetite increafes, they are advifed to omit another fourth part; but if they perceive that their digeftion becomes impaired from the want of this quantity of fpirituous potation, they are advifed to continue as they are, and rather bear the ills they have, than rifk the encounter of greater. At the fame time flefh-meat with or without fpice is recommended, with Peruvian bark and fteel in fmall quantities between their meals, and half a grain of opium or a grain, with five or eight grains of rhubarb at night.

SECT.

SECT. XIII.

OF VEGETABLE ANIMATION.

I. 1. *Vegetables are irritable, mimofa, dionæa mufcipula. Vegetable fecretions.*
2. Vegetable buds are inferior animals, are liable to greater or lefs irritability.
II. *Stamens and piftils of plants fhew marks of fenfibility.*   III. *Vegetables pof-*
fefs fome degree of volition.   IV. *Motions of plants are affociated like thofe of*
animals.   V. 1. *Vegetable ftructure like that of animals, their anthers and ftig-*
mas are living creatures. Male-flowers of Vallifneria.   2. *Whether vegetables*
poffefs ideas? They have organs of fenfe as of touch and fmell, and ideas of exter-
nal things?

I. 1. THE fibres of the vegetable world, as well as thofe of the
animal, are excitable into a variety of motion by the irritations of external objects.   This appears particularly in the mimofa or fenfitive
plant, whofe leaves contract on the flighteft injury ; the dionæa mufcipula, which was lately brought over from the marfhes of America,
prefents us with another curious inftance of vegetable irritability ; its
leaves are armed with fpines on their upper edge, and are fpread on
the ground around the ftem ; when an infect creeps on any of them
in its paffage to the flower or feed, the leaf fhuts up like a fteel rat-
trap, and deftroys its enemy.   See Botanic Garden, Part II. note on
Silene.

The various fecretions of vegetables, as of odour, fruit, gum,
refin, wax, honey, feem brought about in the fame manner as in the
glands of animals : the taftelefs moifture of the earth is converted by
the hop-plant into a bitter juice ; as by the caterpillar in the nut-

fhell

shell the sweet kernel is converted into a bitter powder. While the power of absorption in the roots and barks of vegetables is excited into action by the fluids applied to their mouths like the lacteals and lymphatics of animals.

2. The individuals of the vegetable world may be considered as inferior or less perfect animals; a tree is a congeries of many living buds, and in this respect resembles the branches of coralline, which are a congeries of a multitude of animals. Each of these buds of a tree has its proper leaves or petals for lungs, produces its viviparous or its oviparous offspring in buds or seeds; has its own roots, which extending down the stem of the tree are interwoven with the roots of the other buds, and form the bark, which is the only living part of the stem, is annually renewed, and is superinduced upon the former bark, which then dies, and with its stagnated juices gradually hardening into wood forms the concentric circles, which we see in blocks of timber.

The following circumstances evince the individuality of the buds of trees. First, there are many trees, whose whole internal wood is perished, and yet the branches are vegete and healthy. Secondly, the fibres of the barks of trees are chiefly longitudinal, resembling roots, as is beautifully seen in those prepared barks, that were lately brought from Otaheita. Thirdly, in horizontal wounds of the bark of trees, the fibres of the upper lip are always elongated downwards like roots, but those of the lower lip do not approach to meet them. Fourthly, if you wrap wet moss round any joint of a vine, or cover it with moist earth, roots will shoot out from it. Fifthly, by the inoculation or engrafting of trees many fruits are produced from one stem. Sixthly, a new tree is produced from a branch plucked from an old one, and set in the ground. Whence it appears that the buds of deciduous trees are so many annual plants, that the bark is a contexture of the roots of each individual bud; and that the internal wood

is

is of no other ufe but to fupport them in the air, and that thus they refemble the animal world in their individuality.

The irritability of plants, like that of animals, appears liable to be increafed or decreafed by habit; for thofe trees or fhrubs, which are brought from a colder climate to a warmer, put out their leaves and bloffoms a fortnight fooner than the indigenous ones.

Profeffor Kalm, in his Travels in New York, obferves that the apple-trees brought from England bloffom a fortnight fooner than the native ones. In our country the fhrubs, that are brought a degree or two from the north, are obferved to flourifh better than thofe, which come from the fouth. The Siberian barley and cabbage are faid to grow larger in this climate than the fimilar more fouthern vegetables. And our hoards of roots, as of potatoes and onions, geminate with lefs heat in fpring, after they have been accuftomed to the winter's cold, than in autumn after the fummer's heat.

II. The ftamens and piftils of flowers fhew evident marks of fenfibility, not only from many of the ftamens and fome piftils approaching towards each other at the feafon of impregnation, but from many of them clofing their petals and calyxes during the cold parts of the day. For this cannot be afcribed to irritation, becaufe cold means a defect of the ftimulus of heat; but as the want of accuftomed ftimuli produces pain, as in coldnefs, hunger, and thirft of animals, thefe motions of vegetables in clofing up their flowers muft be afcribed to the difagreeable fenfation, and not to the irritation of cold. Others clofe up their leaves during darknefs, which, like the former, cannot be owing to irritation, as the irritating material is withdrawn.

The approach of the anthers in many flowers to the ftigmas, and of the piftils of fome flowers to the anthers, muft be afcribed to the paffion of love, and hence belongs to fenfation, not to irritation.

III. That the vegetable world poffeffes fome degree of voluntary powers, appears from their neceffity to fleep, which we have fhewn in Sect. XVIII. to confift in the temporary abolition of voluntary

power.

power. This voluntary power feems to be exerted in the circular movement of the tendrils of vines, and other climbing vegetables; or in the efforts to turn the upper furface of their leaves, or their flowers to the light.

IV. The affociations of fibrous motions are obfervable in the vegetable world, as well as in the animal. The divifions of the leaves of the fenfitive plant have been accuftomed to contract at the fame time from the abfence of light; hence if by any other circumftance, as a flight ftroke or injury, one divifion is irritated into contraction, the neighbouring ones contract alfo, from their motions being affociated with thofe of the irritated part. So the various ftamina of the clafs of fyngenefia have been accuftomed to contract together in the evening, and thence if you ftimulate one of them with a pin, according to the experiment of M. Colvolo, they all contract from their acquired affociations.

To evince that the collapfing of the fenfitive plant is not owing to any mechanical vibrations propagated along the whole branch, when a fingle leaf is ftruck with the finger, a leaf of it was flit with fharp fciffors, and fome feconds of time paffed before the plant feemed fenfible of the injury; and then the whole branch collapfed as far as the principal ftem: this experiment was repeated feveral times with the leaft poffible impulfe to the plant.

V. 1. For the numerous circumftances in which vegetable buds are analogous to animals, the reader is referred to the additional notes at the end of the Botanic Garden, Part 1. It is there fhewn, that the roots of vegetables refemble the lacteal fyftem of animals; the fap-veffels in the early fpring, before their leaves expand, are analogous to the placental veffels of the foetus; that the leaves of landplants refemble lungs, and thofe of aquatic plants the gills of fifh; that there are other fyftems of veffels refembling the vena portarum of quadrupeds, or the aorta of fifh; that the digeftive power of vegetables is fimilar to that of animals converting the fluids, which they

abforb,

abforb, into fugar ; that their feeds refemble the eggs of animals, and their buds and bulbs their viviparous offspring. And, laftly, that the anthers and ftigmas are real animals, attached indeed to their parent tree like polypi or coral infects, but capable of fpontaneous motion ; that they are affected with the paffion of love, and furnifhed with powers of reproducing their fpecies, and are fed with honey like the moths and butterflies, which plunder their nectaries. See Botanic Garden, Part I. add. note XXXIX.

The male flowers of vallifneria approach ftill nearer to apparent animality, as they detach themfelves from the parent plant, and float on the furface of the water to the female ones. Botanic Garden, Part II. Art. Vallifneria. Other flowers of the claffes of monecia and diecia, and polygamia, difcharge the fecundating farina, which floating in the air is carried to the ftigma of the female flowers, and that at confiderable diftances. Can this be affected by any fpecific attraction ? or, like the diffufion of the odorous particles of flowers, is it left to the currents of winds, and the accidental mifcarriages of it counteracted by the quantity of its production ?

2. This leads us to a curious enquiry, whether vegetables have ideas of external things ? As all our ideas are originally received by our fenfes, the queftion may be changed to, whether vegetables poffefs any organs of fenfe ? Certain it is, that they poffefs a fenfe of heat and cold, another of moifture and drynefs, and another of light and darknefs ; for they clofe their petals occafionally from the prefence of cold, moifture, or darknefs. And it has been already fhewn, that thefe actions cannot be performed fimply from irritation, becaufe cold and darknefs are negative quantities, and on that account fenfation or volition are implied, and in confequence a fenforium or union of their nerves. So when we go into the light, we contract the iris ; not from any ftimulus of the light on the fine mufcles of the iris, but from its motions being affociated with the fenfation of too much light on the retina : which could not take

P

place

place without a fenforium or center of union of the nerves of the
iris with thofe of vifion.   See Botanic Garden, Part I. Canto 3.
l. 440. note.

Befides thefe organs of fenfe, which diftinguifh cold, moifture,
and darknefs, the leaves of mimofa, and of dionæa, and of drofera,
and the ftamens of many flowers, as of the berbery, and the numerous
clafs of fyngenefia, are fenfible to mechanic impact, that is, they
poffefs a fenfe of touch, as well as a common fenforium; by the
medium of which their mufcles are excited into action. Laftly, in
many flowers the anthers, when mature, approach the ftigma, in
others the female organ approaches to the male.   In a plant of collin-
fonia, a branch of which is now before me, the two yellow ftamens
are about three eights of an inch high, and diverge from each other,
at an angle of about fifteen degrees, the purple ftyle is half an inch,
high, and in fome flowers is now applied to the ftamen on the right
hand, and in others to that of the left; and will, I fuppofe, change
place to-morrow in thofe, where the anthers have not yet effufed their
powder.

I afk, by what means are the anthers in many flowers, and ftigmas
in other flowers, directed to find their paramours? How do either of
them know, that the other exifts in their vicinity? Is this curious kind
of ftorge produced by mechanic attraction, or by the fenfation of love?
The latter opinion is fupported by the ftrongeft analogy, becaufe a
reproduction of the fpecies is the confequence; and then another organ
of fenfe muft be wanted to direct thefe vegetable amourettes to find
each other, one probably analogous to our fenfe of fmell, which in
the animal world directs the new-born infant to its fource of nourifhment, and they may thus poffefs a faculty of perceiving as well
as of producing odours.

Thus, befides a kind of tafte at the extremities of their roots, fimilar to that of the extremities of our lacteal veffels, for the purpofe of
felecting their proper food; and befides different kinds of irritability
refiding

refiding in the various glands, which feparate honey, wax, refin, and other juices from their blood ; vegetable life feems to poffefs an organ of fenfe to diftinguifh the variations of heat, another to diftinguifh the varying degrees of moifture, another of light, another of touch, and probably another analogous to our fenfe of fmell.   To thefe muft be added the indubitable evidence of their paffion of love, and I think we may truly conclude, that they are furnifhed with a common fenforium belonging to each bud, and that they muft occafionally repeat thofe perceptions either in their dreams or waking hours, and confequently poffefs ideas of fo many of the properties of the external world, and of their own exiftence.

SECT. XIV.

OF THE PRODUCTION OF IDEAS.

I. *Of material and immaterial beings. Doctrine of St. Paul.* II. *1. Of the sense of touch. Of solidity. 2. Of figure. Motion. Time. Place. Space. Number. 3. Of the penetrability of matter. 4. Spirit of animation possesses solidity, figure, visibility, &c. Of spirits and angels. 5. The existence of external things.* III. *Of vision.* IV. *Of hearing.* V. *Of smell and taste.* VI. *Of the organ of sense by which we perceive heat and cold, not by the sense of touch.* VII. *Of the sense of extension, the whole of the locomotive muscles may be considered as one organ of sense.* VIII. *Of the senses of hunger, thirst, want of fresh air, suckling children, and lust.* IX. *Of many other organs of sense belonging to the glands. Of painful sensations from the excess of light, pressure, heat, itching, caustics, and electricity.*

I. PHILOSOPHERS have been much perplexed to understand, in what manner we become acquainted with the external world; insomuch that Dr. Berkly even doubted its existence, from having observed (as he thought) that none of our ideas resemble their correspondent objects. Mr. Hume asserts, that our belief depends on the greater distinctness or energy of our ideas from perception; and Mr. Reid has lately contended, that our belief of external objects is an innate principle necessarily joined with our perceptions.

So true is the observation of the famous Malbranch, " that our senses are not given us to discover the essences of things, but to acquaint us with the means of preserving our existence," (L. I. ch. v.) a melancholy reflection to philosophers!

Some philosophers have divided all created beings into material and

immaterial:

immaterial: the former including all that part of being, which obeys the mechanic laws of action and reaction, but which can begin no motion of itfelf; the other is the caufe of all motion, and is either termed the power of gravity, or of fpecific attraction, or the fpirit of animation. This immaterial agent is fuppofed to exift in or with matter, but to be quite diftinct from it, and to be equally capable of exiftence, after the matter, which now poffeffes it, is decompofed.

Nor is this theory ill fupported by analogy, fince heat, electricity, and magnetifm, can be given to or taken from a piece of iron; and muft therefore exift, whether feparated from the metal, or combined with it. From a parity of reafoning, the fpirit of animation would appear to be capable of exifting as well feparately from the body as with it.

I beg to be underftood, that I do not wifh to difpute about words, and am ready to allow, that the powers of gravity, fpecific attraction, electricity, magnetifm, and even the fpirit of animation, may confift of matter of a finer kind; and to believe, with St. Paul and Malbranch, that the ultimate caufe only of all motion is immaterial, that is God. St. Paul fays, " in him we live and move, and have our being;" and, in the 15th chapter to the Corinthians, diftinguifhes between the pfyche or living fpirit, and the pneuma or reviving fpirit. By the words fpirit of animation or fenforial power, I mean only that animal life, which mankind poffeffes in common with brutes, and in fome degree even with vegetables, and leave the confideration of the immortal part of us, which is the object of religion, to thofe who treat of revelation.

II. 1. *Of the Senfe of touch.*

THE firft ideas we become acquainted with, are thofe of the fenfe of touch; for the foetus muft experience fome varieties of agitation,

tation, and exert fome mufcular action, in the womb; and may with great probability be fuppofed thus to gain fome ideas of its own figure, of that of the uterus, and of the tenacity of the fluid, that furrounds it, (as appears from the facts mentioned in the fucceeding Section upon Inftinct.)

Many of the organs of fenfe are confined to a fmall part of the body, as the noftrils, ear, or eye, whilft the fenfe of touch is diffufed over the whole fkin, but exifts with a more exquifite degree of delicacy at the extremities of the fingers and thumbs, and in the lips. The fenfe of touch is thus very commodioufly difpofed for the purpofe of encompaffing fmaller bodies, and for adapting itfelf to the inequalities of larger ones. The figure of fmall bodies feems to be learnt by children by their lips as much as by their fingers; on which account they put every new object to their mouths, when they are fatiated with food, as well as when they are hungry. And puppies feem to learn their ideas of figure principally by the lips in their mode of play.

We acquire our tangible ideas of objects either by the fimple preffure of this organ of touch againft a folid body, or by moving our organ of touch along the furface of it. In the former cafe we learn the length and breadth of the object by the quantity of our organ of touch, that is impreffed by it: in the latter cafe we learn the length and breadth of objects by the continuance of their preffure on our moving organ of touch.

It is hence, that we are very flow in acquiring our tangible ideas, and very flow in recollecting them; for if I now think of the tangible idea of a cube, that is, if I think of its figure, and of the folidity of every part of that figure, I muft conceive myfelf as paffing my fingers over it, and feem in fome meafure to feel the idea, as I formerly did the impreffion, at the ends of them, and am thus very flow in diftinctly recollecting it.

When a body compreffes any part of our fenfe of touch, what happens?

pens? Firſt, this part of our ſenſorium undergoes a mechanical compreſſion, which is termed a ſtimulus; ſecondly, an idea, or contraction of a part of the organ of ſenſe is excited; thirdly, a motion of the central parts, or of the whole ſenſorium, which is termed ſenſation, is produced; and theſe three conſtitute the perception of ſolidity.

2. *Of Figure, Motion, Time, Place, Space, Number.*

No one will deny, that the medulla of the brain and nerves has a certain figure; which, as it is diffuſed through nearly the whole of the body, muſt have nearly the figure of that body. Now it follows, that the ſpirit of animation, or living principle, as it occupies this medulla, and no other part, (which is evinced by a great variety of cruel experiments on living animals,) it follows, that this ſpirit of animation has alſo the ſame figure as the medulla above deſcribed. I appeal to common ſenſe! the ſpirit of animation acts, Where does it act? It acts wherever there is the medulla above mentioned; and that whether the limb is yet joined to a living animal, or whether it be recently detached from it; as the heart of a viper or frog will renew its contractions, when pricked with a pin, for many minutes of time after its exſection from the body.—Does it act any where elſe? —No; then it certainly exiſts in this part of ſpace, and no where elſe; that is, it hath figure; namely, the figure of the nervous ſyſtem, which is nearly the figure of the body. When the idea of ſolidity is excited, as above explained, a part of the extenſive organ of touch is compreſſed by ſome external body, and this part of the ſenſorium ſo compreſſed exactly reſembles *in figure* the figure of the body that compreſſed it. Hence, when we acquire the idea of ſolidity,

dity, we acquire at the fame time the idea of FIGURE ; and this idea of figure, or motion of *a part* of the organ of touch, exactly refembles *in its figure* .the figure of the body that occafions it ; and thus exactly acquaints us with this property of the external world.

Now, as the whole univerfe with all its parts poffeffes a certain form or figure, if any part of it moves, that form or figure of the whole is varied : hence, as MOTION is no other than a perpetual variation of figure, our idea of motion is alfo a real refemblance of the motion that produced it.

It may be faid in objection to this definition of motion, that an ivory globe may revolve on its axis, and that here will be a motion without change of figure. But the figure of the particle x on one fide of this globe is not the *fame* figure as the figure of y on the other fide, any more than the particles themfelves are the fame, though they are *fimilar* figures ; and hence they cannot change place with each other without difturbing or changing the figure of the whole.

Our idea of TIME is from the fame fource, but is more abftracted, as it includes only the comparative velocities of thefe variations of figure ; hence if it be afked, How long was this book in printing ? it may be anfwered, Whilft the fun was paffing through Aries.

Our idea of PLACE includes only the figure of a group of bodies, not the figures of the bodies themfelves. If it be afked where is Nottinghamfhire, the anfwer is, it is furrounded by Derbyfhire, Lincolnfhire, and Leicefterfhire ; hence place is our idea of the figure of one body.furrounded by the figures of other bodies.

The idea of SPACE is a more abftracted idea of place excluding the group of bodies.

The idea of NUMBER includes only the particular arrangements, or diftributions of a group of bodies, and is therefore only a more ab-

ftracted

ftracted idea of the parts of the figure of the group of bodies; thus when I fay England is divided into forty counties, I only fpeak of certain divifions of its figure.

Hence arifes the certainty of the mathematical fciences, as they explain thefe properties of bodies, which are exactly refembled by our ideas of them, whilft we are obliged to collect almoft all our other knowledge from experiment; that is, by obferving the effects exerted by one body upon another.

3. *Of the Penetrability of Matter.*

The impoffibility of two bodies exifting together in the fame fpace cannot be deduced from our idea of folidity, or of figure. As foon as we perceive the motions of objects that furround us, and learn that we poffefs a power to move our own bodies, we experience, that thofe objects, which excite in us the idea of folidity and of figure, oppofe this voluntary movement of our own organs; as whilft I endeavour to comprefs between my hands an ivory ball into a fpheroid. And we are hence taught by experience, that our own body and thofe, which we touch, cannot exift in the fame part of fpace.

But this by no means demonftrates, that no two bodies can exift together in the fame part of fpace. Galilæo in the preface to his works feems to be of opinion, that matter is not impenetrable; Mr. Michel, and Mr. Bofcowich in his Theoria. Philof. Natur. have efpoufed this hypothefis: which has been lately publifhed by Dr. Prieftley, to whom the world is much indebted for fo many important difcoveries in fcience. (Hift. of Light and Colours, p. 391.) The uninterrupted paffage of light through tranfparent bodies, of the electric æther through metallic and aqueous bodies, and of the magnetic effluvia through all bodies, would feem to give fome probability to this

Q   opinion.

opinion. Hence it appears, that beings may exift without poffeffing the property of folidity, as well as they can exift without poffeffing the properties, which excite our fmell or tafte, and can thence occupy fpace without detruding other bodies from it; but we cannot become acquainted with fuch beings by our fenfe of touch, any more than we can with odours or flavours without our fenfes of fmell and tafte.

But that any being can exift without exifting in fpace, is to my ideas utterly incomprehenfible. My appeal is to common fenfe. *To be* implies a when and a where; the one is comparing it with the motions of other beings, and the other with their fituations.

If there was but one object, as the whole creation may be confidered as one object, then I cannot afk where it exifts? for there are no other objects to compare its fituation with. Hence if any one denies, that a being exifts in fpace, he denies, that there are any other beings but that one; for to anfwer the queftion, " Where does it exift?" is only to mention the fituation of the objects that furround it.

In the fame manner if it be afked—" When does a being exift?" The anfwer only fpecifies the fucceffive motions either of itfelf, or of other bodies; hence to fay, a body exifts not in time, is to fay, that there is, or was, no motion in the world.

4. *Of the Spirit of Animation.*

But though there may exift beings in the univerfe, that have not the property of folidity; that is, which can poffefs any part of fpace, at the fame time that it is occupied by other bodies; yet there may be other beings, that can affume this property of folidity, or difrobe themfelves of it occafionally, as we are taught of fpirits, and of angels;

gels; and it would feem, that THE SPIRIT OF ANIMATION muft be endued with this property, otherwife how could it occafionally give motion to the limbs of animals?—or be itfelf ftimulated into motion by the obtrufions of furrounding bodies, as of light, or odour?

If the fpirit of animation was always neceffarily penetrable, it could not influence or be influenced by the folidity of common matter; they would exift together, but could not detrude each other from the part of fpace, where they exift; that is, they could not communicate motion to each other. *No two things can influence or affect each other, which have not fome property common to both of them*; for to influence or affect another body is to give or communicate fome property to it, that it had not before; but how can one body give that to another, which it does not poffefs itfelf?—The words imply, that they muft agree in having the power or faculty of poffeffing fome common property. Thus if one body removes another from the part of fpace, that it poffeffes, it muft have the power of occupying that fpace itfelf: and if one body communicates heat or motion to another, it follows, that they have alike the property of poffeffing heat or motion.

Hence the fpirit of animation at the time it communicates or receives motion from folid bodies, muft itfelf poffefs fome property of folidity. And in confequence at the time it receives other kinds of motion from light, it muft poffefs that property, which light poffeffes, to communicate that kind of motion; and for which no language has a name, unlefs it may be termed Vifibility. And at the time it is ftimulated into other kinds of animal motion by the particles of fapid and odorous bodies affecting the fenfes of tafte and fmell, it muft refemble thefe particles of flavour, and of odour, in poffeffing fome fimilar or correfpondent property; and for which language has no name, unlefs we may ufe the words Saporofity and Odorofity for thofe common properties, which are poffeffed by our organs of tafte and fmell, and by the particles of fapid and odorous bodies; as the

words

words Tangibility and Audibility may exprefs the common property poffeffed by our organs of touch, and of hearing, and by the folid bodies, or their vibrations, which affect thofe organs.

5. Finally, though the figures of bodies are in truth refembled by the figure of the part of the organ of touch, which is ftimulated into motion; and that organ refembles the folid body, which ftimulates it, in its property of folidity; and though the fenfe of hearing refembles the vibrations of external bodies in its capability of being ftimulated into motion by thofe vibrations; and though our other organs of fenfe refemble the bodies, that ftimulate them, in their capability of being ftimulated by them; and we hence become acquainted with thefe properties of the external world; yet as we can repeat all thefe motions of our organs of fenfe by the efforts of volition, or in confequence of the fenfation of pleafure or pain, or by their affociation with other fibrous motions, as happens in our reveries or in fleep, there would ftill appear to be fome difficulty in demonftrating the exiftence of any thing external to us.

In our dreams we cannot determine this circumftance, becaufe our power of volition is fufpended, and the ftimuli of external objects are excluded; but in our waking hours we can compare our ideas belonging to one fenfe with thofe belonging to another, and can thus diftinguifh the ideas occafioned by irritation from thofe excited by fenfation, volition, or affociation. Thus if the idea of the fweetnefs of fugar fhould be excited in our dreams, the whitenefs and hardnefs of it occur at the fame time by affociation; and we believe a material lump of fugar prefent before us. But if, in our waking hours, the idea of the fweetnefs of fugar occurs to us, the ftimuli of furrounding objects, as the edge of the table, on which we prefs, or green colour of the grafs, on which we tread, prevent the other ideas of the hardnefs and whitenefs of the fugar from being exerted by affociation. Or if they fhould occur, we voluntarily compare them with the irritative ideas of the table or grafs above mentioned, and detect

their

their fallacy. We can thus diftinguifh the ideas caufed by the ftimuli of external objects from thofe, which are introduced by affociation, fenfation, or volition; and during our waking hours can thus acquire a knowledge of the external world. Which neverthelefs we cannot do in our dreams, becaufe we have neither perceptions of external bodies, nor the power of volition to enable us to compare them with the ideas of imagination.

III. *Of Vifion.*

OUR eyes obferve a difference of colour, or of fhade, in the prominences and depreffions of objects, and that thofe fhades uniformly vary, when the fenfe of touch obferves any variation. Hence when the retina becomes ftimulated by colours or fhades of light in a certain form, as in a circular fpot; we know by experience, that this is a fign, that a tangible body is before us; and that its figure is refembled by the miniature figure of the part of the organ of vifion, that is thus ftimulated.

Here whilft the ftimulated part of the retina refembles exactly the vifible figure of the whole in miniature, the various kinds of ftimuli from different colours mark the vifible figures of the minuter parts; and by habit we inftantly recall the tangible figures.

Thus when a tree is the object of fight, a part of the retina refembling a flat branching figure is ftimulated by various fhades of colours; but it is by fuggeftion, that the gibbofity of the tree, and the mofs, that fringes its trunk, appear before us. Thefe are ideas of fuggeftion, which we feel or attend to, affociated with the motions of the retina, or irritative ideas, which we do not attend to.

So that though our vifible ideas refemble in miniature the outline of the figure of coloured bodies, in other refpects they ferve only as

a lana language, which by acquired affociations introduce the tangible ideas of bodies. Hence it is, that this fenfe is fo readily deceived by the art of the painter to our amufement and inftruction. The reader will find much very curious knowledge on this fubject in Bifhop Berkley's Effay on Vifion, a work of great ingenuity.

The immediate object however of the fenfe of vifion is light; this fluid, though its velocity is fo great, appears to have no perceptible mechanical impulfe, as was mentioned in the third Section, but feems to ftimulate the retina into animal motion by its tranfmiffion through this part of the fenforium: for though the eyes of cats or other animals appear luminous in obfcure places; yet it is probable, that none of the light, which falls on the retina, is reflected from it, but adheres to or enters into combination with the choroide coat behind it.

The combination of the particles of light with opake bodies, and therefore with the choroide coat of the eye, is evinced from the heat, which is given out, as in other chemical combinations. For the funbeams communicate no heat in their paffage through tranfparent bodies, with which they do not combine, as the air continues cool even in the focus of the largeft burning-glaffes, which in a moment vitrifies a particle of opaque matter.

IV. *Of the Organ of Hearing.*

It is generally believed, that the tympanum of the ear vibrates mechanically, when expofed to audible founds, like the ftrings of one mufical inftrument, when the fame notes are ftruck upon another. Nor is this opinion improbable, as the mufcles and cartilages of the larynx are employed in producing variety of tones by mechanical vibration: fo the mufcles and bones of the ear feem adapted to increafe

or

or diminifh the tenfion of the tympanum for the purpofes of fimilar mechanical vibrations.

But it appears from diffection, that the tympanum is not the immediate organ of hearing, but that like the humours and cornea of the eye, it is only of ufe to prepare the object for the immediate organ. For the portio mollis of the auditory nerve is not fpread upon the tympanum, but upon the veftibulum, and cochlea, and femicircular canals of the ear ; while between the tympanum and the expanfion of the auditory nerve the cavity is faid by Dr. Cotunnus and Dr. Meckel to be filled with water ; as they had frequently obferved by freezing the heads of dead animals before they diffected them; and water being a more denfe fluid than air is much better adapted to the propagation of vibrations. We may add, that even the external opening of the ear is not abfolutely neceffary for the perception of found : for fome people, who from thefe defects would have been completely deaf, have diftinguifhed acute or grave founds by the tremours of a ftick held between their teeth propagated along the bones of the head, (Haller. Phyf. T. V. p. 295.)

Hence it appears, that the immediate organ of hearing is not affected by the particles of the air themfelves, but is ftimulated into animal motion by the vibrations of them. And it is probable from the loofe bones, which are found in the heads of fome fifhes, that the vibrations of water are fenfible to the inhabitants of that element by a fimilar organ.

The motions of the atmofphere, which we become acquainted with by the fenfe of touch, are combined with its folidity, weight, or vis inertiæ; whereas thofe, that are perceived by this organ, depend alone on its elafticity. But though the vibration of the air is the immediate object of the fenfe of hearing, yet the ideas, we receive by this fenfe, like thofe received from light, are only as a language, which by acquired affociations acquaints us with thofe motions of tangible bodies,

3   which

which depend on their elaſticity; and which we had before learned by our ſenſe of touch.

V. *Of Smell and of Taſte.*

The objects of ſmell are diſſolved in the fluid atmoſphere, and thoſe of taſte in the ſaliva, or other aqueous fluid, for the better diffuſing them on their reſpective organs, which ſeem to be ſtimulated into animal motion perhaps by the chemical affinities of theſe particles, which conſtitute the ſapidity and odoroſity of bodies with the nerves of ſenſe, which perceive them.

Mr. Volta has lately obſerved a curious circumſtance relative to our ſenſe of taſte. If a bit of clean lead and a bit of clean ſilver be ſeparately applied to the tongue and palate no taſte is perceived; but by applying them in contact in reſpect to the parts out of the mouth, and nearly ſo in reſpect to the parts, which are immediately applied to the tongue and palate, a ſaline or acidulous taſte is perceived, as of a fluid like a ſtream of electricity paſſing from one of them to the other. This new application of the ſenſe of taſte deſerves further inveſtigation, as it may acquaint us with new properties of matter.

VI. *Of the Senſe of Heat.*

There are many experiments in chemical writers, that evince the exiſtence of heat as a fluid element, which covers and pervades all bodies, and is attracted by the ſolutions of ſome of them, and is detruded from the combination of others. Thus from the combinations of metals with acids, and from thoſe combinations of animal fluids,

which

which are termed fecretions, this fluid matter of heat is given out amongft the neighbouring bodies; and in the folutions of falts in water, or of water in air, it is abforbed from the bodies, that furround them; whilft in its facility in paffing through metallic bodies, and its difficulty in pervading refins and glafs, it refembles the properties of the electric aura; and is like that excited by friction, and feems like that to gravitate amongft other bodies in its uncombined ftate, and to find its equilibrium.

There is no circumftance of more confequence in the animal economy than a due proportion of this fluid of heat; for the digeftion of our nutriment in the ftomach and bowels, and the proper qualities of all our fecreted fluids, as they are produced or prepared partly by animal and partly by chemical procefles, depend much on the quantity of heat; the excefs of which, or its deficiency, alike gives us pain, and induces us to avoid the circumftances that occafion them. And in this the perception of heat effentially differs from the perceptions of the fenfe of touch, as we receive pain from too great preffure of folid bodies, but none from the abfence of it. It is hence probable, that nature has provided us with a fet of nerves for the perception of this fluid, which anatomifts have not yet attended to.

There may be fome difficulty in the proof of this affertion; if we look at a hot fire, we experience no pain of the optic nerve, though the heat along with the light muft be concentrated upon it. Nor does warm water or warm oil poured into the ear give pain to the organ of hearing; and hence as thefe organs of fenfe do not perceive fmall exceffes or deficiences of heat; and as heat has no greater analogy to the folidity or to the figures of bodies, than it has to their colours or vibrations; there feems no fufficient reafon for our afcribing the perception of heat and cold to the fenfe of touch; to which it has generally been attributed, either becaufe it is diffufed beneath the whole fkin like the fenfe of touch, or owing to the inaccuracy of our obfervations, or the defect of our languages.

R

There

There is another circumſtance would induce us to believe, that the perceptions of heat and cold do not belong to the organ of touch; ſince the teeth, which are the leaſt adapted for the perceptions of ſolidity or figure, are the moſt ſenſible to heat or cold; whence we are forewarned from ſwallowing thoſe materials, whoſe degree of coldneſs or of heat would injure our ſtomachs.

The following is an extract from a letter of Dr. R. W. Darwin, of Shrewſbury, when he was a ſtudent at Edinburgh. "I made an experiment yeſterday in our hoſpital, which much favours your opinion, that the ſenſation of heat and of touch depend on different ſets of nerves. A man who had lately recovered from a fever, and was ſtill weak, was ſeized with violent cramps in his legs and feet; which were removed by opiates, except that one of his feet remained inſenſible. Mr. Ewart pricked him with a pin in five or ſix places, and the patient declared he did not feel it in the leaſt, nor was he ſenſible of a very ſmart pinch. I then held a red-hot poker at ſome diſtance, and brought it gradually nearer till it came within three inches, when he aſſerted that he felt it quite diſtinctly. I ſuppoſe ſome violent irritation on the nerves of touch had cauſed the cramps, and had left them paralytic; while the nerves of heat, having ſuffered no increaſed ſtimulus, retained their irritability."

VII. *Of the Senſe of Extenſion.*

THE organ of touch is properly the ſenſe of preſſure, but the muſcular fibres themſelves conſtitute the organ of ſenſe, that feels extenſion. The ſenſe of preſſure is always attended with the ideas of the figure and ſolidity of the object, neither of which accompany our perception of extenſion. The whole ſet of muſcles, whether they are hollow ones, as the heart, arteries, and inteſtines, or longitudinal

ones

ones attached to bones, contract themfelves, whenever they are ftimulated by forcible elongation; and it is obfervable, that the white mufcles, which conftitute the arterial fyftem, feem to be excited into contraction from no other kinds of ftimulus, according to the experiments of Haller. And hence the violent pain in fome inflammations, as in the paronychia, obtains immediate relief by cutting the membrane, that was ftretched by the tumour of the fubjacent parts.

Hence the whole mufcular fyftem may be confidered as one organ of fenfe, and the various attitudes of the body, as ideas belonging to this organ, of many of which we are hourly confcious, while many others, like the irritative ideas of the other fenfes, are performed without our attention.

When the mufcles of the heart ceafe to act, the refluent blood again diftends or elongates them; and thus irritated they contract as before. The fame happens to the arterial fyftem, and I fuppofe to the capillaries, inteftines, and various glands of the body.

When the quantity of urine, or of excrement, diftends the bladder, or rectum, thofe parts contract, and exclude their contents, and many other mufcles by affociation act along with them; but if thefe evacuations are not foon complied-with, pain is produced by a little further extenfion of the mufcular fibres: a fimilar pain is caufed in the mufcles, when a limb is much extended for the reduction of diflocated bones; and in the punifhment of the rack: and in the painful cramps of the calf of the leg, or of other mufcles, for a greater degree of contraction of a mufcle, than the movement of the two bones, to which its ends are affixed, will admit of, muft give fimilar pain to that, which is produced by extending it beyond its due length. And the pain from punctures or incifions arifes from the diftention of the fibres, as the knife paffes through them; for it nearly ceafes as foon as the divifion is completed.

All

All thefe motions of the mufcles, that are thus naturally excited by the ftimulus of diftending bodies, are alfo liable to be called into ftrong action by their catenation, with the irritations or fenfations produced by the momentum of the progreffive particles of blood in the arteries, as in inflammatory fevers, or by acrid fubftances on other fenfible organs, as in the ftrangury, or tenefmus, or cholera.

We fhall conclude this account of the fenfe of extenfion by obferving, that the want of its object is attended with a difagreeable fenfation, as well as the excefs of it. In thofe hollow mufcles, which have been accuftomed to it, this difagreeable fenfation is called faintnefs, emptinefs, and finking; and, when it arifes to a certain degree, is attended with fyncope, or a total quiefcence of all motions, but the internal irritative ones, as happens from fudden lofs of blood, or in the operation of tapping in the dropfy.

VIII. *Of the Appetites of Hunger, Thirft, Heat, Extenfion, the want of frefh Air, animal Love, and the Suckling of Children.*

HUNGER is moft probably perceived by thofe numerous ramifications of nerves that are feen about the upper opening of the ftomach; and thirft by the nerves about the fauces, and the top of the gula. The ideas of thefe fenfes are few in the generality of mankind, but are more numerous in thofe, who by difeafe, or indulgence, defire particular kinds of foods or liquids.

A fenfe of heat has already been fpoken of, which may with propriety be called an appetite, as we painfully defire it, when it is deficient in quantity.

The fenfe of extenfion may be ranked amongft thefe appetites, fince the deficiency of its object gives difagreeable fenfation; when this happens in the arterial fyftem, it is called faintnefs, and feems to

7   bear

bear fome analogy to hunger and to cold; which like it are attended
with emptinefs of a part of the vafcular fyftem.

The fenfe of want of frefh air has not been attended to, but is as
diftinct as the others, and the firft perhaps that we experience
after our nativity; from the want of the object of this fenfe many
difeafes are produced, as the jail-fever, plague, and other epidemic
maladies. Animal love is another appetite, which occurs later in
life, and the females of lactiferous animals have another natural inlet
of pleafure or pain from the fuckling their offspring. The want of
which either owing to the death of their progeny, or to the fafhion of
their country, has been fatal to many of the fex. The males have
alfo pectoral glands, which are frequently turgid with a thin milk at
their nativity, and are furnifhed with nipples, which erect on titillation like thofe of the female; but which feem now to be of no further ufe, owing perhaps to fome change which thefe animals have
undergone in the gradual progreffion of the formation of the earth,
and of all that it inhabit.

Thefe feven laft mentioned fenfes may properly be termed appetites,
as they differ from thofe of touch, fight, hearing, tafte, and fmell,
in this refpect; that they are affected with pain as well by the defect
of their objects as by the excefs of them, which is not fo in the latter.
Thus cold and hunger give us pain, as well as an excefs of heat or
fatiety; but it is not fo with darknefs and filence.

IX. Before we conclude this Section on the organs of fenfe, we
muft obferve, that, as far as we know, there are many more fenfes,
than have been here mentioned, as every gland feems to be influenced
to feparate from the blood, or to abforb from the cavities of the body,
or from the atmofphere, its appropriated fluid, by the ftimulus of
that fluid on the living gland; and not by mechanical capillary abforption, nor by chemical affinity. Hence it appears, that each of
thefe

thefe glands muſt have a peculiar organ to perceive thefe irritations, but as thefe irritations are not fucceeded by fenfation, they have not acquired the names of fenfes.

However when thefe glands are excited into motions ſtronger than uſual, either by the acrimony of their fluids, or by their own irritability being much increaſed, then the fenfation of pain is produced in them as in all the other fenfes of the body ; and thefe pains are all of different kinds, and hence the glands at this time really become each a different organ of fenfe, though thefe different kinds of pain have acquired no names.

Thus a great excefs of light does not give the idea of light but of pain ; as in forcibly opening the eye when it is much inflamed.   The great excefs of preffure or diftention, as when the point of a pin is preffed upon our ſkin, produces pain, (and when this pain of the fenfe of touch is ſlighter, it is termed itching, or tickling,) without any idea of folidity or of figure : an excefs of heat produces ſmarting, of cold another kind of pain ; it is probable by this fenfe of heat the pain produced by cauſtic bodies is perceived, and of electricity, as all thefe are fluids, that permeate, diftend, or decompofe the parts that feel them.

SECT.

SECT. XV.

OF THE CLASSES OF IDEAS.

I. 1. *Ideas received in tribes.*   2. *We combine them further, or abstract from these tribes.*   3. *Complex ideas.*   4. *Compounded ideas.*   5. *Simple ideas, modes, substances, relations, general ideas.*   6. *Ideas of reflexion.*   7. *Memory and imagination imperfectly defined. Ideal presence. Memorandum-rings.*   II. 1. *Irritative ideas. Perception.*   2. *Sensitive ideas, imagination.*   3. *Voluntary ideas, recollection.*   4. *Associated ideas, suggestion.*   III. 1. *Definitions of perception, memory.*   2. *Reasoning, judgment, doubting, distinguishing, comparing.*   3. *Invention.*   4. *Consciousness.*   5. *Identity.*   6. *Lapse of time.*   7. *Free-will.*

I. AS the constituent elements of the material world are only perceptible to our organs of sense in a state of combination; it follows, that the ideas or sensual motions excited by them, are never received singly, but ever with a greater or less degree of combination. So the colours of bodies or their hardnesses occur with their figures: every smell and taste has its degree of pungency as well as its peculiar flavour: and each note in music is combined with the tone of some instrument. It appears from hence, that we can be sensible of a number of ideas at the same time, such as the whiteness, hardness, and coldness, of a snow-ball, and can experience at the same time many irritative ideas of surrounding bodies, which we do not attend to, as mentioned in Section VII. 3. 2. But those ideas which belong to the same sense, seem to be more easily combined into synchronous tribes, than those which were not received by the same sense, as we can

more

more eafily think of the whitenefs and figure of a lump of fugar at the fame time, than the whitenefs and fweetnefs of it.

2. As thefe ideas, or fenfual motions, are thus excited with greater or lefs degrees of combination; fo we have a power, when we repeat them either by our volition or fenfation, to increafe or diminifh this degree of combination, that is, to form compounded ideas from thofe, which were more fimple; and abftract ones from thofe, which were more complex, when they were firft excited; that is, we can repeat a part or the whole of thofe fenfual motions, which did conftitute our ideas of perception; and the repetition of which now conftitutes our ideas of recollection, or of imagination.

3. Thofe ideas, which we repeat without change of the quantity of that combination, with which we firft received them, are called complex ideas, as when you recollect Weftminfter Abbey, or the planet Saturn: but it muft be obferved, that thefe complex ideas, thus re-excited by volition, fenfation, or affociation, are feldom perfect copies of their correfpondent perceptions, except in our dreams, where other external objects do not detract our attention.

4. Thofe ideas, which are more complex than the natural objects that firft excited them, have been called compounded ideas, as when we think of a fphinx, or griffin.

5. And thofe that are lefs complex than the correfpondent natural objects, have been termed abftracted ideas: thus fweetnefs, and whitenefs, and folidity, are received at the fame time from a lump of fugar, yet I can recollect any of thefe qualities without thinking of the others, that were excited along with them.

When ideas are fo far abftracted as in the above example, they have been termed fimple by the writers of metaphyfics, and feem indeed to be more complete repetitions of the ideas or fenfual motions, originally excited by external objects.

Other claffes of thefe ideas, where the abftraction has not been fo great, have been termed, by Mr. Locke, modes, fubftances, and re-

lations,

lations, but they feem only to differ in their degree of abftraction from the complex ideas that were at firft excited; for as thefe complex or natural ideas are themfelves imperfect copies of their correfpondent perceptions, fo thefe abftract or general ideas are only ftill more imperfect copies of the fame perceptions. Thus when I have feen an object but once, as a rhinoceros, my abftract idea of this animal is the fame as my complex one. I may think more or lefs diftinctly of a rhinoceros, but it is the very rhinoceros that I faw, or fome part or property of him, which recurs to my mind.

But when any clafs of complex objects becomes the fubject of converfation, of which I have feen many individuals, as a caftle or an army, fome property or circumftance belonging to it is peculiarly alluded to; and then I feel in my own mind, that my abftract idea of this complex object is only an idea of that part, property, or attitude of it, that employs the prefent converfation, and varies with every fentence that is fpoken concerning it. So if any one fhould fay, " one may fit upon a horfe fafer than on a camel," my abftract idea of the two animals includes only an outline of the level back of the one, and the gibbofity on the back of the other. What noife is that in the ftreet?—Some horfes trotting over the pavement. Here my idea of the horfes includes principally the fhape and motion of their legs. So alfo the abftract ideas of goodnefs and courage are ftill more imperfect reprefentations of the objects they were received from; for here we abftract the material parts, and recollect only the qualities.

Thus we abftract fo much from fome of our complex ideas, that at length it becomes difficult to determine of what perception they partake; and in many inftances our idea feems to be no other than of the found or letters of the word, that ftands for the collective tribe, of which we are faid to have an abftracted idea, as noun, verb, chimæra, apparition.

S

6. Ideas

6. Ideas have been divided into thofe of perception and thofe of reflection, but as whatever is perceived muft be external to the organ that perceives it, all our ideas muft originally be ideas of perception.

7. Others have divided our ideas into thofe of memory, and thofe of imagination; they have faid that a recollection of ideas in the order they were received conftitutes memory, and without that order imagination; but all the ideas of imagination, excepting the few that are termed fimple ideas, are parts of trains or tribes in the order they were received: as if I think of a fphinx, or a griffin, the fair face, bofom, wings, claws, tail, are all complex ideas in the order they were received: and it behoves the writers, who adhere to this definition, to determine, how fmall the trains muft be, that fhall be called imagination; and how great thofe, that fhall be called memory.

Others have thought that the ideas of memory have a greater vivacity than thofe of imagination: but the ideas of a perfon in fleep, or in a waking reverie, where the trains connected with fenfation are uninterrupted, are more vivid and diftinct than thofe of memory, fo that they cannot be diftinguifhed by this criterion.

The very ingenious author of the Elements of Criticifm has defcribed what he conceives to be a fpecies of memory, and calls it ideal prefence; but the inftances he produces are the reveries of fenfation, and are therefore in truth connections of the imagination, though they are recalled in the order they were received.

The ideas connected by affociation are in common difcourfe attributed to memory, as we talk of memorandum-rings, and tie a knot on our handkerchiefs to bring fomething into our minds at a diftance of time. And a fchool-boy, who can repeat a thoufand unmeaning lines in Lilly's Grammar, is faid to have a good memory. But thefe have been already fhewn to belong to the clafs of affociation; and are termed ideas of fuggeftion.

II. Laftly,

II. Laſtly, the method already explained of claſſing ideas into thoſe excited by irritation, ſenſation, volition, or aſſociation, we hope will be found more convenient both for explaining the operations of the mind, and for comparing them with thoſe of the body; and for the illuſtration and the cure of the diſeaſes of both, and which we ſhall here recapitulate.

1. Irritative ideas are thoſe, which are preceded by irritation, which is excited by objects external to the organs of ſenſe: as the idea of that tree, which either I attend to, or which I ſhun in walking near it without attention. In the former caſe it is termed perception, in the latter it is termed ſimply an irritative idea.

2. Senſitive ideas are thoſe, which are preceded by the ſenſation of pleaſure or pain; as the ideas, which conſtitute our dreams or reveries, this is called imagination.

3. Voluntary ideas are thoſe, which are preceded by voluntary exertion, as when I repeat the alphabet backwards: this is called recollection.

4. Aſſociate ideas are thoſe, which are preceded by other ideas or muſcular motions, as when we think over or repeat the alphabet by rote in its uſual order; or ſing a tune we are accuſtomed to; this is called ſuggeſtion.

III. 1. Perceptions ſignify thoſe ideas, which are preceded by irritation and ſucceeded by the ſenſation of pleaſure or pain, for whatever excites our attention intereſts us; that is, it is accompanied with pleaſure or pain; however ſlight may be the degree or quantity of either of them.

The word memory includes two claſſes of ideas, either thoſe which are preceded by voluntary exertion, or thoſe which are ſuggeſted by their aſſociations with other ideas.

2. Reaſoning is that operation of the ſenſorium, by which we excite two or many tribes of ideas; and then re-excite the ideas, in which they differ, or correſpond. If we determine this difference, it

S 2   is

is called judgment; if we in vain endeavour to determine it, it is called doubting.

If we re-excited the ideas, in which they differ, it is called diftinguifhing. If we re-excite thofe in which they correfpond, it is called comparing.

3. Invention is an operation of the fenforium, by which we voluntarily continue to excite one train of ideas, fuppofe the defign of raifing water by a machine; and at the fame time attend to all other ideas, which are connected with this by every kind of catenation; and combine or feparate them voluntarily for the purpofe of obtaining fome end.

For we can create nothing new, we can only combine or feparate the ideas, which we have already received by our perceptions: thus if I wifh to reprefent a monfter, I call to my mind the ideas of every thing difagreeable and horrible, and combine the naftinefs and gluttony of a hog, the ftupidity and obftinacy of an afs, with the fur and awkwardnefs of a bear, and call the new combination Caliban. Yet fuch a monfter may exift in nature, as all his attributes are parts of nature. So when I wifh to reprefent every thing, that is excellent, and amiable; when I combine benevolence with cheerfulnefs, wifdom, knowledge, tafte, wit, beauty of perfon, and elegance of manners, and affociate them in one lady as a pattern to the world, it is called invention; yet fuch a perfon may exift,—fuch a perfon does exift!—It is ——— ———, who is as much a monfter as Caliban.

4. In refpect to confcioufnefs, we are only confcious of our exiftence, when we think about it; as we only perceive the lapfe of time, when we attend to it; when we are bufied about other objects, neither the lapfe of time nor the confcioufnefs of our own exiftence can occupy our attention. Hence, when we think of our own exiftence, we only excite abftracted or reflex ideas (as they are termed), of our principal pleafures or pains, of our defires or averfions, or of the

figure

figure, folidity, colour, or other properties of our bodies, and call that
act of the fenforium a confciousnefs of our exiftence.   Some philofophers, I believe it is Des Cartes, has faid, " I think, therefore I exift."
But this is not right reafoning, becaufe thinking is a mode of exiftence;
and it is thence only faying, " I exift, therefore I exift."   For there
are three modes of exiftence, or in the language of grammarians three
kinds of verbs.   Firft, fimply I am, or exift.   Secondly, I am acting,
or exift in a ftate of activity, as I move.   Thirdly, I am fuffering, or
exift in a ftate of being acted upon, as I am moved.   The when, and
the where, as applicable to this exiftence, depends on the fucceffive
motions of our own or of other bodies ; and on their refpective fituations, as fpoken of Sect. XIV. 2. 5.

5. Our identity is known by our acquired habits or catenated trains
of ideas and mufcular motions; and perhaps, when we compare infancy with old age, in thofe alone can our identity be fuppofed to exift.
For what elfe is there of fimilitude between the firft fpeck of living
entity and the mature man ?—every deduction of reafoning, every fen-
timent or paffion, with every fibre of the corporeal part of our fyftem,
has been fubject almoft to annual mutation ; while fome catenations
alone of our ideas and mufcular actions have continued in part unchanged.

By the facility, with which we can in our waking hours voluntarily
produce certain fucceffive trains of ideas, we know by experience,
that we have before reproduced them ; that is, we are confcious of a
time of our exiftence previous to the prefent time ; that is, of our identity now and heretofore. It is thefe habits of action, thefe catenations
of ideas and mufcular motions, which begin with life, and only terminate with it ; and which we can in fome meafure deliver to our
pofterity ; as explained in Sect. XXXIX.

6. When the progreffive motions of external bodies make a part of
our prefent catenation of ideas, we attend to the lapfe of time ; which
appears the longer, the more frequently we thus attend to it ; as when

we

we expect fomething at a certain hour, which much interefts us, whether it be an agreeable or difagreeable event; or when we count the paffing feconds on a ftop-watch.

When an idea of our own perfon, or a reflex idea of our pleafures and pains, defires and averfions, makes a part of this catenation, it is termed confcioufnefs; and if this idea of confcioufnefs makes a part of a catenation, which we excite by recollection, and know by the facility with which we excite it, that we have before experienced it, it is called identity, as explained above.

7. In refpect to freewill, it is certain, that we cannot will to think of a new train of ideas, without previoufly thinking of the firft link of it; as I cannot will to think of a black fwan, without previoufly thinking of a black fwan. But if I now think of a tail, I can voluntarily recollect all animals, which have tails; my will is fo far free, that I can purfue the ideas linked to this idea of tail, as far as my knowledge of the fubject extends; but to will without motive is to will without defire or averfion; which is as abfurd as to feel without pleafure or pain; they are both folecifms in the terms. So far are we governed by the catenations of motions, which affect both the body and the mind of man, and which begin with our irritability, and end with it.

SECT.

SECT. XVI.

OF INSTINCT.

HAUD EQUIDEM CREDO, QUIA SIT DIVINITUS ILLIS
INGENIUM, AUT RERUM FATO PRUDENTIA MAJOR.

VIRG. GEORG. L. I. 415.

I. *Inftinctive actions defined. Of connate paffions.* II. *Of the fenfations and motions of the fœtus in the womb.* III. *Some animals are more perfectly formed than others before nativity. Of learning to walk.* IV. *Of the fwallowing, breathing, fucking, pecking, and lapping of young animals.* V. *Of the fenfe of fmell, and its ufes to animals. Why cats do not eat their kittens.* VI. *Of the accuracy of fight in mankind, and their fenfe of beauty. Of the fenfe of touch in elephants, monkies, beavers, men.* VII. *Of natural language.* VIII. *The origin of natural language;* 1. *the language of fear;* 2. *of grief;* 3. *Of tender pleafure;* 4. *of ferene pleafure;* 5. *of anger;* 6. *of attention.* IX. *Artificial language of turkies, hens, ducklings, wagtails, cuckoos, rabbits, dogs, and nightingales.* X. *Of mufic; of tooth-edge; of a good ear; of architecture.* XI. *Of acquired knowledge; of foxes, rooks, feildfares, lapwings, dogs, cats, horfes, crows, and pelicans.* XII. *Of birds of paffage, dormice, fnakes, bats, fwallows, quails, ringdoves, ftare, chaffinch, hoopoe, chatterer, hawfinch, crofsbill, rails and cranes.* XIII. *Of birds nefts; of the cuckoo; of fwallows nefts; of the taylor bird.* XIV. *Of the old foldier; of haddocks, cods, and dog fifh; of the remora; of crabs, herrings, and falmon.* XV. *Of fpiders, caterpillars, ants, and the ichneumon.* XVI. 1. *Of locufts, gnats;* 2. *bees;* 3. *dormice, flies, worms, ants, and wafps.* XVII. *Of the faculty that diftinguifhes man from the brutes.*

I. ALL thofe internal motions of animal bodies, which contribute to digeft their aliment, produce their fecretions, repair their injuries, or increafe their growth, are performed without our attention or con-

8   fcioufnefs.

fcioufnefs. They exift as well in our fleep, as in our waking hours, as well in the fœtus during the time of geftation, as in the infant after nativity, and proceed with equal regularity in the vegetable as in the animal fyftem. Thefe motions have been fhewn in a former part of this work to depend on the irritations of peculiar fluids, and as they have never been claffed amongft the inftinctive actions of animals, are precluded from our prefent difquifition.

But all thofe actions of men or animals, that are attended with con-fcioufnefs, and feem neither to have been directed by their appetites, taught by their experience, nor deduced from obfervation or tradition, have been referred to the power of inftinct. And this power has been explained to be a *divine fomething*, a kind of infpiration ; whilft the poor animal, that poffeffes it, has been thought little better than *a machine* !

The *irkfomenefs*, that attends a continued attitude of the body, or the *pains*, that we receive from heat, cold, hunger, or other injuri-ous circumftances, excite us to *general locomotion :* and our fenfes are fo formed and conftituted by the hand of nature, that certain objects prefent us with pleafure, others with pain, and we are induced to ap-proach and embrace thefe, to avoid and abhor thofe, as fuch fenfa-tions direct us.

Thus the palates of fome animals are gratefully affected by the maftication of fruits, others of grains, and others of flefh ; and they are thence inftigated to attain, and to confume thofe materials ; and are furnifhed with powers of mufcular motion, and of digeftion pro-per for fuch purpofes.

Thefe *fenfations* and *defires* conftitute a part of our fyftem, as our *mufcles* and *bones* conftitute another part : and hence they may alike be termed *natural* or *connate* ; but neither of them can properly be termed *inftinctive :* as the word inftinct in its ufual acceptation refers only to the *actions* of animals, as above explained : the origin of thefe *actions* is the fubject of our prefent enquiry.

The

The reader is intreated carefully to attend to this definition of *in-stinctive actions*, left by ufing the word inftinct without adjoining any accurate idea to it, he may not only include the natural defires of love and hunger, and the natural fenfations of pain or pleafure, but the figure and contexture of the body, and the faculty of reafon itfelf under this general term.

II. We experience fome fenfations, and perform fome actions before our nativity; the fenfations of cold and warmth, agitation and reft, fulnefs and inanition, are inftances of the former; and the repeated ftruggles of the limbs of the fœtus, which begin about the middle of geftation, and thofe motions by which it frequently wraps the umbilical chord around its neck or body, and even fometimes ties it on a knot; are inftances of the latter. Smellie's Midwifery, (Vol. I. p. 182).

By a due attention to thefe circumftances many of the actions of young animals, which at firft fight feemed only referable to an inexplicable inftinct, will appear to have been acquired like all other animal actions, that are attended with confcioufnefs, *by the repeated efforts of our mufcles under the conduct of our fenfations or defires.*

The chick in the fhell begins to move its feet and legs on the fixth day of incubation (Mattreican, p. 138); or on the feventh day, (Langley); afterwards they are feen to move themfelves gently in the liquid that furrounds them, and to open and fhut their mouths, (Harvei, de Generat. p. 62, and 197. Form de Poulet. ii. p. 129). Puppies before the membranes are broken, that involve them, are feen to move themfelves, to put out their tongues, and to open and fhut their mouths, (Harvey, Gipfon, Riolan, Haller). And calves lick themfelves and fwallow many of their hairs before their nativity: which however puppies do not, (Swammerden, p 319. Flemyng Phil. Tranf. Ann. 1755. 42). And towards the end of geftation, the fœtus of all animals are proved to drink part of the liquid in which they fwim, (Haller. Phyfiol. T. 8. 204). The white of egg is found

T   in

in the mouth and gizzard of the chick, and is nearly or quite confumed before it is hatched, (Harvei de Generat. 58). And the liquor
amnii is found in the mouth and ftomach of the human fœtus, and of
calves ; and how elfe fhould that excrement be produced in the inteftines of all animals, which is voided in great quantity foon after
their birth ; Gipfon, Med. Effays, Edinb. V. i. 13. Halleri Phyfiolog.
T. 3. p. 318. and T. 8). In the ftomach of a calf the quantity of
this liquid amounted to about three pints, and the hairs amongft it
were of the fame colour with thofe on its fkin, (Blafii Anat. Animal,
p. m. 122). Thefe facts are attefted by many other writers of credit,
befides thofe above mentioned.

III. It has been deemed a furprifing inftance of inftinct, that calves
and chickens fhould be able to walk by a few efforts almoft immediately after their nativity : whilft the human infant in thofe countries
where he is not incumbered with clothes, as in India, is five or fix
months, and in our climate almoft a twelvemonth, before he can
fafely ftand upon his feet.

The ftruggles of all animals in the womb muft refemble their mode
of fwimming, as by this kind of motion they can beft change their
attitude in water. But the fwimming of the calf and chicken refembles their manner of walking, which they have thus in part acquired before their nativity, and hence accomplifh it afterwards with
very few efforts, whilft the fwimming of the human creature refembles that of the frog, and totally differs from his mode of
walking.

There is another circumftance to be attended to in this affair, that
not only the growth of thofe peculiar parts of animals, which are firft
wanted to fecure their fubfiftence, are in general furtheft advanced,
before their nativity : but fome animals come into the world more
completely formed throughout their whole fyftem than others : and
are thence much forwarder in all their habits of motion. Thus the
colt, and the lamb, are much more perfect animals than the blind

3   puppy,

puppy, and the naked rabbit; and the chick of the pheafant, and the partridge, has more perfect plumage, and more perfect eyes, as well as greater aptitude to locomotion, than the callow neftlings of the dove, and of the wren.   The parents of the former only find it neceffary to fhew them their food, and to teach them to take it up ; whilft thofe of the latter are obliged for many days to obtrude it into their gaping mouths.

IV. From the facts mentioned in No. 2. of this Section, it is evinced that the foetus learns to fwallow before its nativity; for it is feen to open its mouth, and its ftomach is found filled with the liquid that furrounds it.   It opens its mouth, either inftigated by hunger, or by the irkfomenefs of a continued attitude of the mufcles of its face ; the liquor amnii, in which it fwims, is agreeable to its palate, as it confifts of a nourifhing material, (Haller Phyf. T. 8. p. 204).   It is tempted to experience its tafte further in the mouth, and by a few efforts learns to fwallow, in the fame manner as we learn all other animal actions, which are attended with confcioufnefs, *by the repeated efforts of our mufcles under the conduct of our fenfations or volitions.*

The infpiration of air into the lungs is fo totally different from that of fwallowing a fluid in which we are immerfed, that it cannot be acquired before our nativity.   But at this time, when the circulation of the blood is no longer continued through the placenta, that fuffocating fenfation, which we feel about the precordia, when we are in want of frefh air, difagreeably affects the infant : and all the mufcles of the body are excited into action to relieve this oppreffion ; thofe of the breaft, ribs, and diaphragm are found to anfwer this purpofe, and thus refpiration is difcovered, and is continued throughout our lives, as often as the oppreffion begins to recur.   Many infants, both of the human creature, and of quadrupeds, ftruggle for a minute after they are born before they begin to breathe, (Haller Phyf. T. 8. p. 400. ib. pt. 2. p. 1).   Mr. Buffon thinks the action of the dry air upon the nerves of fmell of new-born animals, by producing an en-

deavour

deavour to fneeze, may contribute to induce this firft infpiration, and that the rarefaction of the air by the warmth of the lungs contributes to induce expiration, Hift. Nat. Tom. 4. p. 174. Which latter it may effect by producing a difagreeable fenfation by its delay, and a confequent effort to relieve it. Many children fneeze before they refpire, but not all, as far as I have obferved, or can learn from others.

At length, by the direction of its fenfe of fmell, or by the officious care of its mother, the young animal approaches the odoriferous rill of its future nourifhment, already experienced to fwallow. But in the act of fwallowing, it is neceffary nearly to clofe the mouth, whether the creature be immerfed in the fluid it is about to drink, or not : hence, when the child firft attempts to fuck, it does not flightly comprefs the nipple between its lips, and fuck as an adult perfon would do, by abforbing the milk; but it takes the whole nipple into its mouth for this purpofe, compreffes it between its gums, and thus repeatedly chewing (as it were) the nipple, preffes out the milk; exactly in the fame manner as it is drawn from the teats of cows by the hands of the milkmaid. The celebrated Harvey obferves, that the foetus in the womb muft have fucked in a part of its nourifhment, becaufe it knows how to fuck the minute it is born, as any one may experience by putting a finger between its lips, and becaufe in a few days it forgets this art of fucking, and cannot without fome difficulty again acquire it, (Exercit. de Gener. Anim. 48). The fame obfervation is made by Hippocrates.

A little further experience teaches the young animal to fuck by abforption, as well as by compreffion; that is, to open the cheft as in the beginning of refpiration, and thus to rarefy the air in the mouth, that the preffure of the denfer external atmofphere may contribute to force out the milk.

The chick yet in the fhell has learnt to drink by fwallowing a part of the white of the egg for its food; but not having experienced how

to

to take up and fwallow folid feeds, or grains, is either taught by the folicitous induftry of its mother; or by many repeated attempts is enabled at length to diftinguifh and to fwallow this kind of nutriment.

And puppies, though they know how to fuck like other animals from their previous experience in fwallowing, and in refpiration; yet are they long in acquiring the art of lapping with their tongues, which from the flaccidity of their cheeks, and length of their mouths, is afterwards a more convenient way for them to take in water.

V. The fenfes of fmell and tafte in many other animals greatly excel thofe of mankind, for in civilized fociety, as our victuals are generally prepared by others, and are adulterated with falt, fpice, oil, and empyreuma, we do not hefitate about eating whatever is fet before us, and neglect to cultivate thefe fenfes: whereas other animals try every morfel by the fmell, before they take it into their mouths, and by the tafte before they fwallow it: and are led not only each to his proper nourifhment by this organ of fenfe, but it alfo at a maturer age directs them in the gratification of their appetite of love. Which may be further underftood by confidering the fympathies of thefe parts defcribed in Clafs IV. 2. 1. 7. While the human animal is directed to the object of his love by his fenfe of beauty, as mentioned in No. VI. of this Section. Thus Virgil. Georg. III. 250.

> Nonne vides, ut tota tremor pertentat equorum
> Corpora, fi tantum notas odor attulit auras?
> Nonne canis nidum veneris nafutus odore
> Quærit, et erranti trahitur fublambere linguâ?
> Refpuit at guftum cupidus, labiifque retractis
> Elevat os, trepidanfque novis percutitur æftris,
> Inferit et vivum felici vomere femen.—
> Quam tenui filo cæcos adnectit amores
> Docta Venus, vitæque monet renovare favillam!
> Anon.

The

The following curious experiment is related by Galen. "On diffecting a goat great with young I found a brifk embryon, and having
detached it from the matrix, and fnatching it away before it faw its
dam, I brought it into a certain room, where there were many veffels,
fome filled with wine, others with oil, fome with honey, others with
milk, or fome other liquor; and in others were grains and fruits; we
firft obferved the young animal get upon its feet, and walk; then it
fhook itfelf, and afterwards fcratched its fide with one of its feet: then
we faw it fmelling to every one of thefe things, that were fet in the
room; and when it had fmelt to them all, it drank up the milk."
L. 6. de locis. cap. 6.

Parturient quadrupeds, as cats, and bitches, and fows, are led by
their fenfe of fmell to eat the placenta as other common food; why
then do they not devour their whole progeny, as is reprefented in an
antient emblem of TIME? This is faid fometimes to happen in the
unnatural ftate in which we confine fows; and indeed nature would
feem to have endangered her offspring in this nice circumftance! But
at this time the ftimulus of the milk in the tumid teats of the mother
excites her to look out for, and to defire fome unknown circumftance
to relieve her.   At the fame time the fmell of the milk attracts the
exertions of the young animals towards its fource, and thus the delighted mother difcovers a new appetite, as mentioned in Sect. XIV. 8.
and her little progeny are led to receive and to communicate pleafure
by this moft beautiful contrivance.

VI. But though the human fpecies in fome of their fenfations are
much inferior to other animals, yet the accuracy of the fenfe of touch,
which they poffefs in fo eminent a degree, gives them a great fuperi-
ority of underftanding; as is well obferved by the ingenious Mr. Buffon.   The extremities of other animals terminate in horns, and hoofs,
and claws, very unfit for the fenfation of touch; whilft the human
hand is finely adapted to encompafs its object with this organ of
fenfe.

The

The-elephant is indeed endued with a fine fenfe of feeling at the extremity of his probofcis, and hence has acquired much more accurate ideas of touch and of fight than moft other creatures. The two following inftances of the fagacity of thefe animals may entertain the reader, as they were told me by fome gentlemen of diftinct obfervation, and undoubted veracity, who had been much converfant with our eaftern fettlements. Firft, the elephants that are ufed to carry the baggage of our armies, are put each under the care of one of the natives of Indoftan, and whilft himfelf and his wife go into the woods to collect leaves and branches of trees for his food, they fix him to the ground by a length of chain, and frequently leave a child yet unable to walk, under his protection: and the intelligent animal not only defends it, but as it creeps about, when it arrives near the extremity of his chain, he wraps his trunk gently round its body, and brings it again into the centre of his circle. Secondly, the traitor elephants are taught to walk on a narrow path between two pit-falls, which are covered with turf, and then to go into the woods, and to feduce the wild elephants to come that way, who fall into thefe wells, whilft he paffes fafe between them: and it is univerfally obferved, that thofe wild elephants that efcape the fnare, purfue the traitor with the utmoft vehemence, and if they can overtake him, which fometimes happens, they always beat him to death.

The monkey has a hand well enough adapted for the fenfe of touch, which contributes to his great facility of imitation; but in taking objects with his hands, as a ftick or an apple, he puts his thumb on the fame fide of them with his fingers, inftead of counteracting the preffure of his fingers with it : from this neglect he is much flower in acquiring the figures of objects, as he is lefs able to determine the diftances or diameters of their parts, or to diftinguifh their vis inertiæ from their hardnefs. Helvetius adds, that the fhortnefs of his life, his being fugitive before mankind, and his not inhabiting all climates, combine to prevent his improvement. (De l'Efprit. T. i. p.) There

is.

is however at this time an old monkey fhewn in Exeter Change, London, who having loft his teeth, when nuts are given him, takes a ftone into his hand, and cracks them with it one by one; thus ufing tools to effect his purpofe like mankind.

The beaver is another animal that makes much ufe of his hands, and if we may credit the reports of travellers, is poffeffed of amazing ingenuity. This however, M. Buffon affirms, is only where they exift in large numbers, and in countries thinly peopled with men; while in France in their folitary ftate they fhew no uncommon ingenuity.

Indeed all the quadrupeds, that have collar-bones, (claviculæ) ufe their fore-limbs in fome meafure as we ufe our hands, as the cat, fquirrel, tyger, bear and lion; and as they exercife the fenfe of touch more univerfally than other animals, fo are they more fagacious in watching and furprifing their prey. All thofe birds, that ufe their claws for hands, as the hawk, parrot, and cuckoo, appear to be more docile and intelligent; though the gregarious tribes of birds have more acquired knowledge.

Now as the images, that are painted on the retina of the eye, are no other than figns, which recall to our imaginations the objects we had before examined by the organ of touch, as is fully demonftrated by Dr. Berkley in his treatife on vifion; it follows that the human creature has greatly more accurate and diftinct fenfe of vifion than that of any other animal. Whence as he advances to maturity he gradually acquires a fenfe of female beauty, which at this time directs him to the object of his new paffion.

Sentimental love, as diftinguifhed from the animal paffion of that name, with which it is frequently accompanied, confifts in the defire or fenfation of beholding, embracing, and faluting a beautiful object.

The characteriftic of beauty therefore is that it is the object of love; and though many other objects are in common language called beautiful,
ful,

ful, yet they are only called fo metaphorically, and ought to be termed agreeable. A Grecian temple may give us the pleafurable idea of fublimity, a Gothic temple may give us the pleafurable idea of variety, and a modern houfe the pleafurable idea of utility; mufic and poetry may infpire our love by affociation of ideas; but none of thefe, except metaphorically, can be termed beautiful, as we have no wifh to embrace or falute them.

Our perception of beauty confifts in our recognition by the fenfe of vifion of thofe objects, firft, which have before infpired our love by the pleafure, which they have afforded to many of our fenfes; as to our fenfe of warmth, of touch, of fmell, of tafte, hunger and thirft; and, fecondly, which bear any analogy of form to fuch objects.

When the babe, foon after it is born into this cold world, is applied to its mother's bofom; its fenfe of perceiving warmth is firft agreeably affected; next its fenfe of fmell is delighted with the odour of her milk; then its tafte is gratified by the flavour of it; afterwards the appetites of hunger and of thirft afford pleafure by the poffeffion of their objects, and by the fubfequent digeftion of the aliment; and, laftly, the fenfe of touch is delighted by the foftnefs and fmoothnefs of the milky fountain, the fource of fuch variety of happinefs.

All thefe various kinds of pleafure at length become affociated with the form of the mother's breaft; which the infant embraces with its hands, preffes with its lips, and watches with its eyes; and thus acquires more accurate ideas of the form of its mother's bofom, than of the odour and flavour or warmth, which it perceives by its other fenfes. And hence at our maturer years, when any object of vifion is prefented to us, which by its waving or fpiral lines bears any fimilitude to the form of the female bofom, whether it be found in a landfcape with foft gradations of rifing and defcending furface, or in the forms of fome antique vafes, or in other works of the pencil or the chiffel, we feel a general glow of delight, which feems to influence all our fenfes; and, if the object be not too large, we experience an at-

U   traction

traction to embrace it with our arms, and to falute it with our lips, as we did in our early infancy the bofom of our mother. And thus we find, according to the ingenious idea of Hogarth, that the waving lines of beauty were originally taken from the temple of Venus.

This animal attraction is love; which is a fenfation, when the object is prefent; and a defire, when it is abfent. Which conftitutes the pureft fource of human felicity, the cordial drop in the otherwife vapid cup of life, and which overpays mankind for the care and labour, which are attached to the pre-eminence of his fituation above other animals.

It fhould have been obferved, that colour as well as form fometimes enters into our idea of a beautiful object, as a good complexion for inftance, becaufe a fine or fair colour is in general a fign of health, and conveys to us an idea of the warmth of the object; and a pale countenance on the contrary gives an idea of its being cold to the touch.

It was before remarked, that young animals ufe their lips to diftinguifh the forms of things, as well as their fingers, and hence we learn the origin of our inclination to falute beautiful objects with our lips.

VII. There are two ways by which we become acquainted with the paffions of others: firft, by having obferved the effects of them, as of fear or anger, on our own bodies, we know at fight when others are under the influence of thefe affections. So when two cocks are preparing to fight, each feels the feathers rife round his own neck, and knows from the fame fign the difpofition of his adverfary: and children long before they can fpeak, or underftand the language of their parents, may be frightened by an angry countenance, or foothed by fmiles and blandifhments.

Secondly, when we put ourfelves into the attitude that any paffion naturally occafions, we foon in fome degree acquire that paffion; hence when thofe that fcold indulge themfelves in loud oaths, and

violent

violent actions of the arms, they increase their anger by the mode of expressing themselves: and on the contrary the counterfeited smile of pleasure in disagreeable company soon brings along with it a portion of the reality, as is well illustrated by Mr. Burke. (Essay on the Sublime and Beautiful.)

This latter method of entering into the passions of others is rendered of very extensive use by the pleasure we take in imitation, which is every day presented before our eyes, in the actions of children, and indeed in all the customs and fashions of the world. From this our aptitude to imitation, arises what is generally understood by the word sympathy so well explained by Dr. Smith of Glasgow. Thus the appearance of a cheerful countenance gives us pleasure, and of a melancholy one makes us sorrowful. Yawning and sometimes vomiting are thus propagated by sympathy, and some people of delicate fibres, at the presence of a spectacle of misery, have felt pain in the same parts of their own bodies, that were diseased or mangled in the other. Amongst the writers of antiquity Aristotle thought this aptitude to imitation an essential property of the human species, and calls man an imitative animal. Το ζωον μιμωμενον.

These then are the natural signs by which we understand each other, and on this slender basis is built all human language. For without some natural signs, no artificial ones could have been invented or understood, as is very ingeniously observed by Dr. Reid. (Inquiry into the Human Mind.)

VIII. The origin of this universal language is a subject of the highest curiosity, the knowledge of which has always been thought utterly inaccessible. A part of which we shall however here attempt.

Light, sound, and odours, are unknown to the fœtus in the womb, which, except the few sensations and motions already mentioned, sleeps away its time insensible of the busy world. But the moment he arrives into day, he begins to experience many vivid pains and pleasures; these are at the same time attended with certain muscular mo-

tions,

tions, and from this their early, and individual affociation, they acquire habits of occurring together, that are afterwards indiffoluble.

1. *Of Fear.*

As foon as the young animal is born, the firft important fenfations, that occur to him, are occafioned by the oppreffion about his precordia for want of refpiration, and by his fudden tranfition from ninety-eight degrees of heat into fo cold a climate.——He trembles, that is, he exerts alternately all the mufcles of his body, to enfranchife himfelf from the oppreffion about his bofom, and begins to breathe with frequent and fhort refpirations; at the fame time the cold contracts his red fkin, gradually turning it pale; the contents of the bladder and of the bowels are evacuated: and from the experience of thefe firft difagreeable fenfations the paffion of fear is excited, which is no other than the expectation of difagreeable fenfations.   This early affociation of motions and fenfations perfifts throughout life; the paffion of fear produces a cold and pale fkin, with tremblings, quick refpiration, and an evacuation of the bladder and bowels, and thus conftitutes the natural or univerfal language of this paffion.

On obferving a Canary bird this morning, January 28, 1772, at the houfe of Mr. Harvey, near Tutbury, in Derbyfhire, I was told it always fainted away, when its cage was cleaned, and defired to fee the experiment. The cage being taken from the ceiling, and its bottom drawn out, the bird began to tremble, and turned quite white about the root of his bill: he then opened his mouth as if for breath, and refpired quick, ftood ftraighter up on his perch, hung his wings, fpread his tail, clofed his eyes, and appeared quite ftiff and cataleptic for near half an hour, and at length with much trembling and deep refpirations came gradually to himfelf.

2. *Of*

2. Of Grief.

That the internal membrane of the noftrils may be kept always moift, for the better perception of odours, there are two canals, that conduct the tears after they have done their office in moiftening and cleaning the ball of the eye into a fack, which is called the lacrymal fack; and from which there is a duct, that opens into the noftrils: the aperture of this duct is formed of exquifite fenfibility, and when it is ftimulated by odorous particles, or by the drynefs or coldnefs of the air, the fack contracts itfelf, and pours more of its contained moifture on the organ of fmell. By this contrivance the organ is rendered more fit for perceiving fuch odours, and is preferved from being injured by thofe that are more ftrong or corrofive. Many other receptacles of peculiar fluids difgorge their contents, when the ends of their ducts are ftimulated; as the gall bladder, when the contents of the duodenum ftimulate the extremity of the common bile duct: and the falivary glands, when the termination of their ducts in the mouth are excited by the ftimulus of the food we mafticate. Atque veficulæ feminales fuum exprimunt fluidum glande penis fricatâ.

The coldnefs and drynefs of the atmofphere, compared with the warmth and moifture, which the new-born infant had juft before experienced, difagreeably affects the aperture of this lacrymal fack: the tears, that are contained in this fack, are poured into the noftrils, and a further fupply is fecreted by the lacrymal glands, and diffufed upon the eye-balls; as is very vifible in the eyes and noftrils of children foon after their nativity. The fame happens to us at our maturer age, for in fevere frofty weather, fnivelling and tears are produced by the coldnefs and drynefs of the air.

But the lacrymal glands, which feparate the tears from the blood, are fituated on the upper external part of the globes of each eye; and,

when

when a greater quantity of tears are wanted, we contract the fore-head, and bring down the eye-brows, and use many other diftortions of the face, to comprefs thefe glands.

Now as the fuffocating fenfation, that produces refpiration, is removed almoft as foon as perceived, and does not recur again: this difagreeable irritation of the lacrymal ducts, as it muft frequently recur, till the tender organ becomes ufed to variety of odours, is one of the firft pains that is repeatedly attended to: and hence throughout our infancy, and in many people throughout their lives, all difagreeable fenfations are attended with fnivelling at the nofe, a profufion of tears, and fome peculiar diftortions of countenance: according to the laws of early affociation before mentioned, which conftitutes the natural or univerfal language of grief.

You may affure yourfelf of the truth of this obfervation, if you will attend to what paffes, when you read a diftrefsful tale alone; before the tears overflow your eyes, you will invariably feel a titillation at that extremity of the lacrymal duct, which terminates in the noftril, then the compreffion of the eyes fucceeds, and the profufion of tears.

Linnæus afferts, that the female bear fheds tears in grief; the fame has been faid of the hind, and fome other animals.

3. Of Tender Pleafure.

The firft moft lively impreffion of pleafure, that the infant enjoys after its nativity, is excited by the odour of its mother's milk. The organ of fmell is irritated by this perfume, and the lacrymal fack empties itfelf into the noftrils, as before explained, and an increafe of tears is poured into the eyes. Any one may obferve this, when very young infants are about to fuck; for at thofe early periods of life, the

8   fenfation

fenfation affects the organ of fmell, much more powerfully, than after the repeated habits of fmelling has inured it to odours of common ftrength: and in our adult years, the ftronger fmells, though they are at the fame time agreeable to us, as of volatile fpirits, continue to produce an increafed fecretion of tears.

This pleafing fenfation of fmell is followed by the early affection of the infant to the mother that fuckles it, and hence the tender feelings of gratitude and love, as well as of hopelefs grief, are ever after joined with the titillation of the extremity of the lacrymal ducts, and a profufion of tears.

Nor is it fingular, that the lacrymal fack fhould be influenced by pleafing ideas, as the fight of agreeable food produces the fame effect on the falivary glands. Ac dum vidimus infomniis lafcivæ puellæ fimulacrum tenditur penis.

Lambs fhake or wriggle their tails, at the time when they firft fuck, to get free of the hard excrement, which had been long lodged in their bowels. Hence this becomes afterwards a mark of pleafure in them, and in dogs, and other tailed animals. But cats gently extend and contract their paws when they are pleafed, and purr by drawing in their breath, both which refemble their manner of fucking, and thus become their language of pleafure, for thefe animals having collar-bones ufe their paws like hands when they fuck, which dogs and fheep do not.

4. *Of Serene Pleafure.*

In the action of fucking, the lips of the infant are clofed around the nipple of its mother, till he has filled his ftomach, and the pleafure occafioned by the ftimulus of this grateful food fucceeds. Then the fphincter of the mouth, fatigued by the continued action of fucking,

ing, is relaxed; and the antagonist muscles of the face gently acting, produce the smile of pleasure: as cannot but be seen by all who are conversant with children.

Hence this smile during our lives is associated with gentle pleasure; it is visible in kittens, and puppies, when they are played with, and tickled; but more particularly marks the human features. For in children this expression of pleasure is much encouraged, by their imitation of their parents, or friends; who generally address them with a smiling countenance: and hence some nations are more remarkable for the gaiety, and others for the gravity of their looks.

5. *Of Anger.*

The actions that constitute the mode of fighting, are the immediate language of anger in all animals; and a preparation for these actions is the natural language of threatening. Hence the human creature clenches his fist, and sternly surveys his adversary, as if meditating where to make the attack; the ram, and the bull, draws himself some steps backwards, and levels his horns; and the horse, as he fights by striking with his hinder feet, turns his heels to his foe, and bends back his ears, to listen out the place of his adversary, that the threatened blow may not be ineffectual.

6. *Of Attention.*

The eye takes in at once but half our horizon, and that only in the day, and our smell informs us of no very distant objects, hence we confide principally in the organ of hearing to apprize us of danger:

when

when we hear any the fmalleft found, that we cannot immediately account for, our fears are alarmed, we fufpend our fteps, hold every mufcle ftill, open our mouths a little, erect our ears, and liften to gain further information: and this by habit becomes the general language of attention to objects of fight, as well as of hearing; and even to the fucceffive trains of our ideas.

The natural language of violent pain, which is expreffed by writhing the body, grinning, and fcreaming; and that of tumultuous pleafure, expreffed in loud laughter; belong to Section XXXIV. on Difeafes from Volition.

IX. It muft have already appeared to the reader, that all other animals, as well as man, are poffeffed of this natural language of the paffions, expreffed in figns or tones; and we fhall endeavour to evince, that thofe animals, which have preferved themfelves from being enflaved by mankind, and are affociated in flocks, are alfo poffeffed of fome artificial language, and of fome traditional knowledge.

The mother-turkey, when fhe eyes a kite hovering high in air, has either feen her own parents thrown into fear at his prefence, or has by obfervation been acquainted with his dangerous defigns upon her young. She becomes agitated with fear, and ufes the natural language of that paffion, her young ones catch the fear by imitation, and in an inftant conceal themfelves in the grafs.

At the fame time that fhe fhews her fears by her gefture and deportment, fhe ufes a certain exclamation, Koe-ut, Koe-ut, and the young ones afterwards know, when they hear this note, though they do not fee their dam, that the prefence of their adverfary is denounced, and hide themfelves as before.

The wild tribes of birds have very frequent opportunities of knowing their enemies, by obferving the deftruction they make among their progeny, of which every year but a fmall part efcapes to maturity: but to our domeftic birds thefe opportunities fo rarely occur, that their knowledge of their diftant enemies muft frequently be de-

X

livered

livered by tradition in the manner above explained, through many generations.

This note of danger, as well as the other notes of the mother-turkey, when she calls her flock to their flood, or to sleep under her wings, appears to be an artificial language, both as expressed by the mother, and as understood by the progeny. For a hen teaches this language with equal ease to the ducklings, she has hatched from suppositious eggs, and educates as her own offspring: and the wagtails, or hedge-sparrows, learn it from the young cuckoo their foster nurfling, and supply him with food long after he can fly about, whenever they hear his cuckooing, which Linneus tells us, is his call of hunger, (Syst. Nat.) And all our domestic animals are readily taught to come to us for food, when we use one tone of voice, and to fly from our anger, when we use another.

Rabbits, as they cannot easily articulate founds, and are formed into focieties, that live under ground, have a very different method of giving alarm. When danger is threatened, they thump on the ground with one of their hinder feet, and produce a found, that can be heard a great way by animals near the furface of the earth, which would feem to be an artificial fign both from its fingularity and its aptnefs to the fituation of the animal.

The rabbits on the island of Sor, near Senegal, have white flesh, and are well tasted, but do not burrow in the earth, so that we may suspect their digging themselves houses in this cold climate is an acquired art, as well as their note of alarm, (Adanfon's Voyage to Senegal).

The barking of dogs is another curious note of alarm, and would feem to be an acquired language, rather than a natural fign: for "in the island of Juan Fernandes, the dogs did not attempt to bark, till some European dogs were put among them, and then they gradually begun to imitate them, but in a strange manner at first, as if they were learning a thing that was not natural to them," (Voyage to

South

South America by Don G. Juan, and Don Ant. de Ulloa. B. 2. c. 4).

Linnæus alfo obferves, that the dogs of South America do not bark at ftrangers, (Syft. Nat.)　And the European dogs, that have been carried to Guinea, are faid in three or four generations to ceafe to bark, and only howl, like the dogs that are natives of that coaft, (World Difplayed, Vol. XVII. p. 26.)

A circumftance not diffimilar to this, and equally curious, is mentioned by Kircherus. de Mufurgia, in his Chapter de Lufciniis. " That the young nightingales, that are hatched under other birds, never fing till they are inftructed by the company of other nightingales."　And Jonfton affirms, that the nightingales that vifit Scotland, have not the fame harmony as thofe of Italy, (Pennant's Zoology, octavo, p. 255); which would lead us to fufpect that the finging of birds, like human mufic, is an artificial language rather than a natural expreffion of paffion.

X. Our mufic like our language, is perhaps entirely conftituted of artificial tones, which by habit fuggeft certain agreeable paffions. For the fame combination of notes and tones do not excite devotion, love, or poetic melancholy in a native of Indoftan and of Europe. And " the Highlander has the fame warlike ideas annexed to the found of a bagpipe (an inftrument which an Englifhman derides), as the Englifhman has to that of a trumpet or fife," (Dr. Brown's Union of Poetry and Mufic, p. 58.)　So " the mufic of the Turks is very different from the Italian, and the people of Fez and Morocco have again a different kind, which to us appears very rough and horrid, but is highly pleafing to them," (L' Arte Armoniaca a Giorgio Antoniotto).　Hence we fee why the Italian opera does not delight an untutored Englifhman ; and why thofe, who are unaccuftomed to mufic, are more pleafed with a tune, the fecond or third time they hear it, than the firft.　For then the fame melodious train of founds excites the melancholy, they had learned from the fong ; or the fame

X 2　　　　　　　　　　vivid

vivid combination of them recalls all the mirthful ideas of the dance and company.

Even the founds, that were once difagreeable to us, may by habit be affociated with other ideas, fo as to become agreeable. Father Lafitau, in his account of the Iroquois, fays " the mufic and dance of thofe Americans, have fomething in them extremely barbarous, which at firft difgufts. We grow reconciled to them by degrees, and in the end partake of them with pleafure, the favages themfelves are fond of them to diftraction," (Mœurs des Savages, Tom. ii.)

There are indeed a few founds, that we very generally affociate with agreeable ideas, as the whiftling of birds, or purring of animals, that are delighted; and fome others, that we as generally affociate with difagreeable ideas, as the cries of animals in pain, the hifs of fome of them in anger, and the midnight howl of beafts of prey. Yet we receive no terrible or fublime ideas from the lowing of a cow, or the braying of an afs. Which evinces, that thefe emotions are owing to previous affociations. So if the rumbling of a carriage in the ftreet be for a moment miftaken for thunder, we receive a fublime fenfation, which ceafes as foon as we know it is the noife of a coach and fix.

There are other difagreeable founds, that are faid to fet the teeth on edge; which, as they have always been thought a neceffary effect of certain difcordant notes, become a proper fubject of our enquiry. Every one in his childhood has repeatedly bit a part of the glafs or earthen veffel, in which his food has been given him, and has thence had a very difagreeable fenfation in the teeth, which fenfation was defigned by nature to prevent us from exerting them on objects harder than themfelves. The jarring found produced between the cup and the teeth is always attendant on this difagreeable fenfation: and ever after when fuch a found is accidentally produced by the conflict of two hard bodies, we feel by affociation of ideas the concomitant difagreeable fenfation in our teeth.

Others

Others have in their infancy frequently held the corner of a filk handkerchief in their mouth, or the end of the velvet cape of their coat, whilft their companions in play have plucked it from them, and have given another difagreeable fenfation to their teeth, which has afterwards recurred on touching thofe materials. And the fight of a knife drawn along a china plate, though no found is excited by it, and even the imagination of fuch a knife and plate fo fcraped together, I know by repeated experience will produce the fame difagreeable fenfation of the teeth.

Thefe circumftances indifputably prove, that this fenfation of the tooth-edge is owing to affociated ideas; as it is equally excitable by fight, touch, hearing, or imagination.

In refpect to the artificial proportions of found excited by mufical inftruments, thofe, who have early in life affociated them with agreeable ideas, and have nicely attended to diftinguifh them from each other, are faid to have a good ear, in that country where fuch proportions are in fafhion: and not from any fuperior perfection in the organ of hearing, or any inftinctive fympathy between certain founds and paffions.

I have obferved a child to be exquifitely delighted with mufic, and who could with great facility learn to fing any tune that he heard diftinctly, and yet whofe organ of hearing was fo imperfect, that it was neceffary to fpeak louder to him in common converfation than to others.

Our mufic, like our architecture, feems to have no foundation in nature, they are both arts purely of human creation, as they imitate nothing. And the profeffors of them have only claffed thofe circumftances, that are moft agreeable to the accidental tafte of their age, or country; and have called it Proportion. But this proportion muft always fluctuate, as it refts on the caprices, that are introduced into our minds by our various modes of education. And thefe fluctuations

4

of tafte muft become more frequent in the prefent age, where mankind have enfranchifed themfelves from the blind obedience to the
rules of antiquity in perhaps every fcience, but that of architecture.
See Sect. XII. No. 7. 3.

XI. There are many articles of knowledge, which the animals in
cultivated countries feem to learn very early in their lives; either
from each other, or from experience, or obfervation :- one of the moft
general of thefe is to avoid mankind. There is fo great a refemblance
in the natural language of the paffions of all animals, that we generally know, when they are in a pacific, or in a malevolent humour,
they have the fame knowledge of us; and hence we can fcold them
from us by fome tones and geftures, and could poffibly attract them
to us by others, if they were not already apprized of our general ma-
levolence towards them. Mr. Gmelin, Profeffor at Peterfburg, af-
fures us, that in his journey into Siberia, undertaken by order of the
Emprefs of Ruffia, he faw foxes, that exprefled no fear of himfelf or
companions, but permitted him to come quite near them, having
never feen the human creature before. And Mr. Bougainville relates,
that at his arrival at the Malouine, or Falkland's Iflands, which were
not inhabited by men, all the animals came about himfelf and his
people; the fowls fettling upon their heads and fhoulders, and the
quadrupeds running about their feet. From the difficulty of acquiring the confidence of old animals, and the eafe of taming young ones,
it appears that the fear, they all conceive at the fight of mankind, is
an acquired article of knowledge.

This knowledge is more nicely underftood by rooks, who are
formed into focieties, and build, as it were, cities over our heads;
they evidently diftinguifh, that the danger is greater when a man is
armed with a gun. Every one has feen this, who in the fpring of
the year has walked under a rookery with a gun in his hand: the inhabitants of the trees rife on their wings, and fcream to the unfledged
young

young to shrink into their nests from the sight of the enemy.   The vulgar observing this circumstance so uniformly to occur, assert that rooks can smell gun-powder.

The fieldfairs, (turdus pilarus) which breed in Norway, and come hither in the cold season for our winter berries; as they are associated in flocks, and are in a foreign country, have evident marks of keeping a kind of watch, to remark and announce the appearance of danger. On approaching a tree, that is covered with them, they continue fearless till one at the extremity of the bush rising on his wings gives a loud and peculiar note of alarm, when they all immediately fly, except one other, who continues till you approach still nearer, to certify as it were the reality of the danger, and then he also flies off repeating the note of alarm.

And in the woods about Senegal there is a bird called uett-uett by the negroes, and squallers by the French, which, as soon as they see a man, set up a loud scream, and keep flying round him, as if their intent was to warn other birds, which upon hearing the cry immediately take wing.   These birds are the bane of sportsmen, and frequently put me into a passion, and obliged me to shoot them, (Adanson's Voyage to Senegal, 78).   For the same intent the lesser birds of our climate seem to fly after a hawk, cuckoo, or owl, and scream to prevent their companions from being surprised by the general enemies of themselves, or of their eggs and progeny.

But the lapwing, (charadrius pluvialis Lin.) when her unfledged offspring run about the marshes, where they were hatched, not only gives the note of alarm at the approach of men or dogs, that her young may conceal themselves; but flying and screaming near the adversary, she appears more solicitous and impatient, as he recedes from her family, and thus endeavours to mislead him, and frequently succeeds in her design.   These last instances are so apposite to the situation, rather than to the natures of the creatures, that use them; and

are

are fo.fimilar to the actions of men in the fame circumftances, that we cannot but believe, that they proceed from a fimilar principle.

On the northern coaft of Ireland a friend of mine faw above a hundred crows at once preying upon mufcles; each crow took a mufcle up into the air twenty or forty yards high, and let it fall on the ftones, and thus by breaking the fhell, got poffeffion of the animal.—A certain philofopher (I think it was Anaxagoras) walking along the fea-fhore to gather fhells, one of thefe unlucky birds miftaking his bald head for a ftone, dropped a fhell-fifh upon it, and killed at once a philofopher and an oyfter.

Our domeftic animals, that have fome liberty, are alfo poffeffed of fome peculiar traditional knowledge: dogs and cats have been forced into each other's fociety, though naturally animals of a very different kind, and have hence learned from each other to eat the knot-grafs, when they are fick, to promote vomiting. I have feen a cat miftake the blade of barley for this grafs, which evinces it is an acquired knowledge. They have alfo learnt of each other to cover their excrement and urine;—about a fpoonful of water was fpilt upon my hearth from the tea-kettle, and I obferved a kitten cover it with afhes. Hence this muft alfo be an acquired art, as the creature miftook the application of it.

To preferve their fur clean, and efpecially their whifkers, cats wafh their faces, and generally quite behind their ears, every time they eat. As they cannot lick thofe places with their tongues, they firft wet the infide of the leg with faliva, and then repeatedly wafh their faces with it, which muft originally be an effect of reafoning, becaufe a means is ufed to produce an effect; and feems afterwards to be taught or acquired by imitation, like the greateft part of human arts.

Mr. Leonard, a very intelligent friend of mine, faw a cat catch a trout by darting upon it in a deep clear water at the mill at Weaford, near Lichfield. The cat belonged to Mr. Stanley, who had often

feen

feen her catch fifh in the fame manner in fummer, when the mill-pool was drawn fo low, that the fifh could be feen. I have heard of other cats taking fifh in fhallow water, as they ftood on the bank. This feems a natural art of taking their prey in cats, which their acquired delicacy by domeftication has in general prevented them from ufing, though their defire of eating fifh continues in its original ftrength.

Mr. White, in his ingenious Hiftory of Selbourn, was witnefs to a cat's fuckling a young hare, which followed her about the garden, and came jumping to her call of affection. At Elford, near Lichfield, the Rev. Mr. Sawley had taken the young ones out of a hare, which was fhot; they were alive, and the cat, who had juft loft her own kittens, carried them away, as it was fuppofed to eat them; but it prefently appeared, that it was affection not hunger which incited her, as fhe fuckled them, and brought them up as their mother.

Other inftances of the miftaken application of what has been termed inftinct may be obferved in flies in the night, who miftaking a candle for day-light, approach and perifh in the flame. So the putrid fmell of the ftapelia, or carrion-flower, allures the large flefh-fly to depofit its young worms on its beautiful petals, which perifh there for want of nourifhment. This therefore cannot be a neceffary inftinct, becaufe the creature miftakes the application of it.

Though in this country horfes fhew little veftiges of policy, yet in the deferts of Tartary, and Siberia, when hunted by the Tartars they are feen to form a kind of community, fet watches to prevent their being furprifed, and have commanders, who direct, and haften their flight, Origin of Language, Vol. I. p. 212. In this country, where four or five horfes travel in a line, the firft always points his ears forward, and the laft points his backward, while the intermediate ones feem quite carelefs in this refpect; which feems a part of policy to prevent furprife. As all animals depend moft on the ear to apprize them of the approach of danger, the eye taking in only half the horizon at once, and horfes poffefs a great nicety of this fenfe; as ap-

Y   pears

pears from their mode of fighting mentioned No. 8. 5. of this Section, as well as by common obſervation.

There are ſome parts of a horſe, which he cannot conveniently rub, when they itch, as about the ſhoulder, which he can neither bite with his teeth, nor ſcratch with his hind foot; when this part itches, he goes to another horſe, and gently bites him in the part which he wiſhes to be bitten, which is immediately done by his intelligent friend. I once obſerved a young foal thus bite its large mother, who did not chooſe to drop the graſs ſhe had in her mouth, and rubbed her noſe againſt the foal's neck inſtead of biting it; which evinces that ſhe knew the deſign of her progeny, and was not governed by a ne-ceſſary inſtinct to bite where ſhe was bitten.

Many of our ſhrubs, which would otherwiſe afford an agreeable food to horſes, are armed with thorns or prickles, which ſecure them from thoſe animals; as the holly, hawthorn, gooſeberry, gorſe. In the extenſive moorlands of Staffordſhire, the horſes have learnt to ſtamp upon a gorſe-buſh with one of their fore-feet for a minute to-gether, and when the points are broken, they eat it without injury. Which is an art other horſes in the fertile parts of the county do not poſſeſs, and prick their mouths till they bleed, if they are induced by hunger or caprice to attempt eating gorſe.

Swine have a ſenſe of touch as well as of ſmell at the end of their noſe, which they uſe as a hand, both to root up the ſoil, and to turn over and examine objects of food, ſomewhat like the proboſcis of an elephant. As they require ſhelter from the cold in this climate, they have learnt to collect ſtraw in their mouths to make their neſt, when the wind blows cold; and to call their companions by repeated cries to aſſiſt in the work, and add to their warmth by their numerous bed-fellows. Hence theſe animals, which are eſteemed ſo unclean, have alſo learned never to befoul their dens, where they have liberty, with their own excrement; an art, which cows and horſes, which have open hovels to run into, have never acquired. I have obſerved great

ſagacity

fagacity in fwine; but the fhort lives we allow them, and their ge--
neral confinement, prevents their improvement, which might pro--
bably be otherwife greater than that of dogs.

Inftances of the fagacity and knowledge of animals are very nu--
merous to every obferver, and their docility in learning various arts
from mankind, evinces that they may learn fimilar arts from their
own fpecies, and thus be poffeffed of much acquired and traditional
knowledge.

A dog whofe natural prey is fheep, is taught by mankind, not
only to leave them unmolefted, but to guard them; and to hunt, to
fet, or to deftroy other kinds of animals, as birds, or vermin; and in
fome countries to catch fifh, in others to find truffles, and to practife
a great variety of tricks; is it more furprifing that the crows fhould
teach each other, that the hawk can catch lefs birds, by the fuperior
fwiftnefs of his wing, and if two of them follow him, till he fucceeds
in his defign, that they can by force fhare a part of the capture?
This I have formerly obferved with attention and aftonifhment.

There is one kind of pelican mentioned by Mr. Ofbeck, one of
Linnæus's travelling pupils (the pelicanus aquilus), whofe food is
fifh; and which it takes from other birds, becaufe it is not formed to
catch them itfelf; hence it is called by the Englifh a Man-of-war-bird,
Voyage to China, p. 88. There are many other interefting anec-
dotes of the pelican and cormorant, collected from authors of the beft
authority, in a well-managed Natural Hiftory for Children, publifhed
by Mr. Galton. Johnfon. London.

And the following narration from the very accurate Monf. Adanfon, in his Voyage to Senegal, may gain credit with the reader: as his
employment in this country was folely to make obfervations in natural hiftory. On the river Niger, in his road to the ifland Griel, he
faw a great number of pelicans, or wide throats. " They moved
with great ftate like fwans upon the water, and are the largeft bird
next to the oftrich; the bill of the one I killed was upwards of a foot

Y 2   and

and half long, and the bag faftened underneath it held two and twenty-pints of water.   They fwim in flocks, and form a'large circle, which they contract afterwards, driving the fifh before them with their legs: when they fee the fifh in fufficient number confined in this fpace, they plunge their bill wide open into the water, and fhut it again with great quicknefs.' They thus get fifh into their throat-bag, which they eat afterwards on fhore at their leifure." P. 247.

XII. The knowledge and language of thofe birds, that frequently change their climate with the feafons, is ftill more extenfive: as they perform thefe migrations in large focieties, and are lefs fubject to the power of man, than the refident tribes of birds.   They are faid to follow a leader during the day, who is occafionally changed, and to keep a continual cry during the night to keep themfelves together. It is probable that thefe emigrations were at firft undertaken as acci-dent directed, by the more adventurous of their fpecies, and learned from one another like the difcoveries of mankind in navigation.   The following circumftances ftrongly fupport this opinion.

1. Nature has provided thefe animals, in the climates where they are produced, with another refource: when the feafon becomes too cold for their conftitutions, or the food they were fupported with ceafes to be fupplied, I mean that of fleeping.   Dormice, fnakes, and bats, have not the means of changing their country; the two former from the want of wings, and the latter from his being not able to bear the light of the day.   Hence thefe animals are obliged to make ufe of this refource, and fleep during the winter.   And thofe fwallows that have been hatched too late in the year to acquire their full ftrength of pinion, or that have been maimed by accident or difeafe, have been frequently found in the hollows of rocks on the fea coafts, and even under water in this torpid ftate, from which they have been revived by the warmth of a fire.   This torpid ftate of fwal-lows is teftified by innumerable evidences both of antient and modern names.   Ariftotle fpeaking of the fwallows fays, " They pafs into

warmer

warmer climates in winter, if fuch places are at no great diftance; if they are, they bury themfelves in the climates where they dwell," (8. Hift. c. 16.   See alfo Derham's Phyf. Theol. v. ii. p. 177.)

Hence their emigrations cannot depend on a *neceffary* inftinct, as the emigrations themfelves are not *neceffary !*

2. When the weather becomes cold, the fwallows in the neighbourhood affemble in large flocks; that is, the unexperienced attend thofe that have before experienced the journey they are about to undertake: they are then feen fome time to hover on the coaft, till there is calm weather, or a wind, that fuits the direction of their flight. Other birds of paffage have been drowned by thoufands in the fea, or have fettled on fhips quite exhaufted with fatigue. And others, either by miftaking their courfe, or by diftrefs of weather, have arrived in countries where they were never feen before: and thus are evidently fubject to the fame hazards that the human fpecies undergo, in the execution of their artificial purpofes.

3. The fame birds are emigrant from fome countries and not fo from others: the fwallows were feen at Goree in January by an ingenious philofopher of my acquaintance, and he was told that they continued there all the year; as the warmth of the climate was at all feafons fufficient for their own conftitutions, and for the production of the flies that fupply them with nourifhment.   Herodotus fays, that in Libya, about the fprings of the Nile, the fwallows continue all the year. (L. 2.)

Quails (tetrao corturnix, Lin.) are birds of paffage from the coaft of Barbary to Italy, and have frequently fettled in large fhoals on fhips fatigued with their flight. (Ray, Wifdom of God, p. 129.   Derham Phyfic. Theol. v. ii. p. 178.)   Dr. Ruffel, in his Hiftory of Aleppo, obferves that the fwallows vifit that country about the end of February, and having hatched their young difappear about the end of July; and returning again about the beginning of October, continue about a fortnight, and then again difappear. (P. 70.)

When

When my late friend Dr. Chambres, of Derby, was on the ifland of Caprea in the bay of Naples, he was informed that great flights of quails annually fettle on that ifland about the beginning of May, in their paffage from Africa to Europe. And that they always come when the fouth-eaft wind blows, are fatigued when they reft on this ifland, and are taken in fuch amazing quantities and fold to the Continent, that the inhabitants pay the bifhop his ftipend out of the profits arifing from the fale of them.

The flights of thefe birds acrofs the Mediterranean are recorded near three thoufand years ago. " There went forth a wind from the Lord and brought quails from the fea, and let them fall upon the camp, a day's journey round about it, and they were two cubits above the earth," (Numbers, chap. ii. ver. 31.)

In our country, Mr. Pennant informs us, that fome quails migrate, and others only remove from the internal parts of the ifland to the coafts, (Zoology, octavo, 210.) Some of the ringdoves and ftares breed here, others migrate, (ibid. 510, 511.) And the flender billed fmall birds do not all quit thefe kingdoms in the winter, though the difficulty of procuring the worms and infects, that they feed on, fupplies the fame reafon for migration to them all, (ibid. 511.)

Linnæus has obferved, that in Sweden the female chaffinches quit that country in September, migrating into Holland, and leave their mates behind till their return in fpring. Hence he has called them Fringilla cælebs, (Amæn. Acad. ii. 42. iv. 595.) Now in our climate both fexes of them are perennial birds. And Mr. Pennant obferves that the hoopoe, chatterer, hawfinch, and crofsbill, migrate into England fo rarely, and at fuch uncertain times, as not to deferve to be ranked among our birds of paffage, (ibid. 511.)

The water fowl, as geefe and ducks, are better adapted for long migrations, than the other tribes of birds, as, when the weather is calm, they can not only reft themfelves, or fleep upon the ocean, but poffibly procure fome kind of food from it.

Hence

Hence in Siberia, as foon as the lakes are frozen, the water fowl, which are very numerous, all difappear, and are fuppofed to fly to warmer climates, except the rail, which, from its inability for long flights, probably fleeps, like our bat, in their winter. The following account from the Journey of Profeffor Gmelin, may entertain the reader. " In the neighbourhood of Krafnoiark, amongft many other emigrant water fowls, we obferved a great number of rails, which when purfued never took flight, but endeavoured to efcape by running. We enquired how thefe birds, that could not fly, could retire into other countries in the winter, and were told, both by the Tartars and Affanians, that they well knew thofe birds could not alone pafs into other countries: but when the crains (les grues) retire in autumn, each one takes a rail (un rale) upon his back, and carries him to a warmer climate."

Recapitulation.

1. All birds of paffage can exift in the climates, where they are produced.

2. They are fubject in their migrations to the fame accidents and difficulties, that mankind are fubject to in navigation.

3. The fame fpecies of birds migrate from fome countries, and are refident in others.

From all thefe circumftances it appears that the migrations of birds are not produced by a neceffary inftinct, but are accidental improvements, like the arts among mankind, taught by their cotemporaries, or delivered by tradition from one generation of them to another.

XIII. In that feafon of the year which fupplies the nourifhment proper for the expected brood, the birds enter into a contract of marriage, and with joint labour conftruct a bed for the reception of their offspring,

offspring. Their choice of the proper feafon, their contracts of marriage, and the regularity with which they conftruct their nefts, have in all ages excited the admiration of naturalifts ; and have always been attributed to the power of inftinct, which, like the occult qualities of the antient philofophers, prevented all further enquiry. We fhall confider them in their order.

Their Choice of the Seafon.

Our domeftic birds, that are plentifully fupplied throughout the year with their adapted food, and are covered with houfes from the inclemency of the weather, lay their eggs at any feafon: which evinces that the fpring of the year is not pointed out to them by a neceffary inftinct.

Whilft the wild tribes of birds choofe this time of the year from their acquired knowledge, that the mild temperature of the air is more convenient for hatching their eggs, and is foon likely to fupply that kind of nourifhment, that is wanted for their young.

If the genial warmth of the fpring produced the paffion of love, as it expands the foliage of trees, all other animals fhould feel its influence as well as birds: but, the viviparous creatures, as they fuckle their young, that is, as they previoufly digeft the natural food, that it may better fuit the tender ftomachs of their offspring, experience the influence of this paffion at all feafons of the year, as cats and bitches. The graminivorous animals indeed generally produce their young about the time when grafs is fupplied in the greateft plenty; but this is without any degree of exactnefs, as appears from our cows, fheep, and hares, and may be a part of the traditional knowledge, which they learn from the example of their parents.

Their

Their Contracts of Marriage.

Their mutual paffion, and their acquired knowledge, that their joint labour is neceffary to procure fuftenance for their numerous family, induces the wild birds to enter into a contract of marriage, which does not however take place among the ducks, geefe, and fowls, that are provided with their daily food from our barns.

An ingenious philofopher has lately denied, that animals can enter into contracts, and thinks this an effential difference between them and the human creature:—but does not daily obfervation convince us, that they form contracts of friendfhip with each other, and with mankind? When puppies and kittens play together, is there not a tacit contract, that they will not hurt each other? And does not your favorite dog expect you fhould give him his daily food, for his fervices and attention to you? And thus barters his love for your protection? In the fame manner that all contracts are made amongft men, that do not underftand each others arbitrary language.

The Conftruction of their Nefts.

1. They feem to be inftructed how to build their nefts from their obfervation of that, in which they were educated, and from their knowledge of thofe things, that are moft agreeable to their touch in refpect to warmth, cleanlinefs, and ftability. They choofe their fituations from their ideas of fafety from their enemies, and of fhelter from the weather. Nor is the colour of their nefts a circumftance unthought of; the finches, that build in green hedges, cover their habitations with green mofs; the fwallow or martin, that builds

Z   againft

againſt rocks and houſes, covers her's with clay, whilſt the lark chooſes vegetable ſtraw nearly of the colour of the ground ſhe inhabits : by this contrivance, they are all leſs liable to be diſcovered by their adverſaries.

2. Nor are the neſts of the ſame ſpecies of birds conſtructed always of the ſame materials, nor in the ſame form ; which is another circumſtance that aſcertains, that they are led by obſervation.

In the trees before Mr. Levet's houſe in Lichfield, there are annually neſts built by ſparrows, a bird which uſually builds under the tiles of houſes, or the thatch of barns. Not finding ſuch convenient ſituations for their neſts, they build a covered neſt bigger than a man's head, with an opening like a mouth at the ſide, reſembling that of a magpie, except that it is built with ſtraw and hay, and lined with feathers, and ſo nicely managed as to be a defence againſt both wind and rain.

So the jackdaw (corvus monedula) generally builds in church-ſteeples, or under the roofs of high houſes ; but at Selbourn, in Southamptonſhire, where towers and ſteeples are not ſufficiently numerous, theſe ſame birds build in forſaken rabbit burrows. See a curious account of theſe ſubterranean neſts in White's Hiſtory of Selbourn, p. 59. Can the ſkilful change of architecture in theſe birds, and the ſparrows above mentioned be governed by inſtinct ? Then they muſt have two inſtincts, one for common, and the other for extraordinary occaſions.

I have ſeen green worſted in a neſt, which no where exiſts in nature : and the down of thiſtles in thoſe neſts, that were by ſome accident conſtructed later in the ſummer, which material could not be procured for the earlier neſts : in many different climates they cannot procure the ſame materials, that they uſe in ours. And it is well known, that the canary birds, that are propagated in this country, and the finches, that are kept tame, will build their neſts of any flexile

4   materials,

materials, that are given them.   Plutarch, in his Book on Rivers, speaking of the Nile, says, " that the swallows collect a material, when the waters recede, with which they form nests, that are impervious to water."   And in India there is a swallow that collects a glutinous substance for this purpose, whose nest is esculent, and esteemed a principal rarity amongst epicures, (Lin. Syst. Nat.)   Both these must be constructed of very different materials from those used by the swallows of our country.

In India the birds exert more artifice in building their nests on account of the monkeys and snakes: some form their pensile nests in the shape of a purse, deep and open at top; others with a hole in the side; and others, still more cautious, with an entrance at the very bottom, forming their lodge near the summit.   But the taylor-bird will not ever trust its nest to the extremity of a tender twig, but makes one more advance to safety by fixing it to the leaf itself.   It picks up a dead leaf, and sews it to the side of a living one, its slender bill being its needle, and its thread some fine fibres; the lining consists of feathers, gossamer, and down ; its eggs are white, the colour of the bird light yellow, its length three inches, its weight three sixteenths of an ounce ; so that the materials of the nest, and the weight of the bird, are not likely to draw down an habitation so slightly suspended.   A nest of this bird is preserved in the British Museum, (Pennant's Indian Zoology).   This calls to one's mind the Mosaic account of the origin of mankind, the first dawning of art there ascribed to them, is that of sewing leaves together.   For many other curious kinds of nests see Natural History for Children, by Mr. Galton. Johnson. London. Part I. p. 47. Gen. Oriolus.

3. Those birds that are brought up by our care, and have had little communication with others of their own species, are very defective in this acquired knowledge; they are not only very awkward in the construction of their nests, but generally scatter their eggs in various parts of the room or cage, where they are confined, and seldom

produce young ones, till, by failing in their firſt attempt, they have learnt ſomething from their own obſervation.

4. During the time of incubation birds are ſaid in general to turn their eggs every day; ſome cover them, when they leave the neſt, as ducks and geeſe; in ſome the male is ſaid to bring food to the female, that ſhe may have leſs occaſion of abſence, in others he is ſaid to take her place, when ſhe goes in queſt of food; and all of them are ſaid to leave their eggs a ſhorter time in cold weather than in warm. In Senegal the oſtrich ſits on her eggs only during the night, leaving them in the day to the heat of the ſun; but at the Cape of Good Hope, where the heat is leſs, ſhe ſits on them day and night.

If it ſhould be aſked, what induces a bird to ſit weeks on its firſt eggs unconſcious that a brood of young ones will be the product? The anſwer muſt be, that it is the ſame paſſion that induces the human mother to hold her offspring whole nights and days in her fond arms, and preſs it to her boſom, unconſcious of its future growth to ſenſe and manhood, till obſervation or tradition have informed her.

5. And as many ladies are too refined to nurſe their own children, and deliver them to the care and proviſion of others; ſo is there one inſtance of this vice in the feathered world. The cuckoo in ſome parts of England, as I am well informed by a very diſtinct and ingenious gentleman, hatches and educates her own young; whilſt in other parts ſhe builds no neſt, but uſes that of ſome leſſer bird, generally either of the wagtail, or hedge ſparrow, and depoſiting one egg in it, takes no further care of her progeny.

As the Rev. Mr. Stafford was walking in Gloſop Dale, in the Peak of Derbyſhire, he ſaw a cuckoo riſe from its neſt. The neſt was on the ſtump of a tree, that had been ſome time felled, among ſome chips that were in part turned grey, ſo as much to reſemble the colour of the bird, in this neſt were two young cuckoos: tying a ſtring about the leg of one of them, he pegged the other end of it to the ground,

and

and very frequently for many days beheld the old cuckoo feed thefe her young, as he ftood very near them.

Nor is this a new obfervation, though it is entirely overlooked by the modern naturalifts, for Ariftotle fpeaking of the cuckoo, afferts that fhe fometimes builds her neft among broken rocks, and on high mountains, (L. 6. H. c. 1.) but adds in another place that fhe generally poffeffes the neft of another bird, (L. 6. H. c. 7.) And Niphus fays that cuckoos rarely build for themfelves, moft frequently laying their eggs in the nefts of other birds, (Gefner, L. 3. de Cuculo.)

The Philofopher who is acquainted with thefe facts concerning the cuckoo, would feem to have very little *reafon* himfelf, if he could imagine this neglect of her young to be a neceffary *inftinct*!

XIV. The deep receffes of the ocean are inacceffible to mankind, which prevents us from having much knowledge of the arts and government of its inhabitants.

1. One of the baits ufed by the fifherman is an animal called an Old Soldier, his fize and form are fomewhat like the craw-fifh, with this difference, that his tail is covered with a tough membrane inftead of a fhell ; and to obviate this defect, he feeks out the uninhabited fhell of fome dead fifh, that is large enough to receive his tail, and carries it about with him as part of his clothing or armour.

2. On the coafts about Scarborough, where the haddocks, cods, and dog-fifh, are in great abundance, the fifhermen univerfally believe that the dog-fifh make a line, or femicircle, to encompafs a fhoal of haddocks and cod, confining them within certain limits near the fhore, and eating them as occafion requires. For the haddocks and cod are always found near the fhore without any dog-fifh among them, and the dog-fifh further off without any haddocks or cod ; and yet the former are known to prey upon the latter, and in fome years devour fuch immenfe quantities as to render this fifhery more expenfive than profitable.

3. The

3. The remora, when he wifhes to remove his fituation, as he is a very flow fwimmer, is content to take an outfide place on whatever conveyance is going his way; nor can the cunning animal be tempted to quit his hold of a fhip when fhe is failing, not even for the lucre of a piece of pork, left it fhould endanger the lofs of his paffage: at other times he is eafily caught with the hook.

4. The crab-fifh, like many other teftaceous animals, annually changes its fhell; it is then in a foft ftate, covered only with a mucous membrane, and conceals itfelf in holes in the fand or under weeds; at this place a hard fhelled crab always ftands centinel, to prevent the fea infects from injuring the other in its defencelefs ftate; and the fifhermen from his appearance know where to find the foft ones, which they ufe for baits in catching other fifh.

And though the hard fhelled crab, when he is on this duty, advances boldly to meet the foe, and will with difficulty quit the field; yet at other times he fhews great timidity, and has a wonderful fpeed in attempting his efcape; and, if often interrupted, will pretend death like the fpider, and watch an opportunity to fink himfelf into the fand, keeping only his eyes above. My ingenious friend Mr. Burdett, who favoured me with thefe accounts at the time he was furveying the coafts, thinks the commerce between the fexes takes place at this time, and infpires the courage of the creature.

5. The fhoals of herrings, cods, haddocks, and other fifh, which approach our fhores at certain feafons, and quit them at other feafons without leaving one behind; and the falmon, that periodically frequent our rivers, evince, that there are vagrant tribes of fifh, that perform as regular migrations as the birds of paffage already mentioned.

6. There is a cataract on the river Liffey in Ireland about nineteen feet high: here in the falmon feafon many of the inhabitants amufe themfelves in obferving thefe fifh leap up the torrent. They dart themfelves quite out of the water as they afcend, and frequently

fall

fall back many times before they furmount it, and bafkets made of twigs are placed near the edge of the ftream to catch them in their fall.

I have obferved, as I have fat by a fpout of water, which defcends from a ftone trough about two feet into a ftream below, at particular feafons of the year, a great number of little fifh called minums, or pinks, throw themfelves about twenty times their own length out of the water, expecting to get into the trough above.

This evinces that the ftorgee, or attention of the dam to provide for the offspring, is ftrongly exerted amongft the nations of fifh, where it would feem to be the moft neglected; as thefe falmon cannot be fuppofed to attempt fo difficult and dangerous a tafk without being confcious of the purpofe or end of their endeavours.

It is further remarkable, that moft of the old falmon return to the fea before it is proper for the young fhoals to attend them, yet that a few old ones continue in the rivers fo late, that they become perfectly emaciated by the inconvenience of their fituation, and this apparently to guide or to protect the unexperienced brood.

Of the fmaller water animals we have ftill lefs knowledge, who neverthelefs probably poffefs many fuperior arts; fome of thefe are mentioned in Botanic Garden, P. I. Add. Note XXVII. and XXVIII. The nympha of the water-moths of our rivers, which cover themfelves with cafes of ftraw, gravel, and fhell, contrive to make their habitations nearly in equilibrium with the water; when too heavy, they add a bit of wood or ftraw; when too light, a bit of gravel. Edinb. Tranf.

All thefe circumftances bear a near refemblance to the deliberate actions of human reafon.

XV. We have a very imperfect acquaintance with the various tribes of infects: their occupations, manner of life, and even the number of their fenfes, differ from our own, and from each other; but there is reafon to imagine, that thofe which poffefs the fenfe of

touch

touch in the moſt exquiſite degree, and whoſe occupations require the moſt conſtant exertion of their powers, are indued with a greater proportion of knowledge and ingenuity.

The ſpiders of this country manufacture nets of various forms, adapted to various ſituations, to arreſt the flies that are their food ; and ſome of them have a houſe or lodging-place in the middle of the net, well contrived for warmth, ſecurity, or concealment. There is a large ſpider in South America, who conſtructs nets of ſo ſtrong a texture as to entangle ſmall birds, particularly the humming bird. And in Jamaica there is another ſpider, who digs a hole in the earth obliquely downwards, about three inches in length, and one inch in diameter, this cavity ſhe lines with a tough thick web, which when taken out reſembles a leathern purſe : but what is moſt curious, this houſe has a door with hinges, like the operculum of ſome ſea ſhells ; and herſelf and family, who tenant this neſt, open and ſhut the door, whenever they paſs or repaſs. This hiſtory was told me, and the neſt with its operculum ſhewn me by the late Dr. Butt of Bath, who was ſome years phyſician in Jamaica.

The production of theſe nets is indeed a part of the nature or con-formation of the animal, and their natural uſe is to ſupply the place of wings, when ſhe wiſhes to remove to another ſituation. But when ſhe employs them to entangle her prey, there are marks of evident deſign, for ſhe adapts the form of each net to its ſituation, and ſtrengthens thoſe lines, that require it, by joining others to the middle of them, and attaching thoſe others to diſtant objects, with the ſame individual art, that is uſed by mankind in ſupporting the maſts and extending the ſails of ſhips. This work is executed with more ma-thematical exactneſs and ingenuity by the field ſpiders, than by thoſe in our houſes, as their conſtructions are more ſubjected to the injuries of dews and tempeſts.

Beſides the ingenuity ſhewn by theſe little creatures in taking their prey, the circumſtance of their counterfeiting death, when they are

put

put into terror, is truly wonderful; and as foon as the object of terror is removed, they recover and run away. Some beetles are alfo faid to poffefs this piece of hypocrify.

The curious webs, or chords, conftructed by fome young caterpillars to defend themfelves from cold, or from infects of prey; and by filk-worms and fome other caterpillars, when they tranfmigrate into aureliæ or larvæ, have defervedly excited the admiration of the inquifitive. But our ignorance of their manner of life, and even of the number of their fenfes, totally precludes us from underftanding the means by which they acquire this knowledge.

The care of the falmon in choofing a proper fituation for her fpawn, the ftructure of the nefts of birds, their patient incubation, and the art of the cuckoo in depofiting her egg in her neighbour's nurfery, are inftances of great fagacity in thofe creatures: and yet they are much inferior to the arts exerted by many of the infect tribes on fimilar occafions. The hairy excrefcences on briars, the oak apples, the blafted leaves of trees, and the lumps on the backs of cows, are fituations that are rather produced than chofen by the mother infect for the convenience of her offspring. The cells of bees, wafps, fpiders, and of the various coralline infects, equally aftonifh us, whether we attend to the materials or to the architecture.

But the conduct of the ant, and of fome fpecies of the ichneumon fly in the incubation of their eggs, is equal to any exertion of human fcience. The ants many times in a day move their eggs nearer the furface of their habitation, or deeper below it, as the heat of the weather varies; and in colder days lie upon them in heaps for the purpofe of incubation: if their manfion is too dry, they carry them to places where there is moifture, and you may diftinctly fee the little worms move and fuck up the water. When too much moifture approaches their neft, they convey their eggs deeper in the earth, or to fome other place of fafety. (Swammerd, Epil. ad Hift. Infects, p. 153. Phil. Tranf. No. 23. Lowthrop. V. 2. p. 7.)

A a

There

There is one ſpecies of ichneumon-fly, that digs a hole in the earth, and carrying into it two or three living caterpillars, depoſits her eggs, and nicely cloſing up the neſt leaves them there; partly doubtleſs to aſſiſt the incubation, and partly to ſupply food to her future young, (Derham. B. 4. c. 13. Ariſtotle Hiſt. Animal, L. 5. c. 20.)

A friend of mine put about fifty large caterpillars collected from cabbages on ſome bran and a few leaves into a box, and covered it with gauze to prevent their eſcape. After a few days we ſaw, from more than three fourths of them, about eight or ten little caterpillars of the ichneumon-fly come out of their backs, and ſpin each a ſmall cocoon of ſilk, and in a few days the large caterpillars died. This ſmall fly it ſeems lays its egg in the back of the cabbage caterpillar, which when hatched preys upon the material, which is produced there for the purpoſe of making ſilk for the future neſt of the cabbage cater-pillar; of which being deprived, the creature wanders about till it dies, and thus our gardens are preſerved by the ingenuity of this cruel fly. This curious property of producing a ſilk thread, which is com-mon to ſome ſea animals, ſee Botanic Garden, Part I. Note XXVII. and is deſigned for the purpoſe of their transformation as in the ſilk-worm, is uſed for conveying themſelves from higher branches to lower ones of trees by ſome caterpillars, and to make themſelves temporary neſts or tents, and by the ſpider for entangling his prey. Nor is it ſtrange that ſo much knowledge ſhould be acquired by ſuch ſmall ani-mals; ſince there is reaſon to imagine, that theſe inſects have the ſenſe of touch, either in their proboſcis, or their autennæ, to a great degree of perfection; and thence may poſſeſs, as far as their ſphere extends, as accurate knowledge, and as ſubtle invention, as the diſcoverers of human arts.

XVI. 1. If we were better acquainted with the hiſtories of thoſe inſects that are formed into ſocieties, as the bees, waſps, and ants, I make no doubt but we ſhould find, that their arts and improvements are not ſo ſimilar and uniform as they now appear to us, but that they
aroſe

arofe in the fame manner from experience and tradition, as the arts of our own fpecies; though their reafoning is from fewer ideas, is bufied about fewer objects, and is exerted with lefs energy.

There are fome kinds of infects that migrate like the birds before mentioned. The locuft of warmer climates has fometimes come over to England; it is fhaped like a grafshopper, with very large wings, and a body above an inch in length. It is mentioned as coming into Egypt with an eaft wind, " The Lord brought an eaft wind upon the land all that day and night, and in the morning the eaft wind brought the locufts, and covered the face of the earth, fo that the land was dark," Exod. x. 13. The migrations of thefe infects are mentioned in another part of the fcripture, " The locufts have no king, yet go they forth all of them in bands," Prov. xxx. 27.

The accurate Mr. Adanfon, near the river Gambia in Africa, was witnefs to the migration of thefe infects. " About eight in the morning, in the month of February, there fuddenly arofe over our heads a thick cloud, which darkened the air, and deprived us of the rays of the fun. We found it was a cloud of locufts raifed about twenty or thirty fathoms from the ground, and covering an extent of feveral leagues; at length a fhower of thefe infects defcended, and after devouring every green herb, while they refted, again refumed their flight. This cloud was brought by a ftrong eaft-wind, and was all the morning in paffing over the adjacent country." (Voyage to Senegal, 158.)

In this country the gnats are fometimes feen to migrate in clouds, like the mufketoes of warmer climates, and our fwarms of bees frequently travel many miles, and are faid in North America always to fly towards the fouth. The prophet Ifaiah has a beautiful allufion to thefe migrations, " The Lord fhall call the fly from the rivers of Egypt, and fhall hifs for the bee that is in the land of Affyria," Ifa. vii. 18. which has been lately explained by Mr. Bruce, in his travels to difcover the fource of the Nile.

2. I am well informed that the bees that were carried into Barbadoes, and other western iflands, ceafed to lay up any honey after the firft year, as they found it not ufeful to them: and are now become very troublefome to the inhabitants of thofe iflands by infefting their fugarhoufes; but thofe in Jamaica continue to make honey, as the cold north winds, or rainy feafons of that ifland, confine them at home for feveral weeks together. And the bees of Senegal, which differ from thofe of Europe only in fize, make their honey not only fuperior to ours in delicacy of flavour, but it has this fingularity, that it never concretes, but remains liquid as fyrup, (Adanfon). From fome obfervations of Mr. Wildman, and of other people of veracity, it appears, that during the fevere part of the winter feafon for weeks together the bees are quite benumbed and torpid from the cold, and do not confume any of their provifion. This ftate of fleep, like that of fwallows and bats, feems to be the natural refource of thofe creatures in cold climates, and the making of honey to be an artificial improvement.

As the death of our hives of bees appears to be owing to their being kept fo warm, as to require food when their ftock is exhaufted; a very obferving gentleman at my requeft put two hives for many weeks into a dry cellar, and obferved, during all that time, they did not confume any of their provifion, for their weight did not decreafe, as it had done when they were kept in the open air. The fame obfervation is made in the Annual Regifter for 1768, p. 113. And the Rev. Mr. White, in his Method of preferving Bees, adds, that thofe on the north fide of his houfe confumed lefs honey in the winter than thofe on the fouth fide.

There is another obfervation on bees well afcertained, that they at various times, when the feafon begins to be cold, by a general motion of their legs as tl _, hang in clufters produce a degree of warmth, which is eafily perceptible by the hand. Hence by this ingenious exertion,

ertion, they for a long time prevent the torpid state they would naturally fall into.

According to the late observations of Mr. Hunter, it appears that the bee's-wax is not made from the dust of the anthers of flowers, which they bring home on their thighs, but that this makes what is termed bee-bread, and is used for the purpose of feeding the bee-maggots; in the same manner butterflies live on honey, but the previous caterpillar lives on vegetable leaves, while the maggots of large flies require flesh for their food, and those of the ichneumon fly require insects for their food. What induces the bee who lives on honey to lay up vegetable powder for its young? What induces the butterfly to lay its eggs on leaves, when itself feeds on honey? What induces the other flies to seek a food for their progeny different from what they consume themselves? If these are not deductions from their own previous experience or observation, all the actions of mankind must be resolved into instinct.

3. The dormouse consumes but little of its food during the rigour of the season, for they roll themselves up, or sleep, or lie torpid the greatest part of the time; but on warm sunny days experience a short revival, and take a little food, and then relapse into their former state." (Pennant Zoolog. p. 67.) Other animals, that sleep in winter without laying up any provender, are observed to go into their winter beds fat and strong, but return to day-light in the spring season very lean and feeble. The common flies sleep during the winter without any provision for their nourishment, and are daily revived by the warmth of the sun, or of our fires. These whenever they see light endeavour to approach it, having observed, that by its greater vicinity they get free from the degree of torpor, that the cold produces; and are hence induced perpetually to burn themselves in our candles: deceived, like mankind, by the misapplication of their knowledge. Whilst many of the subterraneous insects, as the common worms, seem to retreat so deep into the earth as not to be enlivened or awak-

ened

ened by the difference of our winter days; and ftop up their holes with leaves or ftraws, to prevent the frofts from injuring them, or the centipes from devouring them. The habits of peace, or the ftratagems of war, of thefe fubterranean nations are covered from our view; but a friend of mine prevailed on a diftreffed worm to enter the hole of another worm on a bowling-green, and he prefently returned much wounded about his head. And I once faw a worm rife haftily out of the earth into the funfhine, and obferved a centipes hanging at its tail; the centipes nimbly quitted the tail, and feizing the worm about its middle cut it in half with its forceps, and preyed upon one part, while the other efcaped. Which evinces they have defign in ftopping the mouths of their habitations.

4. The wafp of this country fixes his habitation under ground, that he may not be affected with the various changes of our climate; but in Jamaica he hangs it on the bough of a tree, where the feafons are lefs fevere. He weaves a very curious paper of vegetable fibres to cover his neft, which is conftructed on the fame principle with that of the bee, but with a different material; but as his prey confifts of flefh, fruits, and infects, which are perifhable commodities, he can lay up no provender for the winter.

M. de la Loubiere, in his relation of Siam, fays, " That in a part of that kingdom, which lies open to great inundations, all the ants make their fettlements upon trees; no ants' nefts are to be feen any where elfe." Whereas in our country the ground is their only fituation. From the fcriptural account of thefe infects, one might be led to fufpect, that in fome climates they lay up a provifion for the winter. Origen affirms the fame, (Cont. Celf. L. 4.) But it is generally believed that in this country they do not, (Prov. vi. 6. xxx. 25.) The white ants of the coaft of Africa make themfelves pyramids eight or ten feet high, on a bafe of about the fame width, with a fmooth furface of rich clay, exceffively hard and well built, which appear at a diftance like an affemblage of the huts of the negroes, (Adanfon).

The

The hiftory of thefe have been lately well defcribed in the Philofoph. Tranfactions, under the name of termes, or termites. Thefe differ very much from the nefts of our large ant; but the real hiftory of this creature, as well as of the wafp, is yet very imperfectly known.

Wafps are faid to catch large fpiders, and to cut off their legs, and carry their mutilated bodies to their young, Dict. Raifon. Tom. I. p. 152.

One circumftance I fhall relate which fell under my own eye, and fhewed the power of reafon in a wafp, as it is exercifed among men. A wafp, on a gravel walk, had caught a fly nearly as large as himfelf; kneeling on the ground I obferved him feparate the tail and the head from the body part, to which the wings were attached. He then took the body part in his paws, and rofe about two feet from the ground with it; but a gentle breeze wafting the wings of the fly turned him round in the air, and he fettled again with his prey upon the gravel. I then diftinctly obferved him cut off with his mouth, firft one of the wings, and then the other, after which he flew away with it unmolefted by the wind.

Go, thou fluggard, learn arts and induftry from the bee, and from the ant!

Go, proud reafoner, and call the worm thy fifter!

XVII. *Conclufion.*

It was before obferved how much the fuperior accuracy of our fenfe of touch contributes to increafe our knowledge; but it is the greater energy and activity of the power of volition (as explained in the former Sections of this work) that marks mankind, and has given him the empire of the world.

There

There is a criterion by which we may diftinguifh our voluntary acts or thoughts from thofe that are excited by our fenfations: "The former are always employed about the *means* to acquire pleafureable objects, or to avoid painful ones: while the latter are employed about the *poffeffion* of thofe that are already in our power."

If we turn our eyes upon the fabric of our fellow animals, we find they are fupported with bones, covered with fkins, moved by mufcles; that they poffefs the fame fenfes, acknowledge the fame appetites, and are nourifhed by the fame aliment with ourfelves; and we fhould hence conclude from the ftrongeft analogy, that their internal faculties were alfo in fome meafure fimilar to our own.

Mr. Locke indeed publifhed an opinion, that other animals poffeffed no abftract or general ideas, and thought this circumftance was the barrier between the brute and the human world. But thefe abftracted ideas have been fince demonftrated by Bifhop Berkley, and allowed by Mr. Hume, to have no exiftence in nature, not even in the mind of their inventor, and we are hence neceffitated to look for fome other mark of diftinction.

The ideas and actions of brutes, like thofe of children, are almoft perpetually produced by their prefent pleafures, or their prefent pains; and, except in the few inftances that have been mentioned in this Section, they feldom bufy themfelves about the *means* of procuring future blifs, or of avoiding future mifery.

Whilft the acquiring of languages, the making of tools, and the labouring for money; which are all only the *means* of procuring pleafure: and the praying to the Deity, as another *means* to procure happinefs, are characteriftic of human nature.

SECT.

SECT. XVII.

THE CATENATION OF MOTIONS.

I.   1. *Catenations of animal motion.*   2. *Are produced by irritations, by senfations, by volitions.*   3. *They continue fome time after they have been excited. Caufe of catenation.*   4. *We can then exert our attention on other objeEts.*   5. *Many catenations of motions go on together.*   6. *Some links of the catenations of motions may be left out without difuniting the chain.*   7. *Interrupted circles of motion continue confufedly till they come to the part of the circle, where they were difturbed.*   8. *Weaker catenations are diffevered by ftronger.*   9. *Then new catenations take place.*   10. *Much effort prevents their reuniting. Impediment of fpeech.*   11. *Trains more eafily diffevered than circles.*   12. *Sleep deftroys volition and external ftimulus.*   II. *Inftances of various catenations in a young lady playing on the harpfichord.*   III. 1. *What catenations are the ftrongeft.*   2. *Irritations joined with affociations form ftrongeft connexions. Vital motions.*   3. *New links with increafed force, cold fits of fever produced.*   4. *New links with decreafed force. Cold bath.*   5. *Irritation joined with fenfation. Inflammatory fever. Why children cannot tickle themfelves.*   6. *Volition joined with fenfation. Irritative ideas of found become fenfible.*   7. *Ideas of imagination diffevered by irritations, by volition, produEtion of furprife.*

I.   1. TO inveftigate with precifion the catenations of animal motions, it would be well to attend to the manner of their produEtion; but we cannot begin this difquifition early enough for this purpofe, as the catenations of motion feem to begin with life, and are only extinguifhable with it.   We have fpoken of the power of irritation, of fenfation, of volition, and of affociation, as preceding the fibrous

B b   motions;

motions; we now ftep forwards, and confider, that converfely they are in their turn preceded by thofe motions; and that all the fucceffive trains or circles of our actions are compofed of this twofold concatenation. Thofe we fhall call trains of action, which continue to proceed without any ftated repetitions; and thofe circles of action, when the parts of them return at certain periods, though the trains, of which they confift, are not exactly fimilar. The reading an epic poem is a train of actions; the reading a fong with a chorus at equal diftances in the meafure conftitutes fo many circles of action.

2. Some catenations of animal motion are produced by reiterated fucceffive irritations, as when we learn to repeat the alphabet in its order by frequently reading the letters of it. Thus the vermicular motions of the bowels were originally produced by the fucceffive irritations of the paffing aliment; and the fucceffion of actions of the auricles and ventricles of the heart was originally formed by fucceffive ftimulus of the blood, thefe afterwards become part of the diurnal circles of animal actions, as appears by the periodical returns of hunger, and the quickened pulfe of weak people in the evening.

Other catenations of animal motion are gradually acquired by fucceffive agreeable fenfations, as in learning a favourite fong or dance; others by difagreeable fenfations, as in coughing or nictitation; thefe become affociated by frequent repetition, and afterwards compofe parts of greater circles of action like thofe above mentioned.

Other catenations of motions are gradually acquired by frequent voluntary repetitions; as when we deliberately learn to march, read, fence, or any mechanic art, the motions of many of our mufcles become gradually linked together in trains, tribes, or circles of action. Thus when any one at firft begins to ufe the tools in turning wood or metals in a lathe, he wills the motions of his hand or fingers, till at length thefe actions become fo connected with the effect, that he feems only to will the point of the chiffel. Thefe are caufed by volition,

7

lition, connected by affociation like thofe above defcribed, and after-wards become parts of our diurnal trains or circles of action.

3. All thefe catenations of animal motions are liable to proceed fome time after they are excited, unlefs they are difturbed or impeded by other irritations, fenfations, or volitions; and in many inftances in fpite of our endeavours to ftop them; and this property of animal motions is probably the caufe of their catenation. Thus when a child revolves fome minutes on one foot, the fpectra of the ambient objects appear to circulate round him fome time after he falls upon the ground. Thus the palpitation of the heart continues fome time after the object of fear, which occafioned it, is removed. The blufh of fhame, which is an excefs of fenfation, and the glow of anger, which is an excefs of volition, continue fome time, though the affected per-fon finds, that thofe emotions were caufed by miftaken facts, and en-deavours to extinguifh their appearance. See Sect. XII. 1. 5.

4. When a circle of motions becomes connected by frequent repe-titions as above, we can exert our attention ftrongly on other objects, and the concatenated circle of motions will neverthelefs proceed in due order; as whilft you are thinking on this fubject, you ufe variety of mufcles in walking about your parlour, or in fitting at your writ-ing-table.

5. Innumerable catenations of motions may proceed at the fame time, without incommoding each other. Of thefe are the motions of the heart and arteries; thofe of digeftion and glandular fecretion; of the ideas, or fenfual motions; thofe of progreffion, and of fpeak-ing; the great annual circle of actions fo apparent in birds in their times of breeding and moulting; the monthly circles of many female animals; and the diurnal circles of fleeping and waking, of fulnefs and inanition.

6. Some links of fucceffive trains or of fynchronous tribes of action may be left out without disjoining the whole. Such are our ufual trains of recollection; after having travelled through an entertaining

country, and viewed many delightful lawns, rolling rivers, and echoing rocks ; in the recollection of our journey we leave out the many diftricts, that we croffed, which were marked with no peculiar pleasure. Such also are our complex ideas, they are catenated tribes of ideas, which do not perfectly refemble their correfpondent perceptions, becaufe fome of the parts are omitted.

7. If an interrupted circle of actions is not entirely diffevered, it will continue to proceed confufedly, till it comes to the part of the circle, where it was interrupted.

The vital motions in a fever from drunkennefs, and in other periodical difeafes, are inftances of this circumftance. The accidental inebriate does not recover himfelf perfectly till about the fame hour on the fucceeding day. The accuftomed drunkard is difordered, if he has not his ufual potation of fermented liquor. So if a confiderable part of a connected tribe of action be difturbed, that whole tribe goes on with confufion, till the part of the tribe affected regains its accuftomed catenations. So vertigo produces vomiting, and a great fecretion of bile, as in fea-ficknefs, all thefe being parts of the tribe of irritative catenations.

8. Weaker catenated trains may be diffevered by the fudden exertion of the ftronger. When a child firft attempts to walk acrofs a room, call to him, and he inftantly falls upon the ground. So while I am thinking over the virtues of my friends, if the tea-kettle fpurt out fome hot water on my ftocking; the fudden pain breaks the weaker chain of ideas, and introduces a new group of figures of its own. This circumftance is extended to fome unnatural trains of action, which have not been confirmed by long habit ; as the hiccough, or an ague-fit, which are frequently curable by furprife. A young lady about eleven years old had for five days had a contraction of one mufcle in her fore arm, and another in her arm, which occurred four or five times every minute ; the mufcles were feen to leap, but without bending the arm. To counteract this new morbid habit, an iffue

3   was

was placed over the convulfed mufcle of her arm, and an adhefive plafter wrapped tight like a bandage over the whole fore arm, by which the new motions were immediately deftroyed, but the means were continued fome weeks to prevent a return.

9. If any circle of actions is diffevered, either by omiffion of fome of the links, as in fleep, or by infertion of other links, as in furprife, new catenations take place in a greater or lefs degree. The laft link of the broken chain of actions becomes connected with the new motion which has broken it, or with that which was neareft the link omitted; and thefe new catenations proceed inftead of the old ones. Hence the periodic returns of ague-fits, and the chimeras of our dreams.

10. If a train of actions is diffevered, much effort of volition or fenfation will prevent its being reftored. Thus in the common impediment of fpeech, when the affociation of the motions of the mufcles of enunciation with the idea of the word to be fpoken is difordered, the great voluntary efforts, which diftort the countenance, prevent the rejoining of the broken affociations. See No. II. 10. of this Section. It is thus likewife obfervable in fome inflammations of the bowels, the too ftrong efforts made by the mufcles to carry forwards the offending material fixes it more firmly in its place, and prevents the cure. So in endeavouring to recal to our memory fome particular word of a fentence, if we exert ourfelves too ftrongly about it, we are lefs likely to regain it.

11. Catenated trains or tribes of action are eafier diffevered than catenated circles of action. Hence in epileptic fits the fynchronous connected tribes of action, which keep the body erect, are diffevered, but the circle of vital motions continues undifturbed.

12. Sleep deftroys the power of volition, and precludes the ftimuli of external objects, and thence diffevers the trains, of which thefe are a part; which confirms the other catenations, as thofe of the vital

motions,

motions, fecretions, and abforptions; and produces the new trains of ideas, which conftitute our dreams.

II. 1. All the preceding circumftances of the catenations of animal motions will be more clearly underftood by the following example of a perfon learning mufic; and when we recollect the variety of mechanic arts, which are performed by affociated trains of mufcular actions catenated with the effects they produce, as in knitting, netting, weaving; and the greater variety of affociated trains of ideas caufed or catenated by volitions or fenfations, as in our hourly modes of reafoning, or imagining, or recollecting, we fhall gain fome idea of the innumerable catenated trains and circles of action, which form the tenor of our lives, and which began, and will only ceafe entirely with them.

2. When a young lady begins to learn mufic, fhe voluntarily applies herfelf to the characters of her mufic-book, and by many repetitions endeavours to catenate them with the proportions of found, of which they are fymbols. The ideas excited by the mufical characters are flowly connected with the keys of the harpfichord, and much effort is neceffary to produce every note with the proper finger, and in its due place and time; till at length a train of voluntary exertions becomes catenated with certain irritations. As the various notes by frequent repetitions become connected in the order, in which they are produced, a new catenation of fenfitive exertions becomes mixed with the voluntary ones above defcribed; and not only the mufical fymbols of crotchets and quavers, but the auditory notes and tones at the fame time, become fo many fucceffive or fynchronous links in this circle of catenated actions.

At length the motions of her fingers become catenated with the mufical characters; and thefe no fooner ftrike the eye, than the finger preffes down the key without any voluntary attention between them; the activity of the hand being connected with the irritation of the

figure

figure or place of the mufical fymbol on the retina; till at length by
frequent repetitions of the fame tune the movements of her fingers in
playing, and the mufcles of the larynx in finging, become affociated
with each other, and form part of thofe intricate trains and circles of
catenated motions, according with the fecond article of the preceding
propofitions in No. 1. of this Section.

3. Befides the facility, which by habit attends the execution of
this mufical performance, a curious circumftance occurs, which is,
that when our young mufician has began a tune, fhe finds herfelf inclined to continue it; and that even when fhe is carelefsly finging
alone without attending to her own fong; according with the third
preceding article.

4. At the fame time that our young performer continues to play
with great exactnefs this accuftomed tune, fhe can bend her mind,
and that intenfely, on fome other object, according with the fourth
article of the preceding propofitions.

The manufcript copy of this work was lent to many of my friends
at different times for the purpofe of gaining their opinions and criti-
cifms on many parts of it, and I found the following anecdote written with a pencil oppofite to this page, but am not certain by whom.
" I remember feeing the pretty young actrefs, who fucceeded Mrs.
Arne in the performance of the celebrated Padlock, rehearfe the mufical parts at her harpfichord under the eye of her mafter with great
tafte and accuracy; though I obferved her countenance full of emotion, which I could not account for; at laft fhe fuddenly burft into
tears; for fhe had all this time been eyeing a beloved canary bird,
fuffering great agonies, which at that inftant fell dead from its
perch."

5. At the fame time many other catenated circles of action are
going on in the perfon of our fair mufician, as well as the motions of
her fingers, fuch as the vital motions, refpiration, the movements of

her

her eyes and eyelids, and of the intricate muscles of vocality, according with the fifth preceding article.

6. If by any strong impression on the mind of our fair musician she should be interrupted for a very inconsiderable time, she can still continue her performance, according to the sixth article.

7. If however this interruption be greater, though the chain of actions be not dissevered, it proceeds confusedly, and our young performer continues indeed to play, but in a hurry without accuracy and elegance, till she begins the tune again, according to the seventh of the preceding articles.

8. But if this interruption be still greater, the circle of actions becomes entirely dissevered, and she finds herself immediately under the necessity to begin over again to recover the lost catenation, according to the eighth preceding article.

9. Or in trying to recover it she will sing some dissonant notes, or strike some improper keys, according to the ninth preceding article.

10. A very remarkable thing attends this breach of catenation, if the performer has forgotten some word of her song, the more energy of mind she uses about it, the more distant is she from regaining it; and artfully employs her mind in part on some other object, or endeavours to dull its perceptions, continuing to repeat, as it were inconsciously, the former part of the song, that she remembers, in hopes to regain the lost connexion.

For if the activity of the mind itself be more energetic, or takes its attention more, than the connecting word, which is wanted; it will not perceive the slighter link of this lost word; as who listens to a feeble sound, must be very silent and motionless; so that in this case the very vigour of the mind itself seems to prevent it from regaining the lost catenation, as well as the too great exertion in endeavouring to regain it, according to the tenth preceding article.

We

We frequently experience, when we are doubtful about the spelling of a word, that the greater voluntary exertion we use, that is the more intensely we think about it, the further are we from regaining the lost association between the letters of it, but which readily recurs when we have become careless about it. In the same manner, after having for an hour laboured to recollect the name of some absent person, it shall seem, particularly after sleep, to come into the mind as it were spontaneously; that is, the word we are in search of, was joined to the preceding one by association; this association being dissevered, we endeavour to recover it by volition; this very action of the mind strikes our attention more, than the faint link of association, and we find it impossible by this means to retrieve the lost word. After sleep, when volition is entirely suspended, the mind becomes capable of perceiving the fainter link of association, and the word is regained.

On this circumstance depends the impediment of speech before mentioned; the first syllable of a word is causable by volition, but the remainder of it is in common conversation introduced by its associations with this first syllable acquired by long habit. Hence when the mind of the stammerer is vehemently employed on some idea of ambition of shining, or fear of not succeeding, the associations of the motions of the muscles of articulation with each other become dissevered by this greater exertion, and he endeavours in vain by voluntary efforts to rejoin the broken association. For this purpose he continues to repeat the first syllable, which is causable by volition, and strives in vain, by various distortions of countenance, to produce the next links, which are subject to association. See Class IV. 3. 1. 1.

11. After our accomplished musician has acquired great variety of tunes and songs, so that some of them begin to cease to be easily recollected, she finds progressive trains of musical notes more frequently forgotten, than those which are composed of reiterated circles, according with the eleventh preceding article.

C c

12. To

12. To finiſh our example with the preceding articles we muſt at length ſuppoſe, that our fair performer falls aſleep over her harpſichord; and thus by the ſuſpenſion of volition, and the excluſion of external ſtimuli, ſhe diſſevers the trains and circles of her muſical exertions.

III. 1. Many of theſe circumſtances of catenations of motions receive an eaſy explanation from the four following conſequences to the ſeventh law of animal cauſation in Sect. IV. Theſe are, firſt, that thoſe ſucceſſions or combinations of animal motions, whether they were united by cauſation, aſſociation, or catenation, which have been moſt frequently repeated, acquire the ſtrongeſt connection. Secondly, that of theſe, thoſe, which have been leſs frequently mixed with other trains or tribes of motion, have the ſtrongeſt connection. Thirdly, that of theſe, thoſe, which were firſt formed, have the ſtrongeſt connection. Fourthly, that if an animal motion be excited by more than one cauſation, aſſociation, or catenation, at the ſame time, it will be performed with greater energy.

2. Hence alſo we underſtand, why the catenations of irritative motions are more ſtrongly connected than thoſe of the other claſſes, where the quantity of unmixed repetition has been equal; becauſe they were firſt formed. Such are thoſe of the ſecerning and abſorbent ſyſtems of veſſels, where the action of the gland produces a fluid, which ſtimulates the mouths of its correſpondent abſorbents. The aſſociated motions ſeem to be the next moſt ſtrongly united, from their frequent repetition; and where both theſe circumſtances unite, as in the vital motions, their catenations are indiſſoluble but by the deſtruction of the animal.

3. Where a new link has been introduced into a circle of actions by ſome accidental defect of ſtimulus; if that defect of ſtimulus be repeated at the ſame part of the circle a ſecond or a third time, the defective motions thus produced, both by the repeated defect of ſti-

8

mulus

mulus and by their catenation with the parts of the circle of actions, will be performed with lefs and lefs energy. Thus if any perfon is expofed to cold at a certain hour to-day, fo long as to render fome part of the fyftem for a time torpid; and is again expofed to it at the fame hour to-morrow, and the next day; he will be more and more affected by it, till at length a cold fit of fever is completely formed, as happens at the beginning of many of thofe fevers, which are called nervous or low fevers. Where the patient has flight periodical fhiverings and palenefs for many days before the febrile paroxifm is completely formed.

4. On the contrary, if the expofure to cold be for fo fhort a time, as not to induce any confiderable degree of torpor or quiefcence, and is repeated daily as above mentioned, it lofes its effect more and more at every repetition, till the conftitution can bear it without inconvenience, or indeed without being confcious of it. As in walking into the cold air in frofty weather. The fame rule is applicable to increafed ftimulus, as of heat, or of vinous fpirit, within certain limits, as is applied in the two laft paragraphs to Deficient Stimulus, as is further explained in Sect. XXXVI. on the Periods of Difeafes.

5. Where irritation coincides with fenfation to produce the fame catenations of motion, as in inflammatory fevers, they are excited with ftill greater energy than by the irritation alone. So when children expect to be tickled in play, by a feather lightly paffed over the lips, or by gently vellicating the foles of their feet, laughter is moft vehemently excited; though they can ftimulate thefe parts with their own fingers unmoved. Here the pleafureable idea of playfulnefs coincides with the vellication; and there is no voluntary exertion ufed to diminifh the fenfation, as there would be, if a child fhould endeavour to tickle himfelf. See Sect. XXXIV. 1. 4.

6. And laftly, the motions excited by the junction of voluntary exertion with irritation are performed with more energy, than thofe

C c 2

by

by irritation fingly; as when we liften to fmall noifes, as to the tick-
ing of a watch in the night, we perceive the moft weak founds, that
are at other times unheeded.   So when we attend to the irritative
ideas of found in our ears, which are generally not attended to, we
can hear them; and can fee the fpectra of objects, which remain in
the eye, whenever we pleafe to exert our voluntary power in aid of
thofe weak actions of the retina, or of the auditory nerve.

7. The temporary catenations of ideas, which are caufed by the
fenfations of pleafure or pain, are eafily differered either by irritations,
as when a fudden noife difturbs a day-dream; or by the power of volition, as when we awake from fleep.   Hence in our waking hours,
whenever an idea occurs, which is incongruous to our former experience, we inftantly differer the train of imagination by the power
of volition, and compare the incongruous idea with our previous
knowledge of nature, and reject it.   This operation of the mind has
not yet acquired a fpecific name, though it is exerted every minute
of our waking hours; unlefs it may be termed INTUITIVE ANA-
LOGY.   It is an act of reafoning of which we are unconfcious except
from its effects in preferving the congruity of our ideas, and bears the
fame relation to the fenforial power of volition, that irritative ideas,
of which we are inconfcious except by their effects, do to the fenforial power of irritation; as the former is produced by volition without
our attention to it, and the latter by irritation without our attention
to them.

If on the other hand a train of imagination or of voluntary ideas are
excited with great energy, and paffing on with great vivacity, and become differered by fome violent ftimulus, as the difcharge of a piftol
near one's ear, another circumftance takes place, which is termed
SURPRISE; which by exciting violent irritation, and violent fenfation,
employs for a time the whole fenforial energy, and thus differers the
paffing trains of ideas, before the power of volition has time to compare

pare them with the ufual phenomena of nature.   In this cafe fear is generally the companion of furprife, and adds to our embarraffment, as every one experiences in fome degree when he hears a noife in the dark, which he cannot inftantly account for.   This catenation of fear with furprife is owing to our perpetual experience of injuries from external bodies in motion, unlefs we are upon our guard againft them. See Sect. XVIII. 17. and XIX. 2.

Many other examples of the catenations of animal motions are explained in Sect. XXXVI. on the Periods of Difeafes.

SECT.

SECT. XVIII.

OF SLEEP.

1. Volition is suspended in sleep. 2. Sensation continues. Dreams prevent delirium and inflammation. 3. Nightmare. 4. Ceaseless flow of ideas in dreams. 5. We seem to receive them by the senses. Optic nerve perfectly sensible in sleep. Eyes less dazzled after dreaming of visible objects. 6. Reverie, belief. 7. How we distinguish ideas from perceptions. 8. Variety of scenery in dreams, excellence of the sense of vision. 9. Novelty of combination in dreams. 10. Distinctness of imagery in dreams. 11. Rapidity of transaction in dreams. 12. Of measuring time. Of dramatic time and place. Why a dull play induces sleep, and an interesting one reverie. 13. Consciousness of our existence and identity in dreams. 14. How we awake sometimes suddenly, sometimes frequently. 15. Irritative motions continue in sleep, internal irritations are succeeded by sensation. Sensibility increases during sleep, and irritability. Morning dreams. Why epilepsies occur in sleep. Ecstacy of children. Case of convulsions in sleep. Cramp, why painful. Asthma. Morning sweats. Increase of heat. Increase of urine in sleep. Why more liable to take cold in sleep. Catarrh from thin night-caps. Why we feel chilly at the approach of sleep, and at waking in the open air. 16. Why the gout commences in sleep. Secretions are more copious in sleep, young animals and plants grow more in sleep. 17. Inconsistency of dreams. Absence of surprise in dreams. 18. Why we forget some dreams and not others. 19. Sleep-talkers awake with surprise. 20. Remote causes of sleep. Atmosphere with less oxygene. Compression of the brain in spina bifida. By whirling on an horizontal wheel. By cold. 21. Definition of sleep.

I. THERE are four situations of our system, which in their moderate degrees are not usually termed diseases, and yet abound with many very curious and instructive phenomena; these are sleep, reverie,

verie, vertigo, drunkennefs.   Thefe we fhall previoufly confider, before we ftep forwards to develop the caufes and cures of difeafes with the modes of the operation of medicines.

As all thofe trains and tribes of animal motion, which are fubjected to volition, were the laft that were caufed, their connection is weaker than that of the other claffes; and there is a peculiar circumftance attending this caufation, which is, that it is entirely fufpended during fleep; whilft the other claffes of motion, which are more immediately neceffary to life, as thofe caufed by internal ftimuli, for inftance the pulfations of the heart and arteries, or thofe catenated with pleafurable fenfation, as the powers of digeftion, continue to ftrengthen their habits without interruption.   Thus though man in his fleeping ftate is a much lefs perfect animal, than in his waking hours; and though he confumes more than one third of his life in this his irrational fituation; yet is the wifdom of the Author of nature manifeft even in this feeming imperfection of his work!

The truth of this affertion with refpect to the large mufcles of the body, which are concerned in locomotion, is evident; as no one in perfect fanity walks about in his fleep, or performs any domeftic offices: and in refpect to the mind, we never exercife our reafon or recollection in dreams; we may fometimes feem diftracted between contending paffions, but we never compare their objects, or deliberate about the acquifition of thofe objects, if our fleep is perfect.   And though many fynchronous tribes or fucceffive trains of ideas may reprefent the houfes or walks, which have real exiftence, yet are they here introduced by their connection with our fenfations, and are in truth ideas of imagination, not of recollection.

2. For our fenfations of pleafure and pain are experienced with great vivacity in our dreams; and hence all that motley group of ideas, which are caufed by them, called the ideas of imagination, with their various affociated trains, are in a very vivid manner acted over in the fenforium; and thefe fometimes call into action the larger mufcles,

which

which have been much affociated with them; as appears from the muttering fentences, which fome people utter in their dreams, and from the obfcure barking of fleeping dogs, and the motions of their feet and noftrils.

This perpetual flow of the trains of ideas, which conftitute our dreams, and which are caufed by painful or pleafureable fenfation, might at firft view be conceived to be an ufelefs expenditure of fenforial power. But it has been fhewn, that thofe motions, which are perpetually excited, as thofe of the arterial fyftem by the ftimulus of the blood, are attended by a great accumulation of fenforial power, after they have been for a time fufpended; as the hot-fit of fever is the confequence of the cold one. Now as thefe trains of ideas caufed by fenfation are perpetually excited during our waking hours, if they were to be fufpended in fleep like the voluntary motions, (which are exerted only by intervals during our waking hours,) an accumulation of fenforial power would follow; and on our awaking a delirium would fupervene, fince thefe ideas caufed by fenfation would be produced with fuch energy, that we fhould miftake the trains of imagination for ideas excited by irritation; as perpetually happens to people debilitated by fevers on their firft awaking; for in thefe fevers with debility the general quantity of irritation being diminifhed, that of fenfation is increafed. In like manner if the actions of the ftomach, inteftines, and various glands, which are perhaps in part at leaft caufed by or catenated with agreeable fenfation, and which perpetually exift during our waking hours, were like the voluntary motions fufpended in our fleep; the great accumulation of fenforial power, which would necefarily follow, would be liable to excite inflammation in them.

3. When by our continued pofture in fleep, fome uneafy fenfations are produced, we either gradually awake by the exertion of volition, or the mufcles connected by habit with fuch fenfations alter the pofition of the body; but where the fleep is uncommonly profound,

5   found,

found, and thofe uneafy fenfations great, the difeafe called the incu-
bus, or nightmare, is produced. Here the defire of moving the
body is painfully exerted, but the power of moving it, or volition, is
incapable of action, till we awake. Many lefs difagreeable ftruggles
in our dreams, as when we wifh in vain to fly from terrifying objects,
conftitute a flighter degree of this difeafe. In awaking from the
nightmare I have more than once obferved, that there was no diforder
in my pulfe; nor do I believe the refpiration is laborious, as fome
have affirmed. It occurs to people whofe fleep is too profound, and
fome difagreeable fenfation exifts, which at other times would have
awakened them, and have thence prevented the difeafe of nightmare;
as after great fatigue or hunger with too large a fupper and wine,
which occafion our fleep to be uncommonly profound. See No. 14,
of this Section.

4. As the larger mufcles of the body are much more frequently
excited by volition than by fenfation, they are but feldom brought
into action in our fleep: but the ideas of the mind are by habit much
more frequently connected with fenfation than with volition; and
hence the ceafelefs flow of our ideas in dreams. Every one's experience will teach him this truth, for we all daily exert much voluntary
mufcular motion: but few of mankind can bear the fatigue of much
voluntary thinking.

5. A very curious circumftance attending thefe our fleeping imaginations is, that we feem to receive them by the fenfes. The mufcles, which are fubfervient to the external organs of fenfe, are connected with volition, and ceafe to act in fleep; hence the eyelids are
clofed, and the tympanum of the ear relaxed; and it is probable a
fimilarity of voluntary exertion may be neceffary for the perceptions
of the other nerves of fenfe; for it is obferved that the papillæ of the
tongue can be feen to become erected, when we attempt to tafte any
thing extremely grateful. Hewfon Exper. Enquir. V. 2. 186. Albini
Annot. Acad. L. i. c. 15. Add to this, that the immediate organs

D d

of

of fenfe have no objects to excite them in the darknefs and filence of
the night; but their nerves of fenfe neverthelefs continue to poffefs
their perfect activity fubfervient to all their numerous fenfitive connections.. This vivacity of our nerves of fenfe during the time of
fleep is evinced by a circumftance, which almoft every one muft at
fome time or other have experienced; that is, if we fleep in the day-
light, and endeavour to fee fome object in our dream, the light is exceedingly painful to our eyes; and after repeated ftruggles we lament
in our fleep, that we cannot fee it.   In this cafe I apprehend the eye-
lid is in fome degree opened by the vehemence of our fenfations; and,
the iris being dilated, the optic nerve fhews as great or greater fenfibility than in our waking hours: See No. 15. of this Section.

When we are forcibly waked at midnight from profound fleep, our
eyes are much dazzled with the light of the candle for a minute or
two, after there has been fufficient time allowed for the contraction
of the iris; which is owing to the accumulation of fenforial power in
the organ of vifion during its ftate of lefs activity.   But when we
have dreamt much of vifible objects, this accumulation of fenforial
power in the organ of vifion is leffened or prevented, and we awake
in the morning without being dazzled with the light, after the iris
has had time to contract itfelf.   This is a matter of great curiofity,
and may be thus tried by any one in the day-light.   Clofe your eyes,
and cover them with your hat; think for a minute on a tune, which
you are accuftomed to, and endeavour to fing it with as little activity
of mind as poffible.   Suddenly uncover and open your eyes, and in
one fecond of time the iris will contract itfelf, but you will perceive
the day more luminous for feveral feconds, owing to the accumulation
of fenforial power in the optic nerve.

Then again clofe and cover your eyes, and think intenfely on a
cube of ivory two inches diameter, attending firft to the north and
fouth fides of it, and then to the other four fides of it; then get a
clear image in your mind's eye of all the fides of the fame cube co-

loured

loured red; and then of it coloured green; and then of it coloured blue; laftly, open your eyes as in the former experiment, and after the firft fecond of time allowed for the contraction of the iris, you will not perceive any increafe of the light of the day, or dazzling; becaufe now there is no accumulation of fenforial power in the optic nerve; that having been expended by its action in thinking over vifible objects.

This experiment is not eafy to be made at firft, but by a few patient trials the fact appears very certain; and fhews clearly, that our ideas of imagination are repetitions of the motions of the nerve, which were originally occafioned by the ftimulus of external bodies; becaufe they equally expend the fenforial power in the organ of fenfe. See Sect. III. 4. which is analogous to our being as much fatigued by thinking as by labour.

6. Nor is it in our dreams alone, but even in our waking reveries, and in great efforts of invention, fo great is the vivacity of our ideas, that we do not for a time diftinguifh them from the real prefence of fubftantial objects; though the external organs of fenfe are open, and furrounded with their ufual ftimuli. Thus whilft I am thinking over the beautiful valley, through which I yefterday travelled, I do not perceive the furniture of my room: and there are fome, whofe waking imaginations are fo apt to run into perfect reverie, that in their common attention to a favourite idea they do not hear the voice of the companion, who accofts them, unlefs it is repeated with unufual energy.

This perpetual miftake in dreams and reveries, where our ideas of imagination are attended with a belief of the prefence of external objects, evinces beyond a doubt, that all our ideas are repetitions of the motions of the nerves of fenfe, by which they were acquired; and that this belief is not, as fome late philofophers contend, an inftinct neceffarily connected only with our perceptions.

7. A

7. A curious queſtion demands our attention in this place; as we do not diſtinguiſh in our dreams and reveries between our perceptions of external objeſts, and our ideas of them in their abſence, how do we diſtinguiſh them at any time? In a dream, if the ſweetneſs of ſugar occurs to my imagination, the whiteneſs and hardneſs of it, which were ideas uſually connected with the ſweetneſs, immediately follow in the train; and I believe a material lump of ſugar preſent before my ſenſes: but in my waking hours, if the ſweetneſs occurs to my imagination, the ſtimulus of the table to my hand, or of the window to my eye, prevents the other ideas of the hardneſs and whiteneſs of the ſugar from ſucceeding; and hence I perceive the fallacy, and diſbelieve the exiſtence of objects correſpondent to thoſe ideas, whoſe tribes or trains are broken by the ſtimulus of other ob-jects. And further in our waking hours, we frequently exert our volition in comparing preſent appearances with ſuch, as we have uſually obſerved; and thus correct the errors of one ſenſe by our ge-neral knowledge of nature by intuitive analogy. See Sect. XVII. 3. 7. Whereas in dreams the power of volition is ſuſpended, we can recollect and compare our preſent ideas with none of our acquired knowledge, and are hence incapable of obſerving any abſurdities in them.

By this criterion we diſtinguiſh our waking from our ſleeping hours, we can voluntarily recollect our ſleeping ideas, when we are awake, and compare them with our waking ones; but we cannot in our ſleep *voluntarily* recollect our waking ideas at all.

8. The vaſt variety of ſcenery, novelty of combination, and dif-tinctneſs of imagery, are other curious circumſtances of our ſleeping imaginations. The variety of ſcenery ſeems to ariſe from the ſupe-rior activity and excellence of our ſenſe of viſion; which in an inſtant unfolds to the mind extenſive fields of pleaſurable ideas; while the other ſenſes collect their objects ſlowly, and with little combination;

3   add

add to this, that the ideas, which this organ prefents us with, are more frequently connected with our fenfation than thofe of any other.

9. The great novelty of combination is owing to another circum-ftance; the trains of ideas, which are carried on in our waking thoughts, are in our dreams diffevered in a thoufand places by the fufpenfion of volition, and the abfence of irritative ideas, and are hence perpetually falling into new catenations. As explained in Sect. XVI. 1. 9. For the power of volition is perpetually exerted during our waking hours in comparing our paffing trains of ideas with our acquired knowledge of nature, and thus forms many intermediate links in their catenation. And the irritative ideas excited by the ftimulus of the objects, with which we are furrounded, are every moment in-truded upon us, and form other links of our unceafing catenations of ideas.

10. The abfence of the ftimuli of external bodies, and of volition, in our dreams renders the organs of fenfe liable to be more ftrongly af-fected by the powers of fenfation, and of affociation. For our defires or averfions, or the obtrufions of furrounding bodies, diffever the fenfitive and affociate tribes of ideas in our waking hours by intro-ducing thofe of irritation and volition amongft them. Hence pro-ceeds the fuperior diftinctnefs of pleafureable or painful imagery in our fleep; for we recal the figure and the features of a long loft friend, whom we loved, in our dreams with much more accuracy and viva-city than in our waking thoughts. This circumftance contributes to prove, that our ideas of imagination are reiterations of thofe motions of our organs of fenfe, which were excited by external objects; be-caufe while we are expofed to the ftimuli of prefent objects, our ideas of abfent objects cannot be fo diftinctly formed.

11. The rapidity of the fucceffion of tranfactions in our dreams is almoft inconceivable; infomuch that, when we are accidentally awakened by the jarring of a door, which is opened into our bed-chamber,

chamber, we fometimes dream a whole hiftory of thieves or fire in the very inftant of awaking.

During the fufpenfion of volition we cannot compare our other ideas with thofe of the parts of time in which they exift; that is, we cannot compare the imaginary fcene, which is before us, with thofe changes of it, which precede or follow it; becaufe this act of comparing requires recollection or voluntary exertion. Whereas in our waking hours, we are perpetually making this comparifon, and by that means our waking ideas are kept confiftent with each other by intuitive analogy; but this comparifon retards the fucceffion of them, by occafioning their repetition. Add to this, that the tranfactions of our dreams confift chiefly of vifible ideas, and that a whole hiftory of thieves and fire may be *beheld* in an inftant of time like the figures in a picture.

12. From this incapacity of attending to the parts of time in our dreams, arifes our ignorance of the length of the night; which, but from our conftant experience to the contrary, we fhould conclude was but a few minutes, when our fleep is perfect. The fame happens in our reveries; thus when we are poffeffed with vehement joy, grief, or anger, time appears fhort, for we exert no volition to compare the prefent fcenery with the paft or future; but when we are compelled to perform thofe exercifes of mind or body, which are unmixed with paffion, as in travelling over a dreary country, time appears long; for our defire to finifh our journey occafions us more frequently to compare our prefent fituation with the parts of time or place, which are before and behind us.

So when we are enveloped in deep contemplation of any kind, or in reverie, as in reading a very interefting play or romance, we meafure time very inaccurately; and hence, if a play greatly affects our paffions, the abfurdities of paffing over many days or years, and of perpetual changes of place, are not perceived by the audience; as is experienced by every one, who reads or fees fome plays of the im-

mortal

mortal Shakefpear; but it is neceffary for inferior authors to obferve thofe rules of the πιθανον and πρεπον inculcated by Ariftotle, becaufe their works do not intereft the paffions fufficiently to produce complete reverie.

Thofe works, however, whether a romance or a fermon, which do not intereft us fo much as to induce reverie, may neverthelefs incline us to fleep. For thofe pleafureable ideas, which are prefented to us, and are too gentle to excite laughter, (which is attended with interrupted voluntary exertions, as explained Sect. XXXIV. 1. 4.) and which are not accompanied with any other emotion, which ufually excites fome voluntary exertion, as anger, or fear, are liable to produce fleep; which confifts in a fufpenfion of all voluntary power. But if the ideas thus prefented to us, and intereft our attention, are accompanied with fo much pleafureable or painful fenfation as to excite our voluntary exertion at the fame time, reverie is the confequence. Hence an interefting play produces reverie, a tedious one produces fleep: in the latter we become exhaufted by attention, and are not excited to any voluntary exertion, and therefore fleep; in the former we are excited by fome emotion, which prevents by its pain the fufpenfion of volition, and in as much as it interefts us, induces reverie, as explained in the next Section.

But when our fleep is imperfect, as when we have determined to rife in half an hour, time appears longer to us than in moft other fituations. Here our folicitude not to overfleep the determined time induces us in this imperfect fleep to compare the quick changes of imagined fcenery with the parts of time or place, they would have taken up, had they real exiftence; and that more frequently than in our waking hours; and hence the time appears longer to us: and I make no doubt, but the permitted time appears long to a man going to the gallows, as the fear of its quick lapfe will make him think frequently about it.

13. As

13. As we gain our knowledge of time by comparing the prefent fcenery with the paft and future, and of place by comparing the fituations of objects with each other; fo we gain our idea of confcioufnefs by comparing ourfelves with the fcenery around us; and of identity by comparing our prefent confcioufnefs with our paft confcioufnefs: as we never think of time or place, but when we make the comparifons above mentioned, fo we never think of confcioufnefs, but when we compare our own exiftence with that of other objects; nor of identity, but when we compare our prefent and our paft confcioufnefs. Hence the confcioufnefs of our own exiftence, and of our identity, is owing to a voluntary exertion of our minds: and on that account in our complete dreams we neither meafure time, are furprifed at the fudden changes of place, nor attend to our own exiftence, or identity; becaufe our power of volition is fufpended. But all thefe circumftances are more or lefs obfervable in our incomplete ones; for then we attend a little to the lapfe of time, and the changes of place, and to our own exiftence; and even to our identity of perfon; for a lady feldom dreams, that fhe is a foldier; nor a man, that he is brought to bed.

14. As long as our fenfations only excite their fenfual motions, or ideas, our fleep continues found; but as foon as they excite defires or averfions, our fleep becomes imperfect; and when that defire or averfion is fo ftrong, as to produce voluntary motions, we begin to awake; the larger mufcles of the body are brought into action to remove that irritation or fenfation, which a continued pofture has caufed; we ftretch our limbs, and yawn, and our fleep is thus broken by the accumulation of voluntary power.

Sometimes it happens, that the act of waking is fuddenly produced, and this foon after the commencement of fleep; which is occafioned by fome fenfation fo difagreeable, as inftantaneoufly to excite the power of volition; and a temporary action of all the voluntary

motions

motions suddenly succeeds, and we start awake. This is sometimes accompanied with loud noise in the ears, and with some degree of fear; and when it is in great excess, so as to produce continued convulsive motions of those muscles, which are generally subservient to volition, it becomes epilepsy: the fits of which in some patients generally commence during sleep. This differs from the night-mare described in No. 3. of this Section, because in that the disagreeable sensation is not so great as to excite the power of volition into action; for as soon as that happens, the disease ceases.

Another circumstance, which sometimes awakes people soon after the commencement of their sleep, is where the voluntary power is already so great in quantity as almost to prevent them from falling asleep, and then a little accumulation of it soon again awakens them; this happens in cases of insanity, or where the mind has been lately much agitated by fear or anger. There is another circumstance in which sleep is likewise of short duration, which arises from great debility, as after great over-fatigue, and in some fevers, where the strength of the patient is greatly diminished, as in these cases the pulse intermits or flutters, and the respiration is previously affected, it seems to originate from the want of some voluntary efforts to facilitate respiration, as when we are awake. And is further treated of in Vol. II. Class I. 2. 1. 2. on the Diseases of the Voluntary Power. Art. Somnus interruptus.

15. We come now to those motions which depend on irritation. The motions of the arterial and glandular systems continue in our sleep, proceeding slower indeed, but stronger and more uniformly, than in our waking hours, when they are incommoded by external stimuli, or by the movements of volition; the motions of the muscles subservient to respiration continue to be stimulated into action, and the other internal senses of hunger, thirst, and lust, are not only occasionally excited in our sleep, but their irritative motions are succeeded by their usual sensations, and make a part of the farrago of our dreams. These

E e   sensations

senfations of the want of air, of hunger, thirst, and lust, in our dreams,
contribute to prove, that the nerves of the external senses are also
alive and excitable in our sleep; but as the stimuli of external objects
are either excluded from them by the darkness and silence of the
night, or their access to them is prevented by the suspension of volition, these nerves of sense fall more readily into their connexions
with sensation and with association; because much sensorial power,
which during the day was expended in moving the external organs of
sense in consequence of irritation from external stimuli, or in consequence of volition, becomes now in some degree accumulated, and
renders the internal or immediate organs of sense more easily excitable
by the other sensorial powers. Thus in respect to the eye, the irritation from external stimuli, and the power of volition during our waking hours, elevate the eye-lids, adapt the aperture of the iris to the
quantity of light, the focus of the cryftalline humour, and the angle
of the optic axifes to the diftance of the object, all which perpetual
activity during the day expends much sensorial power, which is saved
during our sleep.

Hence it appears, that not only those parts of the system, which
are always excited by internal stimuli, as the stomach, intestinal canal,
bile-ducts, and the various glands, but the organs of sense also may
be more violently excited into action by the irritation from internal
stimuli, or by sensation, during our sleep than in our waking hours;
because during the suspension of volition, there is a greater quantity
of the spirit of animation to be expended by the other sensorial powers.
On this account our irritability to internal stimuli, and our sensibility
to pain or pleasure, is not only greater in sleep, but increases as our
sleep is prolonged. Whence digestion and secretion are performed
better in sleep, than in our waking hours, and our dreams in the
morning have greater variety and vivacity, as our sensibility increases,
than at night when we first lie down. And hence epileptic fits,
which are always occasioned by some disagreeable sensation, so frequently

quently attack thofe, who are fubject to them, in their fleep; becaufe at this time the fyftem is more excitable by painful fenfation in confequence of internal ftimuli; and the power of volition is then fuddenly exerted to relieve this pain, as explained Sect. XXXIV. 1. 4.

There is a difeafe, which frequently affects children in the cradle, which is termed ecftafy, and feems to confift in certain exertions to relieve painful fenfation, in which the voluntary power is not fo far excited as totally to awaken them, and yet is fufficient to remove the difagreeable fenfation, which excites it; in this cafe changing the pofture of the child frequently relieves it.

I have at this time under my care an elegant young man about twenty-two years of age, who feldom fleeps more than an hour without experiencing a convulfion fit; which ceafes in about half a minute without any fubfequent ftupor. Large dofes of opium only prevented the paroxyfms, fo long as they prevented him from fleeping by the intoxication, which they induced. Other medicines had no effect on him. He was gently awakened every half hour for one night, but without good effect, as he foon flept again, and the fit returned at about the fame periods of time, for the accumulated fenforial power, which occafioned the increafed fenfibility to pain, was not thus exhaufted. This cafe evinces, that the fenfibility of the fyftem to internal excitation increafes, as our fleep is prolonged; till the pain thus occafioned produces voluntary exertion; which, when it is in its ufual degree, only awakens us; but when it is more violent, it occafions convulfions.

The cramp in the calf of the leg is another kind of convulfion, which generally commences in fleep, occafioned by the continual increafe of irritability from internal ftimuli, or of fenfibility, during that ftate of our exiftence. The cramp is a violent exertion to relieve pain, generally either of the fkin from cold, or of the bowels, as in fome diarrhoeas, or from the mufcles having been previoufly overftretched, as in walking up or down fteep hills. But in thefe

E e 2   convulfions

convulfions of the mufcles, which form the calf of the leg, the contraction is fo violent as to occafion another pain in confequence of
their own too violent contraction; as foon as the original pain, which
caufed the contraction, is removed.   And hence the cramp, or fpafm,
of thefe mufcles is continued without intermiffion by this new pain,
unlike the alternate convulfions and remiffions in epileptic fits.   The
reafon, that the contraction of thefe mufcles of the calf of the leg is
more violent during their convulfion than that of others, depends on
the weaknefs of their antagonift mufcles; for after thefe have been
contracted in their ufual action, as at every ftep in walking, they are
again extended, not, as moft other mufcles are, by their antagonifts,
but by the weight of the whole body on the balls of the toes; and
that weight applied to great mechanical advantage on the heel, that is,
on the other end of the bone of the foot, which thus acts as a lever.

Another difeafe, the periods of which generally commence during
our fleep, is the afthma.   Whatever may be the remote caufe of paroxyfms of afthma, the immediate caufe of the convulfive refpiration,
whether in the common afthma, or in what is termed the convulfive
afthma, which are perhaps only different degrees of the fame difeafe,
muft be owing to violent voluntary exertions to relieve pain, as in other
convulfions; and the increafe of irritability to internal ftimuli, or of
fenfibility, during fleep muft occafion them to commence at this time.

Debilitated people, who have been unfortunately accuftomed to
great ingurgitation of fpirituous potation, frequently part with a great
quantity of water during the night, but with not more than ufual in
the day-time.   This is owing to a beginning torpor of the abforbent
fyftem, and precedes anafarca, which commences in the day, but is
cured in the night by the increafe of the irritability of the abforbent
fyftem during fleep, which thus imbibes from the cellular membrane
the fluids, which had been accumulated there during the day; though
it is poffible the horizontal pofition of the body may contribute fomething to this purpofe, and alfo the greater irritability of fome branches

of

of the abforbent veffels, which open their mouths in the cells of the cellular membrane, than that of other branches.

As foon as a perfon begins to fleep, the irritability and fenfibility of the fyftem begins to increafe, owing to the fufpenfion of volition and the exclufion of external ftimuli.   Hence the actions of the veffels in obedience to internal ftimulation become ftronger and more energetic, though lefs frequent in refpect to number.   And as many of the fe- cretions are increafed, fo the heat of the fyftem is gradually increafed, and the extremities of feeble people, which had been cold during the day, become warm.   Till towards morning many people become fo warm, as to find it neceffary to throw off fome of their bed-clothes, as foon as they awake ; and in others fweats are fo liable to occur to- wards morning during their fleep.

Thus thofe, who are not accuftomed to fleep in the open air, are very liable to take cold, if they happen to fall afleep on a garden bench, or in a carriage with the window open.   For as the fyftem is warmer during fleep, as above explained, if a current of cold air affects any part of the body, a torpor of that part is more effectually pro- duced, as when a cold blaft of air through a key-hole or cafement falls upon a perfon in a warm room.   In thofe cafes the affected part poffeffes lefs irritability in refpect to heat from its having previoufly been expofed to a greater ftimulus of heat, as in the warm room, or during fleep ; and hence, when the ftimulus of heat is diminifhed, a torpor is liable to enfue; that is, we take cold.   Hence people who fleep in the open air, generally feel chilly both at the approach of fleep, and on their awaking ; and hence many people are perpetually fubject to catarrhs if they fleep in a lefs warm head-drefs, than that which they wear in the day.

16. Not only the fenforial powers of irritation and of fenfation, but that of affociation alfo appear to act with greater vigour during the fufpenfion of volition in fleep.   It will be fhewn in another place, that the gout generally firft attacks the liver, and that afterwards an

3

inflammation

inflammation of the ball of the great toe commences by affociation, and that of the liver ceafes. Now as this change or metaftafis of the activity of the fyftem generally commences in fleep, it follows, that thefe affociations of motion exift with greater energy at that time; that is, that the fenforial faculty of affociation, like thofe of irritation and of fenfation, becomes in fome meafure accumulated during the fufpenfion of volition.

Other affociate tribes and trains of motions, as well as the irritative and fenfitive ones, appear to be increafed in their activity during the fufpenfion of volition in fleep. As thofe which contribute to circulate the blood, and to perform the various fecretions; as well as the affociate tribes and trains of ideas, which contribute to furnifh the perpetual ftreams of our dreaming imaginations.

In fleep the fecretions have generally been fuppofed to be diminifhed, as the expectorated mucus in coughs, the fluids difcharged in diarrhœas, and in falivation, except indeed the fecretion of fweat, which is often vifibly increafed. This error feems to have arifen from attention to the excretions rather than to the fecretions. For the fecretions, except that of fweat, are generally received into refervoirs, as the urine into the bladder, and the mucus of the inteftines and lungs into their refpective cavities; but thefe refervoirs do not exclude thefe fluids immediately by their ftimulus, but require at the fame time fome voluntary efforts, and therefore permit them to remain during fleep. And as they thus continue longer in thofe receptacles in our fleeping hours, a greater part is abforbed from them, and the remainder becomes thicker, and fometimes in lefs quantity, though at the time it was fecreted the fluid was in greater quantity than in our waking hours. Thus the urine is higher coloured after long fleep; which fhews, that a greater quantity has been fecreted, and that more of the aqueous and faline part has been reabforbed, and the earthy part left in the bladder; hence thick urine in fevers fhews

only

only a greater action of the veffels which fecrete it in the kidneys, and of thofe which abforb it from the bladder.

The fame happens to the mucus expectorated in coughs, which is thus thickened by abforption of its aqueous and faline parts; and the fame of the feces of the inteftines. From hence it appears, and from what has been faid in No. 15 of this Section concerning the increafe of irritability and of fenfibility during fleep, that the fecretions are in general rather increafed than diminifhed during thefe hours of our exiftence; and it is probable that nutrition is almoft entirely performed in fleep; and that young animals grow more at this time than in their waking hours, as young plants have long fince been obferved to grow more in the night, which is their time of fleep.

17. Two other remarkable circumftances of our dreaming ideas are their inconfiftency, and the total abfence of furprife. Thus we feem to be prefent at more extraordinary metamorphofes of animals or trees, than are to be met with in the fables of antiquity; and appear to be tranfported from place to place, which feas divide, as quickly as the changes of fcenery are performed in a play-houfe; and yet are not fenfible of their inconfiftency, nor in the leaft degree affected with furprife.

We muft confider this circumftance more minutely. In our waking trains of ideas, thofe that are inconfiftent with the ufual order of nature, fo rarely have occurred to us, that their connexion is the flighteft of all others: hence, when a confiftent train of ideas is exhaufted, we attend to the external ftimuli, that ufually furround us, rather than to any inconfiftent idea, which might otherwife prefent itfelf: and if an inconfiftent idea fhould intrude itfelf, we immediately compare it with the preceding one, and voluntarily reject the train it would introduce; this appears further in the Section on Reverie, in which ftate of the mind external ftimuli are not attended to, and yet the ftreams of ideas are kept confiftent by the efforts of volition. But as our faculty of volition is fufpended, and all external ftimuli are excluded

cluded in fleep, this flighter connexion of ideas takes place; and the train is faid to be inconfiftent; that is, diffimilar to the ufual order of nature.

But, when any confiftent train of fenfitive or voluntary ideas is flowing along, if any external ftimulus affects us fo violently, as to intrude irritative ideas forcibly into the mind, it difunites the former train of ideas, and we are affected with furprife. Thefe ftimuli of unufual energy or novelty not only difunite our common trains of ideas, but the trains of mufcular motions alfo, which have not been long eftablifhed by habit, and difturb thofe that have. Some people become motionlefs by great furprife, the fits of hiccup and of ague have been often removed by it, and it even affects the movements of the heart, and arteries; but in our fleep, all external ftimuli are excluded, and in confequence no furprife can exift. See Section XVII. 3. 7.

18. We frequently awake with pleafure from a dream, which has delighted us, without being able to recollect the tranfactions of it; unlefs perhaps at a diftance of time, fome analogous idea may introduce afrefh this forgotten train: and in our waking reveries we fometimes in a moment lofe the train of thought, but continue to feel the glow of pleafure, or the depreffion of fpirits, it occafioned: whilft at other times we can retrace with eafe thefe hiftories of our reveries and dreams.

The above explanation of furprife throws light upon this fubject. When we are fuddenly awaked by any violent ftimulus, the furprife totally difunites the trains of our fleeping ideas from thefe of our waking ones; but if we gradually awake, this does not happen; and we readily unravel the preceding trains of imagination.

19. There are various degrees of furprife; the more intent we are upon the train of ideas, which we are employed about, the more violent muft be the ftimulus that interrupts them, and the greater is the degree of furprife. I have obferved dogs, who have flept by the fire,

                                                                    and

and by their obfcure barking and ftruggling have appeared very intent
on their prey, that fhewed great furprife for a few feconds after their
awaking by looking eagerly around them ; which they did not do at
other times of waking. And an intelligent friend of mine has remarked, that his lady, who frequently fpeaks much and articularly
in her fleep, could never recollect her dreams in the morning, when
this happened to her : but that when fhe did not fpeak in her fleep,
fhe could always recollect them.

Hence, when our fenfations act fo ftrongly in fleep as to influence
the larger mufcles, as in thofe, who talk or ftruggle in their dreams ;
or in thofe, who are affected with complete reverie (as defcribed in
the next Section), great furprife is produced, when they awake ; and
thefe as well as thofe, who are completely drunk or delirious, totally
forget afterwards their imaginations at thofe times.

20. As the immediate caufe of fleep confifts in the fufpenfion of volition, it follows, that whatever diminifhes the general quantity of
fenforial power, or derives it from the faculty of volition, will conftitute a remote caufe of fleep ; fuch as fatigue from mufcular or mental exertion, which diminifhes the general quantity of fenforial power ;
or an increafe of the fenfitive motions, as by attending to foft mufic,
which diverts the fenforial power from the faculty of volition ; or
laftly, by increafe of the irritative motions, as by wine, or food, or
warmth ; which not only by their expenditure of fenforial power diminifh the quantity of volition ; but alfo by their producing pleafureable fenfations (which occafion other mufcular or fenfual motions in
confequence), doubly decreafe the voluntary power, and thus more
forceably produce fleep. See Sect. XXXIV. 1. 4.

Another method of inducing fleep is delivered in a very ingenious
work lately publifhed by Dr Beddoes. Who, after lamenting that
opium frequently occafions reftleffnefs, thinks, " that in moft cafes
it would be better to induce fleep by the abftraction of ftimuli, than
by exhaufting the excitability ;" and adds, " upon this principle we

F f   could

could not have a better foporific than an atmofphere with a diminifhed proportion of oxygene air, and that common air might be admitted
after the patient was afleep." (Obferv. on Calculus, &c. by Dr. Beddoes. Murray.)   If it fhould be found to be true, that the excitability
of the fyftem depends on the quantity of oxygene abforbed by the
lungs in refpiration according to the theory of Dr. Beddoes, and of
M. Girtanner, this idea of fleeping in an atmofphere with lefs oxygene
in its compofition might be of great fervice in epileptic cafes, and in
cramp, and even in fits of the afthma, where their periods commence
from the increafe of irritability during fleep.

Sleep is likewife faid to be induced by mechanic preffure on
the brain in the cafes of fpina bifida.   Where there has been a defect of one of the vertebræ of the back, a tumour is protruded in confequence; and, whenever this tumour has been compreffed by the hand,
fleep is faid to be induced, becaufe the whole of the brain both within
the head and fpine becomes compreffed by the retroceffion of the fluid
within the tumour.   But by what means a compreffion of the brain
induces fleep has not been explained, but probably by diminifhing the
fecretion of fenforial power, and then the voluntary motions become
fufpended previoufly to the irritative ones, as occurs in moft dying
perfons.

Another way of procuring fleep mechanically was related to me by
Mr. Brindley, the famous canal engineer, who was brought up to
the bufinefs of a mill-wright; he told me, that he had more than
once feen the experiment of a man extending himfelf acrofs the large
ftone of a corn-mill, and that by gradually letting the ftone whirl,
the man fell afleep, before the ftone had gained its full velocity, and
he fuppofed would have died without pain by the continuance or increafe of the motion.   In this cafe the centrifugal motion of the head
and feet muft accumulate the blood in both thofe extremities of the
body, and thus comprefs the brain.

Laftly, we fhould mention the application of cold; which, when

3

in a lefs degree, produces watchfulnefs by the pain it occafions, and
the tremulous convulfions of the fubcutaneous mufcles; but when it
is applied in great degree, is faid to produce fleep. To explain this
effect it has been faid, that as the veffels of the fkin and extremities
become firft torpid by the want of the ftimulus of heat, and as thence
lefs blood is circulated through them, as appears from their palenefs,
a greater quantity of blood poured upon the brain produces fleep by
its compreffion of that organ. But I fhould rather imagine, that the
fenforial power becomes exhaufted by the convulfive actions in confequence of the pain of cold, and of the voluntary exercife previoufly
ufed to prevent it, and that the fleep is only the beginning to die, as
the fufpenfion of voluntary power in lingering deaths precedes for many
hours the extinction of the irritative motions.

21. The following are the characteriftic circumftances attending
perfect fleep.

1. The power of volition is totally fufpended.

2. The trains of ideas caufed by fenfation proceed with greater facility and vivacity; but become inconfiftent with the ufual order of
nature. The mufcular motions caufed by fenfation continue; as
thofe concerned in our evacuations during infancy, and afterwards in
digeftion, and in priapifmus.

3. The irritative mufcular motions continue, as thofe concerned in
the circulation, in fecretion, in refpiration. But the irritative fenfual
motions, or ideas, are not excited; as the immediate organs of fenfe
are not ftimulated into action by external objects, which are excluded
by the external organs of fenfe; which are not in fleep adapted to their
reception by the power of volition, as in our waking hours.

4. The affociate motions continue; but their firft link is not excited into action by volition, or by external ftimuli. In all refpects,
except thofe above mentioned, the three laft fenforial powers are
fomewhat increafed in energy during the fufpenfion of volition, owing
to the confequent accumulation of the fpirit of animation.

F f 2

SECT.

SECT. XIX.

OF REVERIE.

1. *Various degrees of reverie.* 2. *Sleep-walkers. Cafe of a young lady. Great furprife at awaking. And total forgetfulnefs of what paffed in reverie.* 3. *No fufpenfion of volition in reverie.* 4. *Senfitive motions continue, and are confiftent.* 5. *Irritative motions continue, but are not fucceeded by fenfation.* 6. *Volition neceffary for the perception of feeble impreffions.* 7. *Affociated motions continue.* 8. *Nerves of fenfe are irritable in fleep, but not in reverie.* 9. *Somnambuli are not afleep. Contagion received but once.* 10. *Definition of reverie.*

1. WHEN we are employed with great fenfation of pleafure, or with great efforts of volition, in the purfuit of fome interefting train of ideas, we ceafe to be confcious of our exiftence, are inattentive to time and place, and do not diftinguifh this train of fenfitive and voluntary ideas from the irritative ones excited by the prefence of external objects, though our organs of fenfe are furrounded with their accuftomed ftimuli, till at length this interefting train of ideas becomes exhaufted, or the appulfes of external objects are applied with unufual violence, and we return with furprife, or with regret, into the common track of life. This is termed reverie or ftudium.

In fome conftitutions thefe reveries continue a confiderable time, and are not to be removed without greater difficulty, but are experienced in a lefs degree by us all; when we attend earneftly to the ideas excited by volition or fenfation, with their affociated connexions, but are at the fame time confcious at intervals of the ftimuli of furrounding bodies. Thus in being prefent at a play, or in reading a romance,

some

some persons are so totally absorbed as to forget their usual time of sleep, and to neglect their meals; while others are said to have been so involved in voluntary study as not to have heard the discharge of artillery; and there is a story of an Italian politician, who could think so intensely on other subjects, as to be insensible to the torture of the rack.

From hence it appears, that these catenations of ideas and muscular motions, which form the trains of reverie, are composed both of voluntary and sensitive associations of them; and that these ideas differ from those of delirium or of sleep, as they are kept consistent by the power of volition; and they differ also from the trains of ideas belonging to insanity, as they are as frequently excited by sensation as by volition. But lastly, that the whole sensorial power is so employed on these trains of complete reverie, that like the violent efforts of volition, as in convulsions or insanity; or like the great activity of the irritative motions in drunkenness; or of the sensitive motions in delirium; they preclude all sensation consequent to external stimulus.

2. Those persons, who are said to walk in their sleep, are affected with reverie to so great a degree, that it becomes a formidable disease; the essence of which consists in the inaptitude of the mind to attend to external stimuli. Many histories of this disease have been published by medical writers; of which there is a very curious one in the Lausanne Transactions. I shall here subjoin an account of such a case, with its cure, for the better illustration of this subject.

A very ingenious and elegant young lady, with light eyes and hair, about the age of seventeen, in other respects well, was suddenly seized soon after her usual menstruation with this very wonderful malady. The disease began with vehement convulsions of almost every muscle of her body, with great but vain efforts to vomit, and the most violent hiccoughs, that can be conceived: these were succeeded in about an hour with a fixed spasm; in which one hand was applied to her head, and the other to support it: in about half an hour these

ceased,

ceafed, and the reverie began fuddenly, and was at firft manifeft by the look of her eyes and countenance, which feemed to exprefs attention. Then fhe converfed aloud with imaginary perfons with her eyes open, and could not for about an hour be brought to attend to the ftimulus of external objects by any kind of violence, which it was proper to ufe: thefe fymptoms returned in this order every day for five or fix weeks.

These converfations were quite confiftent, and we could underftand, what fhe fuppofed her imaginary companions to anfwer, by the continuation of her part of the difcourfe. Sometimes fhe was angry, at other times fhewed much wit and vivacity, but was moft frequently inclined to melancholy. In thefe reveries fhe fometimes fung over fome mufic with accuracy, and repeated whole pages from the Englifh poets. In repeating fome lines from Mr. Pope's works fhe had forgot one word, and began again, endeavouring to recollect it; when fhe came to the forgotten word, it was fhouted aloud in her ear, and this repeatedly, to no purpofe; but by many trials fhe at length regained it herfelf.

These paroxyfms were terminated with the appearance of inexpreffible furprife, and great fear, from which fhe was fome minutes in recovering herfelf, calling on her fifter with great agitation, and very frequently underwent a repetition of convulfions, apparently from the pain of fear. See Sect. XVII. 3. 7.

After having thus returned for about an hour every day for two or three weeks, the reveries feemed to become lefs complete, and fome of their circumftances varied; fo that fhe could walk about the room in them without running againft any of the furniture; though thefe motions were at firft very unfteady and tottering. And afterwards fhe once drank a difh of tea, when the whole apparatus of the teatable was fet before her; and expreffed fome fufpicion, that a medicine was put into it, and once feemed to fmell of a tuberofe, which was in flower in her chamber, and deliberated aloud about breaking it

from

from the ftem, faying, " it would make her fifter fo charmingly
angry." At another time in her melancholy moments fhe heard the
found of a passing bell, " I wifh I was dead," fhe cried, liftening to
the bell, and then taking off one of her fhoes, as fhe fat upon the
bed, " I love the colour black," fays fhe, " a little wider, and a
little longer, even this might make me a coffin!' —Yet it is evident,
fhe was not fenfible at this time, any more than formerly, of feeing
or hearing any perfon about her; indeed when great light was thrown
upon her, by opening the fhutters of the window, her trains of ideas
feemed lefs melancholy; and when I have forcibly held her hands, or
covered her eyes, fhe appeared to grow impatient, and would fay, fhe
could not tell what to do, for fhe could neither fee nor move. In all
thefe circumftances her pulfe continued unaffected as in health. And
when the paroxyfm was over, fhe could never recollect a fingle idea
of what had paffed in it.

This aftonifhing difeafe, after the ufe of many other medicines and
applications in vain, was cured by very large dofes of opium given
about an hour before the expected returns of the paroxyfms; and after
a few relapfes, at the intervals of three or four months, entirely difappeared. But fhe continued at times to have other fymptoms of
epilepfy.

3. We fhall only here confider, what happened during the time of
her reveries, as that is our prefent fubject; the fits of convulfion belong to another part of this treatife. Sect. XXXIV. 44.

There feems to have been no fufpenfion of volition during the fits
of reverie, becaufe fhe endeavoured to regain the loft idea in repeating
the lines of poetry, and deliberated about breaking the tuberofe, and
fufpected the tea to have been medicated.

4. The ideas and mufcular movements depending on fenfation were
exerted with their ufual vivacity, and were kept from being inconfiftent by the power of volition, as appeared from her whole converfation, and was explained in Sect. XVII. 3. 7. and XVIII. 16.

5. The

5. The ideas and motions dependant on irritation during the first weeks of her disease, whilst the reverie was complete, were never succeeded by the sensation of pleasure or pain; as she neither saw, heard, nor felt any of the surrounding objects. Nor was it certain that any irritative motions succeeded the stimulus of external objects, till the reverie became less complete, and then she could walk about the room without running against the furniture of it. Afterwards, when the reverie became still less complete from the use of opium, some few irritations were at times succeeded by her attention to them. As when she smelt at a tuberose, and drank a dish of tea, but this only when she seemed voluntarily to attend to them.

6. In common life when we listen to distant sounds, or wish to distinguish objects in the night, we are obliged strongly to exert our volition to dispose the organs of sense to perceive them, and to suppress the other trains of ideas, which might interrupt these feeble sensations. Hence in the present history the strongest stimuli were not perceived, except when the faculty of volition was exerted on the organ of sense; and then even common stimuli were sometimes perceived: for her mind was so strenuously employed in pursuing its own trains of voluntary or sensitive ideas, that no common stimuli could so far excite her attention as to disunite them; that is, the quantity of volition or of sensation already existing was greater than any, which could be produced in consequence of common degrees of stimulation. But the few stimuli of the tuberose, and of the tea, which she did perceive, were such, as accidentally coincided with the trains of thought, which were passing in her mind; and hence did not disunite those trains, and create surprise. And their being perceived at all was owing to the power of volition preceding or coinciding with that of irritation.

This explication is countenanced by a fact mentioned concerning a somnambulist in the Lausanne Transactions, who sometimes opened his eyes for a short time to examine, where he was, or where his ink-

pot

pot ftood, and then fhut them again, dipping his pen into the pot
every now and then, and writing on, but never opening his eyes afterwards, although he wrote on from line to line regularly, and cor-
rected fome errors of the pen, or in fpelling: fo much eafier was it
to him to refer to his ideas of the pofitions of things, than to his perceptions of them.

7. The affociated motions perfifted in their ufual channel, as appeared by the combinations of her ideas, and the ufe of her mufcles,
and the equality of her pulfe; for the natural motions of the arterial
fyftem, though originally excited like other motions by ftimulus,
feem in part to continue by their affociation with each other. As the
heart of a viper pulfates long after it is cut out of the body, and removed from the ftimulus of the blood.

8. In the fection on fleep, it was obferved that the nerves of fenfe
are equally alive and fufceptible to irritation in that ftate, as when we
are awake; but that they are fecluded from ftimulating objects, or
rendered unfit to receive them: but in complete reverie the reverfe
happens, the immediate organs of fenfe are expofed to their ufual ftimuli; but are either not excited into action at all, or not into fo great
action, as to produce attention or fenfation.

The total forgetfulnefs of what paffes in reveries; and the furprife
on recovering from them, are explained in Section XVIII. 19. and in
Section XVII. 3. 7.

9. It appears from hence, that reverie is a difeafe of the epileptic
or cataleptic kind, fince the paroxyfms of this young lady always began and frequently terminated with convulfions; and though in its
greateft degree it has been called fomnambulation, or fleep-walking,
it is totally different from fleep; becaufe the effential character of
fleep confifts in the total fufpenfion of volition, which in reverie is
not affected; and the effential character of reverie confifts not in the
abfence of thofe irritative motions of our fenfes, which are occafioned
by the ftimulus of external objects, but in their never being produc-

G g   tive

tive of fenfation. So that during a fit of reverie that ftrange event happens to the whole fyftem of nerves, which occurs only to fome particular branches of them in thofe, who are a fecond time expofed to the action of contagious matter. If the matter of the fmall-pox be inferted into the arm of one, who has previoufly had that difeafe, it will ftimulate the wound, but the general fenfation or inflammation of the fyftem does not follow, which conftitutes the difeafe. See Sect. XII. 7. 6. XXXIII. 2. 8.

10. The following is the definition or character of complete reverie. 1. The irritative motions occafioned by internal ftimuli continue, thofe from the ftimuli of external objects are either not produced at all, or are never fucceeded by fenfation or attention, unlefs they are at the fame time excited by volition. 2. The fenfitive motions continue, and are kept confiftent by the power of volition. 3. The voluntary motions continue undifturbed. 4. The affociate motions continue undifturbed.

Two other cafes of reverie are related in Section XXXIV. 3. which further evince, that reverie is an effort of the mind to relieve fome painful fenfation, and is hence allied to convulfion, and to infanity.

SECT.

SECT. XX.

OF VERTIGO.

1. IN learning to walk we judge of the diſtances of the objects, which we approach, by the eye; and by obſerving their perpendicularity determine our own. This circumſtance not having been attended to by the writers on viſion, the diſeaſe called vertigo or dizzineſs has been little underſtood.

When

When any perfon lofes the power of mufcular action, whether he is erect or in a fitting pofture, he finks down upon the ground; as is feen in fainting fits, and other inftances of great debility. Hence it follows, that fome exertion of mufcular power is neceffary to preferve our perpendicular attitude. This is performed by proportionally exerting the antagonift mufcles of the trunk, neck, and limbs; and if at any time in our locomotions we find ourfelves inclining to one fide, we either reftore our equilibrium by the efforts of the mufcles on the other fide, or by moving one of our feet extend the bafe, which we reft upon, to the new center of gravity.

But the moft eafy and habitual manner of determining our want of perpendicularity, is by attending to the apparent motion of the objects within the fphere of diftinct vifion; for this apparent motion of objects, when we incline from our perpendicularity, or begin to fall, is as much greater than the real motion of the eye, as the diameter of the fphere of diftinct vifion is to our perpendicular height.

Hence no one, who is hood-winked, can walk in a ftraight line for a hundred fteps together; for he inclines fo greatly, before he is warned of his want of perpendicularity by the fenfe of touch, not having the apparent motions of ambient objects to meafure this inclination by, that he is neceffitated to move one of his feet outwards, to the right or to the left, to fupport the new centre of gravity, and thus errs from the line he endeavours to proceed in.

For the fame reafon many people become dizzy, when they look from the fummit of a tower, which is raifed much above all other objects, as thefe objects are out of the fphere of diftinct vifion, and they are obliged to balance their bodies by the lefs accurate feelings of their mufcles.

There is another curious phenomenon belonging to this place, if the circumjacent vifible objects are fo fmall, that we do not diftinguifh their minute parts; or fo fimilar, that we do not know them from each other; we cannot determine our perpendicularity by them. Thus

in

in a room hung with a paper, which is coloured over with similar
small black lozenges or rhomboids, many people become dizzy; for
when they begin to fall, the next and the next lozenge succeeds upon
the eye; which they mistake for the first, and are not aware, that
they have any apparent motion.   But if you fix a sheet of paper, or
draw any other figure, in the midst of these lozenges, the charm
ceases, and no dizziness is perceptible.—The same occurs, when we
ride over a plain covered with snow without trees or other eminent
objects.

2. But after having compared visible objects at rest with the sense
of touch, and learnt to distinguish their shapes and shades, and to
measure our want of perpendicularity by their apparent motions, we
come to consider them in real motion.   Here a new difficulty occurs,
and we require some experience to learn the peculiar mode of motion
of any moving objects, before we can make use of them for the purposes of determining our perpendicularity.   Thus some people become
dizzy at the sight of a whirling wheel, or by gazing on the fluctuations of a river, if no steady objects are at the same time within the
sphere of their distinct vision; and when a child first can stand erect
upon his legs, if you gain his attention to a white handkerchief steadily extended like a sail, and afterwards make it undulate, he instantly
loses his perpendicularity, and tumbles on the ground.

3. A second difficulty we have to encounter is to distinguish our
own real movements from the apparent motions of objects.   Our
daily practice of walking and riding on horseback soon instructs us
with accuracy to discern these modes of motion, and to ascribe the
apparent motions of the ambient objects to ourselves; but those,
which we have not acquired by repeated habit, continue to confound
us.   So as we ride on horseback the trees and cottages, which occur
to us, appear at rest; we can measure their distances with our eye,
and regulate our attitude by them; yet if we carelessly attend to diftant hills or woods through a thin hedge, which is near us, we ob-

ferve

ferve the jumping and progreſſive motions of them; as this is increaſed by the paralax of theſe objects; which we have not habituated
ourſelves to attend to. When firſt an European mounts an elephant
ſixteen feet high, and whoſe mode of motion he is not accuſtomed to,
the objects ſeem to undulate, as he paſſes, and he frequently becomes
ſick and vertiginous, as I am well informed. Any other unuſual movement of our bodies has the ſame effect, as riding backwards in a coach,
ſwinging on a rope, turning round ſwiftly on one leg, ſcating on the
ice, and a thouſand others. So after a patient has been long confined
to his bed, when he firſt attempts to walk, he finds himſelf vertiginous, and is obliged by practice to learn again the particular modes
of the apparent motions of objects, as he walks by them.

4. A third difficulty, which occurs to us in learning to balance
ourſelves by the eye, is, when both ourſelves and the circumjacent
objects are in real motion. Here it is neceſſary, that we ſhould be
habituated to both theſe modes of motion in order to preſerve our perpendicularity. Thus on horſeback we accurately obſerve another
perſon, whom we meet, trotting towards us, without confounding
his jumping and progreſſive motion with our own, becauſe we have
been accuſtomed to them both; that is, to undergo the one, and to
ſee the other at the ſame time. But in riding over a broad and fluc-
tuating ſtream, though we are well experienced in the motions of our
horſe, we are liable to become dizzy from our inexperience in that of
the water. And when firſt we go on ſhip-board, where the movements of ourſelves, and the movements of the large waves are both
new to us, the vertigo is almoſt unavoidable with the terrible ſickneſs,
which attends it. And this I have been aſſured has happened to ſeveral from being removed from a large ſhip into a ſmall one; and
again from a ſmall one into a man of war.

5. From the foregoing examples it is evident, that, when we are
ſurrounded with unuſual motions, we loſe our perpendicularity: but
there are ſome peculiar circumſtances attending this effect of moving

5   objects,

objects, which we come now to mention, and shall hope from the recital of them to gain some insight into the manner of their production.

When a child moves round quick upon one foot, the circumjacent objects become quite indistinct, as their distance increases their apparent motions; and this great velocity confounds both their forms, and their colours, as is seen in whirling round a many coloured wheel; he then loses his usual method of balancing himself by vision, and begins to stagger, and attempts to recover himself by his muscular feelings. This staggering adds to the instability of the visible objects by giving a vibratory motion besides their rotatory one. The child then drops upon the ground, and the neighbouring objects seem to continue for some seconds of time to circulate around him, and the earth, under him appears to librate like a balance. In some seconds of time these sensations of a continuation of the motion of objects vanish; but if he continues turning round somewhat longer, before he falls, sickness and vomiting are very liable to succeed. But none of these circumstances affect those who have habituated themselves to this kind of motion, as the dervises in Turkey, amongst whom these swift gyrations are a ceremony of religion.

In an open boat passing from Leith to Kinghorn in Scotland, a sudden change of the wind shook the undistended sail, and stopt our boat; from this unusual movement the passengers all vomited except myself. I observed, that the undulation of the ship, and the instability of all visible objects, inclined me strongly to be sick; and this continued or increased, when I closed my eyes, but as often as I bent my attention with energy on the management and mechanism of the ropes and sails, the sickness ceased; and recurred again, as often as I relaxed this attention; and I am assured by a gentleman of observation and veracity, that he has more than once observed, when the vessel has been in immediate danger, that the sea-sickness of the passengers

has

has instantaneously ceased, and recurred again, when the danger was over.

Those, who have been upon the water in a boat or ship so long, that they have acquired the necessary habits of motion upon that unstable element, at their return on land frequently think in their reveries, or between sleeping and waking, that they observe the room, they sit in, or some of its furniture, to librate like the motion of the vessel. This I have experienced myself, and have been told, that after long voyages, it is some time before these ideas entirely vanish. The same is observable in a less degree after having travelled some days in a stage coach, and particularly when we lie down in bed, and compose ourselves to sleep; in this case it is observable, that the rattling noise of the coach, as well as the undulatory motion, haunts us. The drunken vertigo, and the vulgar custom of rocking children, will be considered in the next Section.

6. The motions, which are produced by the power of volition, may be immediately stopped by the exertion of the same power on the antagonist muscles; otherwise these with all the other classes of motion continue to go on, some time after they are excited, as the palpitation of the heart continues after the object of fear, which occasioned it, is removed. But this circumstance is in no class of motions more remarkable than in those dependent on irritation; thus if any one looks at the sun, and then covers his eyes with his hand, he will for many seconds of time, perceive the image of the sun marked on his retina: a similar image of all other visible objects would remain some time formed on the retina, but is extinguished by the perpetual change of the motions of this nerve in our attention to other objects. To this must be added, that the longer time any movements have continued to be excited without fatigue to the organ, the longer will they continue spontaneously, after the excitement is withdrawn: as the taste of tobacco in the mouth after a person has been smoaking it.

This

This tafte remains fo ftrong, that if a perfon continues to draw air through a tobacco pipe in the dark, after having been fmoking fome time, he cannot diftinguifh whether his pipe be lighted or not.

From thefe two confiderations it appears, that the dizzinefs felt in the head, after feeing objects in unufual motion, is no other than a continuation of the motions of the optic nerve excited by thofe objects, and which engage our attention. Thus on turning round on one foot, the vertigo continues for fome feconds of time after the perfon is fallen on the ground; and the longer he has continued to revolve, the longer will continue thefe fucceffive motions of the parts of the optic nerve.

Any one, who ftands alone on the top of a high tower, if he has not been accuftomed to balance himfelf by objects placed at fuch diftances and with fuch inclinations, begins to ftagger, and endeavours to recover himfelf by his mufcular feelings. During this time the apparent motion of objects at a diftance below him is very great, and the impreffions of thefe apparent motions continue a little time after he has experienced them; and he is perfuaded to incline the contrary way to counteract their effects; and either immediately falls, or applying his hands to the building, ufes his mufcular feelings to preferve his perpendicular attitude, contrary to the erroneous perfuafions of his eyes. Whilft the perfon, who walks in the dark, ftaggers, but without dizzinefs; for he neither has the fenfation of moving objects to take off his attention from his mufcular feelings, nor has he the fpectra of thofe motions continued on his retina to add to his confufion. It happens indeed fometimes to one ftanding on a tower, that the idea of his not having room to extend his bafe by moving one of his feet outwards, when he begins to incline, fuperadds fears to his other inconveniences; which like furprife, joy, or any great degree of fenfation, enervates him in a moment, by employing the whole fenforial power, and by thus breaking all the affociated trains and tribes of motion.

H h

7. The

7. The irritative ideas of objects, whilst we are awake, are perpetually present to our sense of sight; as we view the furniture of our rooms, or the ground, we tread upon, throughout the whole day without attending to it. And as our bodies are never at perfect rest during our waking hours, these irritative ideas of objects are attended perpetually with irritative ideas of their apparent motions. The ideas of apparent motions are always irritative ideas, because we never attend to them, whether we attend to the objects themselves, or to their real motions, or to neither. Hence the ideas of the apparent motions of objects are a complete circle of irritative ideas, which continue throughout the day.

Also during all our waking hours, there is a perpetual confused sound of various bodies, as of the wind in our rooms, the fire, distant conversations, mechanic business; this continued buzz, as we are seldom quite motionless, changes its loudness perpetually, like the sound of a bell; which rises and falls as long as it continues, and seems to pulsate on the ear. This any one may experience by turning himself round near a waterfall; or by striking a glass bell, and then moving the direction of its mouth towards the ears, or from them, as long as its vibrations continue. Hence this undulation of indistinct sound makes another concomitant circle of irritative ideas, which continues throughout the day.

We hear this undulating sound, when we are perfectly at rest ourselves, from other sonorous bodies besides bells; as from two organ-pipes, which are nearly but not quite in unison, when they are sounded together. When a bell is struck, the circular form is changed into an eliptic one; the longest axis of which, as the vibrations continue, moves round the periphery of the bell; and when either axis of this elipse is pointed towards our ears, the sound is louder; and less when the intermediate parts of the elipse are opposite to us. The vibrations of the two organ-pipes may be compared to Nonius's rule; the sound is louder, when they coincide, and less at the intermediate

times.

times.   But, as the found of bells is the moſt familiar of thoſe founds, which have a conſiderable battement, the vertiginous patients, who attend to the irritative circles of founds above deſcribed, generally compare it to the noiſe of bells.

The periſtaltic motions of our ſtomach and inteſtines, and the ſecretions of the various glands, are other circles of irritative motions, ſome of them more or leſs complete, according to our abſtinence or ſatiety.

So that the irritative ideas of the apparent motions of objects, the irritative battements of founds, and the movements of our bowels and glands compoſe a great circle of irritative tribes of motion : and when one conſiderable part of this circle of motions becomes interrupted, the whole proceeds in confuſion, as deſcribed in Section XVII. 1. 7. on Catenation of Motions.

8. Hence a violent vertigo, from whatever cauſe it happens, is generally attended with undulating noiſe in the head, perverſions of the motions of the ſtomach and duodenum, unuſual excretion of bile and gaſtic juice, with much pale urine, ſometimes with yellowneſs of the ſkin, and a diſordered ſecretion of almoſt every gland of the body, till at length the arterial ſyſtem is affected, and fever ſucceeds.

Thus bilious vomitings accompany the vertigo occaſioned by the motion of a ſhip; and when the brain is rendered vertiginous by a paralytic affection of any part of the body, a vomiting generally enſues, and a great diſcharge of bile: and hence great injuries of the head from external violence are ſucceeded with bilious vomitings, and ſometimes with abſceſſes of the liver.   And hence, when a patient is inclined to vomit from other cauſes, as in ſome fevers, any motions of the attendants in his room, or of himſelf when he is raiſed or turned in his bed, preſently induces the vomiting by ſuperadding a degree of vertigo.

9. And converſely it is very uſual with thoſe, whoſe ſtomachs are affected from internal cauſes, to be afflicted with vertigo, and noiſe

H h 2   in

in the head; such is the vertigo of drunken people, which continues, when their eyes are closed, and themselves in a recumbent posture, as well as when they are in an erect posture, and have their eyes open. And thus the irritation of a stone in the bile-duct, or in the ureter, or an inflammation of any of the intestines, are accompanied with vomitings and vertigo.

In these cases the irritative motions of the stomach, which are in general not attended to, become so changed by some unnatural stimulus, as to become uneasy, and excite our sensation or attention. And thus the other irritative trains of motions, which are associated with it, become disordered by their sympathy. The same happens, when a piece of gravel sticks in the ureter, or when some part of the intestinal canal becomes inflamed. In these cases the irritative muscular motions are first disturbed by unusual stimulus, and a disordered action of the sensual motions, or dizziness ensues. While in sea-sickness the irritative sensual motions, as vertigo, precedes; and the disordered irritative muscular motions, as those of the stomach in vomiting, follow.

10. When these irritative motions are disturbed, if the degree be not very great, the exertion of voluntary attention to any other object, or any sudden sensation, will disjoin these new habits of motion. Thus some drunken people have become sober immediately, when any accident has strongly excited their attention; and sea-sickness has vanished, when the ship has been in danger. Hence when our attention to other objects is most relaxed, as just before we fall asleep, or between our reveries when awake, these irritative ideas of motion and sound are most liable to be perceived; as those, who have been at sea, or have travelled long in a coach, seem to perceive the vibrations of the ship, or the rattling of the wheels, at these intervals; which cease again, as soon as they exert their attention. That is, at those intervals they attend to the apparent motions, and to the battement of sounds of the bodies around them, and for a moment mistake them

7   for

for thofe real motions of the fhip, and noife of wheels, which they had lately been accuftomed to : or at thefe intervals of reverie, or on the approach of fleep, thefe fuppofed motions or founds may be produced entirely by imagination.

We may conclude from this account of vertigo, that fea-ficknefs is not an effort of nature to relieve herfelf, but a neceffary confequence of the affociations or catenations of animal motions. And may thence infer, that the vomiting, which attends the gravel in the ureter, inflammations of the bowels, and the commencement of fome fevers, has a fimilar origin, and is not always an effort of the vis medicatrix naturæ. But where the action of the organ is the immediate confequence of the ftimulating caufe, it is frequently exerted to diflodge that ftimulus, as in vomiting up an emetic drug ; at other times, the action of an organ is a general effort to relieve pain, as in convulfions of the locomotive mufcles ; other actions drink up and carry on the fluids, as in abforption and fecretion; all which may be termed efforts of nature to relieve, or to preferve herfelf.

11. The cure of vertigo will frequently depend on our previoufly inveftigating the caufe of it, which from what has been delivered above may originate from the diforder of any part of the great tribes of irritative motions, and of the affociate motions catenated with them.

Many people, when they arrive at fifty or fixty years of age, are affected with flight vertigo; which is generally but wrongly afcribed to indigeftion, but in reality arifes from a beginning defect of their fight; as about this time they alfo find it neceffary to begin to ufe fpectacles, when they read fmall prints, efpecially in winter, or by candle light, but are yet able to read without them during the fummer days, when the light is ftronger. Thefe people do not fee objects fo diftinctly as formerly, and by exerting their eyes more than ufual, they perceive the apparent motions of objects, and confound them with the real motions of them; and therefore cannot accurately balance themfelves fo as eafily to preferve their perpendicularity by them.

That

That is, the apparent motions of objects, which are at reft, as we move by them, fhould only excite irritative ideas: but as thefe are now become lefs diftinct, owing to the beginning imperfection of our fight, we are induced *voluntarily* to attend to them; and then thefe apparent motions become fucceeded by fenfation; and thus the other parts of the trains of irritative ideas, or irritative mufcular motions, become difordered, as explained above. In thefe cafes of flight vertigo I have always promifed my patients, that they would get free from it in two or three months, as they fhould acquire the habit of balancing their bodies by lefs diftinct objects, and have feldom been miftaken in my prognoftic.

There is an auditory vertigo, which is called a noife in the head, explained in No. 7. of this fection, which alfo is very liable to affect people in the advance of life, and is owing to their hearing lefs perfectly than before. This is fometimes called a ringing, and fometimes a finging, or buzzing, in the ears, and is occafioned by our firft experiencing a difagreeable fenfation from our not being able diftinctly to hear the founds, we ufed formerly to hear diftinctly. And this difagreeable fenfation excites defire and confequent volition; and when we voluntarily attend to fmall indiftinct founds, even the whifpering of the air in a room, and the pulfations of the arteries of the ear are fucceeded by fenfation; which minute founds ought only to have produced irritative fenfual motions, or unperceived ideas. See Section XVII. 3. 6. Thefe patients after a while lofe this auditory vertigo, by acquiring a new habit of not attending voluntarily to thefe indiftinct founds, but contenting themfelves with the lefs accuracy of their fenfe of hearing.

Another kind of vertigo begins with the difordered action of fome irritative mufcular motions, as thofe of the ftomach from intoxication, or from emetics; or thofe of the ureter, from the ftimulus of a ftone lodged in it; and it is probable, that the difordered motions of fome of the great congeries of glands, as of thofe which form the liver, or of
                                                                    the

the inteftinal canal, may occafion vertigo in confequence of their motions being affociated or catenated with the great circles of irritative motions; and from hence it appears, that the means of cure muft be adapted to the caufe.

To prevent fea-ficknefs it is probable, that the habit of fwinging for a week or two before going on fhipboard might be of fervice. For the vertigo from failure of fight, fpectacles may be ufed. For the auditory vertigo, æther may be dropt into the ear to ftimulate the part, or to diffolve ear-wax, if fuch be a part of the caufe. For the vertigo arifing from indigeftion, the peruvian bark and a blifter are recommended. And for that owing to a ftone in the ureter, venefection, cathartics, opiates, fal foda aerated.

12. Definition of vertigo. 1. Some of the irritative fenfual, or mufcular motions, which were ufually not fucceeded by fenfation, are in this difeafe fucceeded by fenfation; and the trains or circles of motions, which were ufually catenated with them, are interrupted, or inverted, or proceed in confufion. 2. The fenfitive and voluntary motions continue undifturbed. 3. The affociate trains or circles of motions continue; but their catenations with fome of the irritative motions are difordered, or inverted, or diffevered.

SECT.

SECT. XXI.

OF DRUNKENNESS.

1. Sleep from satiety of hunger. From rocking children. From uniform sounds. 2. Intoxication from common food after fatigue and inanition. 3. From wine or opium. Chilness after meals. Vertigo. Why pleasure is produced by intoxication, and by swinging and rocking children. And why pain is relieved by it. 4. Why drunkards stagger and stammer, and are liable to weep. 5. And become delirious, sleepy, and stupid. 6. Or make pale urine and vomit. 7. Objects are seen double. 8. Attention of the mind diminishes drunkenness. 9. Disordered irritative motions of all the senses. 10. Diseases from drunkenness. 11. Definition of drunkenness.

1. IN the state of nature when the sense of hunger is appeased by the stimulus of agreeable food, the business of the day is over, and the human savage is at peace with the world, he then exerts little attention to external objects, pleasing reveries of imagination succeed, and at length sleep is the result : till the nourishment which he has procured, is carried over every part of the system to repair the injuries of action, and he awakens with fresh vigour, and feels a renewal of his sense of hunger.

The juices of some bitter vegetables, as of the poppy and the lauro-cerasus, and the ardent spirit produced in the fermentation of the sugar found in vegetable juices, are so agreeable to the nerves of the stomach, that, taken in a small quantity, they instantly pacify the sense of hunger ; and the inattention to external stimuli with the reveries

veries of imagination, and fleep, fucceeds, in the fame manner as when the ftomach is filled with other lefs intoxicating food.

This inattention to the irritative motions occafioned by external ftimuli is a very important circumftance in the approach of fleep, and is produced in young children by rocking their cradles: during which all vifible objects become indiftinct to them. An uniform foft repeated found, as the murmurs of a gentle current, or of bees, are faid to produce the fame effect, by prefenting indiftinct ideas of inconfequential founds, and by thus ftealing our attention from other objects, whilft by their continued reiterations they become familiar themfelves, and we ceafe gradually to attend to any thing, and fleep enfues.

2. After great fatigue or inanition, when the ftomach is fuddenly filled with flefh and vegetable food, the inattention to external ftimuli, and the reveries of imagination, become fo confpicuous as to amount to a degree of intoxication. The fame is at any time produced by fuperadding a little wine or opium to our common meals; or by taking thefe feparately in confiderable quantity; and this more efficacioufly after fatigue or inanition; becaufe a lefs quantity of any ftimulating material will excite an organ into energetic action, after it has lately been torpid from defect of ftimulus; as objects appear more luminous, after we have been in the dark; and becaufe the fufpenfion of volition, which is the immediate caufe of fleep, is fooner induced, after a continued voluntary exertion has in part exhaufted the fenforial power of volition; in the fame manner as we cannot contract a fingle mufcle long together without intervals of inaction.

3. In the beginning of intoxication we are inclined to fleep, as mentioned above, but by the excitement of external circumftances, as of noife, light, bufinefs, or by the exertion of volition, we prevent the approaches of it, and continue to take into our ftomach greater quantities of the inebriating materials. By thefe means the irritative movements of the ftomach are excited into greater action than is na-

I. i.   tural;

tural; and in confequence all the irritative tribes and trains of motion, which are catenated with them, become fufceptible of ftronger
action from their accuftomed ftimuli; becaufe thefe motions are excited both by their ufual irritation, and by their affociation with the
increafed actions of the ftomach and lacteals. Hence the fkin glows,
and the heat of the body is increafed, by the more energetic action of
the whole glandular fyftem; and pleafure is introduced in confequence
of thefe increafed motions from internal ftimulus. According to
Law, 5. Sect. IV. on Animal Caufation.

From this great increafe of irritative motions from internal ftimulus, and the increafed fenfation introduced into the fyftem in confequence; and fecondly, from the increafed fenfitive motions in confequence of this additional quantity of fenfation, fo much fenforial power
is expended, that the voluntary power becomes feebly exerted, and
the irritation from the ftimulus of external objects is lefs forcible;
the external parts of the eye are not therefore voluntarily adapted to
the diftances of objects, whence the apparent motions of thofe objects
either are feen double, or become too indiftinct for the purpofe of balancing the body, and vertigo is induced.

Hence we become acquainted with that very curious circumftance,
why the drunken vertigo is attended with an increafe of pleafure; for
the irritative ideas and motions occafioned by internal ftimulus, that
were not attended to in our fober hours, are now juft fo much increafed as to be fucceeded by pleafureable fenfation, in the fame manner as the more violent motions of our organs are fucceeded by painful
fenfation. And hence a greater quantity of pleafureable fenfation is
introduced into the conftitution; which is attended in fome people
with an increafe of benevolence and good humour.

If the apparent motions of objects is much increafed, as when we
revolve on one foot, or are fwung on a rope, the ideas of thefe apparent motions are alfo attended to, and are fucceeded with pleafureable fenfation, till they become familiar to us by frequent ufe. Hence
                                                            children

children are at firft delighted with thefe kinds of exercife, and with
riding, and failing, and hence rocking young children inclines them
to fleep. For though in the vertigo from intoxication the irritative
ideas of the apparent motions of objects are indiftinct from their decreafe of energy: yet in the vertigo occafioned by rocking or fwinging the irritative ideas of the apparent motions of objects are increafed
in energy, and hence they induce pleafure into the fyftem, but are
equally indiftinct, and in confequence equally unfit to balance ourfelves by. This addition of pleafure precludes defire or averfion, and
in confequence the voluntary power is feebly exerted, and on this account rocking young children inclines them to fleep.

In what manner opium and wine act in relieving pain is another
article, that well deferves our attention. There are many pains that
originate from defect as well as from excefs of ftimulus; of thefe are
thofe of the fix appetites of hunger, thirft, luft, the want of heat, of
diftention, and of frefh air. Thus if our cutaneous capillaries ceafe to
act from the diminifhed ftimulus of heat, when we are expofed to
cold weather, or our ftomach is uneafy for want of food; thefe are
both pains from defect of ftimulus, and in confequence opium, which
ftimulates all the moving fyftem into increafed action, muft relieve
them. But this is not the cafe in thofe pains, which arife from excefs
of ftimulus, as in violent inflammations: in thefe the exhibition of
opium is frequently injurious by increafing the action of the fyftem
already too great, as in inflammation of the bowels mortification is
often produced by the ftimulus of opium. Where, however, no fuch
bad confequences follow; the ftimulus of opium, by increafing all the
motions of the fyftem, expends fo much of the fenforial power, that
the actions of the whole fyftem foon become feebler, and in confequence thofe which produced the pain and inflammation.

4. When intoxication proceeds a little further, the quantity of
pleafureable fenfation is fo far increafed, that all defire ceafes, for there
is no pain in the fyftem to excite it. Hence the voluntary exertions

are

are diminifhed, ftaggering and ftammering fucceed; and the trains of ideas become more and more inconfiftent from this defect of voluntary exertion, as explained in the fections on fleep and reverie, whilft thofe paffions which are unmixed with volition are more vividly felt, and fhewn with lefs referve; hence pining love, or fuperftitious fear, and the maudling tear dropped on the remembrance of the moft trifling diftrefs.

5. At length all thefe circumftances are increafed; the quantity of pleafure introduced into the fyftem by the increafed irritative mufcular motions of the whole fanguiferous, and glandular, and abforbent fyftems, becomes fo great, that the organs of fenfe are more forcibly excited into action by this internal pleafureable fenfation, than by the irritation from the ftimulus of external objects. Hence the drunkard ceafes to attend to external ftimuli, and as volition is now alfo fufpended, the trains of his ideas become totally inconfiftent as in dreams, or delirium: and at length a ftupor fucceeds from the great exhauftion of fenforial power, which probably does not even admit of dreams, and in which, as in apoplexy, no motions continue but thofe from internal ftimuli, from fenfation, and from affociation.

6. In other people a paroxyfm of drunkennefs has another termination; the inebriate, as foon as he begins to be vertiginous, makes pale urine in great quantities and very frequently, and at length becomes fick, vomits repeatedly, or purges, or has profufe fweats, and a temporary fever enfues with a quick ftrong pulfe. This in fome hours is fucceeded by fleep; but the unfortunate bacchanalian does not perfectly recover himfelf till about the fame time of the fucceeding day, when his courfe of inebriation began. As fhewn in Sect. XVII. 1. 7. on Catenation. The temporary fever with ftrong pulfe is owing to the fame caufe as the glow on the fkin mentioned in the third paragraph of this Section: the flow of urine and ficknefs arifes from the whole fyftem of irritative motions being thrown into confufion by their affociations with each other; as in fea-ficknefs, mentioned in Sect. XX. 4.

On

on Vertigo; and which is more fully explained in Section XXIX. on Diabetes.

7. In this vertigo from internal caufes we fee objects double, as two candles inftead of one, which is thus explained. Two lines drawn through the axes of our two eyes meet at the object we attend to: this angle of the optic axes increafes or diminifhes with the lefs or greater diftances of objects. All objects before or behind the place where this angle is formed, appear double; as any one may obferve by holding up a pen between his eyes and the candle; when he looks attentively at a fpot on the pen, and carelefsly at the candle, it will appear double; and the reverfe when he looks attentively at the candle and carelefsly at the pen; fo that in this cafe the mufcles of the eye, like thofe of the limbs, ftagger and are difobedient to the expiring efforts of volition. Numerous objects are indeed fometimes feen by the inebriate, occafioned by the refractions made by the tears, which ftand upon his eye-lids.

8. This vertigo alfo continues, when the inebriate lies in his bed, in the dark, or with his eyes clofed; and this more powerfully than when he is erect, and in the light. For the irritative ideas of the apparent motions of objects are now excited by irritation from internal ftimulus, or by affociation with other irritative motions; and the inebriate, like one in a dream, believes the objects of thefe irritative motions to be prefent, and feels himfelf vertiginous. I have obferved in this fituation, fo long as my eyes and mind were intent upon a book, the ficknefs and vertigo ceafed, and were renewed again the moment I difcontinued this attention; as was explained in the preceding account of fea-ficknefs. Some drunken people have been known to become fober inftantly from fome accident, that has ftrongly excited their attention, as the pain of a broken bone, or the news of their houfe being on fire.

9. Sometimes the vertigo from internal caufes, as from intoxication, or at the beginning of fome fevers, becomes fo univerfal, that

the irritative motions which belong to other organs of fenfe are fucceeded by fenfation or attention, as well as thofe of the eye. The
vertiginous noife in the ears has been explained in Section XX. on
Vertigo. The tafte of the faliva, which in general is not attended
to, becomes perceptible, and the patients complain of a bad tafte in
their mouth.

The common fmells of the furrounding air fometimes excite the attention of thefe patients, and bad fmells are complained of, which to
other people are imperceptible. The irritative motions that belong
to the fenfe of preffure, or of touch, are attended to, and the patient
conceives the bed to librate, and is fearful of falling out of it. The
irritative motions belonging to the fenfes of diftention, and of heat,
like thofe above mentioned, become attended to at this time: hence
we feel the pulfation of our arteries all over us, and complain of heat,
or of cold, in parts of the body where there is no accumulation or diminution of actual heat. All which are to be explained, as in the laft
paragraph, by the irritative ideas belonging to the various fenfes being
now excited by internal ftimuli, or by their affociations with other irritative motions. And that the inebriate, like one in a dream, believes the external objects, which ufually caufed thefe irritative ideas,
to be now prefent.

10. The difeafes in confequence of frequent inebriety, or of daily
taking much vinous fpirit without inebriety, confift in the paralyfis,
which is liable to fucceed violent ftimulation. Organs, whofe actions
are affociated with others, are frequently more affected than the organ,
which is ftimulated into too violent action. See Sect. XXIV. 2. 8.
Hence in drunken people it generally happens, that the fecretory veffels
of the liver become firft paralytic, and a torpor with confequent gall-
ftones or fchirrus of this vifcus is induced with concomitant jaundice;
otherwife it becomes inflamed in confequence of previous torpor, and
this inflammation is frequently transferred to a more fenfible part, which
is affociated with it, and produces the gout, or the rofy eruption of

7

the

the face, or fome other leprous eruption on the head, or arms, or legs. Sometimes the ftomach is firft affected, and paralyfis of the lacteal fyftem is induced; whence a total abhorrence from flefh-food, and general emaciation. In others the lymphatic fyftem is affected with paralyfis, and dropfy is the confequence. In fome inebriates the torpor of the liver produces pain without apparent fchirrus, or gall-ftones, or inflammation, or confequent gout, and in thefe epilepfy or infanity are often the confequence. All which will be more fully treated of in the courfe of the work.

I am well aware, that it is a common opinion, that the gout is as frequently owing to gluttony in eating, as to intemperance in drinking fermented or fpirituous liquors. To this I anfwer, that I have feen no perfon afflicted with the gout, who has not drank freely of fermented liquor, as wine and water, or fmall beer; though as the difpofition to all the difeafes, which have originated from intoxication, is in fome degree hereditary, a lefs quantity of fpirituous potation will induce the gout in thofe, who inherit the difpofition from their parents. To which I muft add, that in young people the rheumatifm is frequently miftaken for the gout.

Spice is feldom taken in fuch quantity as to do any material injury to the fyftem, flefh-meats as well as vegetables are the natural diet of mankind; with thefe a glutton may be crammed up to the throat, and fed fat like a ftalled ox; but he will not be difeafed, unlefs he adds fpirituous or fermented liquor to his food. This is well known in the diftilleries, where the fwine, which are fattened by the fpirituous fediments of barrels, acquire difeafed livers. But mark what happens to a man, who drinks a quart of wine or of ale, if he has not been habituated to it. He lofes the ufe both of his limbs and of his underftanding! He becomes a temporary idiot, and has a temporary ftroke of the palfy! And though he flowly recovers after fome hours, is it not reafonable to conclude, that a perpetual repetition of fo powerful a poifon muft at length permanently affect him?—If a perfon accidentally

tally becomes intoxicated by eating a few muſhrooms of a peculiar kind, a general alarm is excited, and he is ſaid to be poiſoned, and emetics are exhibited; but ſo familiariſed are we to the intoxication from vinous ſpirit, that it occaſions laughter rather than alarm.

There is however conſiderable danger in too haſtily diſcontinuing the uſe of ſo ſtrong a ſtimulus, left the torpor of the ſyſtem, or para- lyſis, ſhould ſooner be induced by the omiſſion than by the continu- ance of this habit, when unfortunately acquired. A golden rule for determining the quantity, which may with ſafety be diſcontinued, is delivered in Sect. XII. 7. 8.

11. Definition of drunkenneſs. Many of the irritative motions are much increaſed in energy by internal ſtimulation.

2. A great additional quantity of pleaſureable ſenſation is occaſioned by this increaſed exertion of the irritative motions. And many ſenſi- tive motions are produced in conſequence of this increaſed ſenſation.

3. The aſſociated trains and tribes of motions, catenated with the increaſed irritative and ſenſitive motions, are diſturbed, and proceed in confuſion.

4. The faculty of volition is gradually impaired, whence proceeds the inſtability of locomotion, inaccuracy of perception, and incon- ſiſtency of ideas; and is at length totally ſuſpended, and a temporary apoplexy ſucceeds.

SECT. XXII.

OF PROPENSITY TO MOTION, REPETITION AND IMITATION.

I. *Accumulation of senforial power in hemiplagia, in sleep, in cold fit of fever, in the locomotive muscles, in the organs of sense. Produces propensity to action.* II. *Repetition by three sensorial powers. In rhimes and alliterations, in music, dancing, architecture, landscape-painting, beauty.* III. 1. *Perception consists in imitation. Four kinds of imitation.* 2. *Voluntary. Dogs taught to dance.* 3. *Sensitive. Hence sympathy, and all our virtues. Contagious matter of venereal ulcers, of hydrophobia, of jail-fever, of small-pox, produced by imitation, and the sex of the embryon.* 4. *Irritative imitation.* 5. *Imitations resolvable into associations.*

I. 1. IN the hemiplagia, when the limbs on one side have lost their power of voluntary motion, the patient is for many days perpetually employed in moving those of the other. 2. When the voluntary power is suspended during sleep, there commences a ceaseless flow of sensitive motions, or ideas of imagination, which compose our dreams. 3. When in the cold fit of an intermittent fever some parts of the system have for a time continued torpid, and have thus expended less than their usual expenditure of sensorial power; a hot fit succeeds, with violent action of those vessels, which had previously been quiescent. All these are explained from an accumulation of sensorial power during the inactivity of some part of the system.

Besides the very great quantity of sensorial power perpetually produced and expended in moving the arterial, venous, and glandular

K k   systems,

fyftems, with the various organs of digeftion, as defcribed in Section XXXII. 3. 2. there is alfo a conftant expenditure of it by the action of our locomotive mufcles and organs of fenfe. Thus the thicknefs of the optic nerves, where they enter the eye, and the great expanfion of the nerves of touch beneath the whole of the cuticle, evince the great confumption of fenforial power by thefe fenfes. And our perpetual mufcular actions in the common offices of life, and in conftantly preferving the perpendicularity of our bodies during the day, evince a confiderable expenditure of the fpirit of animation by our locomotive mufcles. It follows, that if the exertion of thefe organs of fenfe and mufcles be for a while intermitted, that fome quantity of fenforial power muft be accumulated, and a propenfity to activity of fome kind enfue from the increafed excitability of the fyftem. Whence proceeds the irkfomenefs of a continued attitude, and of an indolent life.

However fmall this hourly accumulation of the fpirit of animation may be, it produces a propenfity to fome kind of action; but it neverthelefs requires either defire or averfion, either pleafure or pain, or fome external ftimulus, or a previous link of affociation, to excite the fyftem into activity; thus it frequently happens, when the mind and body are fo unemployed as not to poffefs any of the three firft kinds of ftimuli, that the laft takes place, and confumes the fmall but perpetual accumulation of fenforial power. Whence fome indolent people repeat the fame verfe for hours together, or hum the fame tune. Thus the poet:

> Onward he trudged, not knowing what he fought,
> And whiftled, as he went, for want of thought.

II. The repetitions of motions may be at firft produced either by volition, or by fenfation, or by irritation, but they foon become eafier to perform than any other kinds of action, becaufe they foon become affociated together, according to Law the feventh, Section IV. on

Animal

Animal Caufation.   And becaufe their frequency of repetition, if as much fenforial power be produced during every reiteration as is expended, adds to the facility of their production.

If a ftimulus be repeated at uniform intervals of time, as defcribed in Sect. XII. 3. 3. the action, whether of our mufcles or organs of fenfe, is produced with ftill greater facility or energy; becaufe the fenforial power of affociation, mentioned above, is combined with the fenforial power of irritation ; that is, in common language, the acquired habit affifts the power of the ftimulus.

This not only obtains in the annual, lunar, and diurnal catenations of animal motions, as explained in Sect. XXXVI. which are thus performed with great facility and energy ; but in every lefs circle of actions or ideas, as in the burthen of a fong, or the reiterations of a dance.   To the facility and diftinctnefs, with which we hear founds at repeated intervals, we owe the pleafure, which we receive from mufical time, and from poetic time ; as defcribed in Botanic Garden, P. 2. Interlude 3.   And to this the pleafure we receive from the rhimes and alliterations of modern verfification ; the fource of which without this key would be difficult to difcover.   And to this likewife fhould be afcribed the beauty of the duplicature in the perfect tenfe of the Greek verbs, and of fome Latin ones, as tango tetegi, mordeo momordi.

There is no variety of notes referable to the gamut in the beating of the drum, yet if it be performed in mufical time, it is agreeable to our ears ; and therefore this pleafureable fenfation muft be owing to the repetition of the divifions of the founds at certain intervals of time, or mufical bars.   Whether thefe times or bars are diftinguifhed by a paufe, or by an emphafis, or accent, certain it is, that this diftinction is perpetually repeated ; otherwife the ear could not determine inftantly, whether the fucceffions of found were in common or in triple time. In common time there is a divifion between every two crotchets, or

K k 2   other

other notes of equivalent time; though the bar in written mufic is put after every fourth crotchet, or notes equivalent in time; in triple time the divifion or bar is after every three crotchets, or notes equivalent; fo that in common time the repetition recurs more frequently than in triple time. The grave or heroic verfes of the Greek and Latin poets are written in common time; the French heroic verfes, and Mr. Anftie's humorous verfes in his bath guide, are written in the fame time as the Greek and Latin verfes, but are one bar fhorter. The Englifh grave or heroic verfes are meafured by triple time, as Mr. Pope's tranflation of Homer.

But befides thefe little circles of mufical time, there are the greater returning periods, and the ftill more diftant chorufes, which, like the rhimes at the ends of verfes, owe their beauty to repetition; that is, to the facility and diftinctnefs with which we perceive founds, which we expect to perceive, or have perceived before; or in the language of this work, to the greater eafe and energy with which our organ is excited by the combined fenforial powers of affociation and irritation, than by the latter fingly.

A certain uniformity or repetition of parts enters the very compofition of harmony. Thus two octaves neareft to each other in the fcale commence their vibrations together after every fecond vibration of the higher one. And where the firft, third, and fifth compofe a chord the vibrations concur or coincide frequently, though lefs fo than in the two octaves. It is probable that thefe chords bear fome analogy to a mixture of three alternate colours in the fun's fpectrum feparated by a prifm.

The pleafure we receive from a melodious fucceffion of notes referable to the gamut is derived from another fource, viz. to the pandiculation or counteraction of antagonift fibres. See Botanic Garden, P. 2. Interlude 3. If to thefe be added our early affociations of agreeable ideas with certain proportions of found, I fuppofe, from thefe

three

three fources fprings all the delight of mufic, fo celebrated by ancient authors, and fo enthufiastically cultivated at prefent. See Sect. XVI. No. 10. on Inftinct.

This kind of pleafure arifing from repetition, that is from the facility and diftinctnefs, with which we perceive and underftand repeated fenfations, enters into all the agreeable arts ; and when it is carried to excefs is termed formality. The art of dancing like that of mufic depends for a great part of the pleafure, it affords, on repetition ; architecture, efpecially the Grecian, confifts of one part being a repetition of another ; and hence the beauty of the pyramidal outline in landfcape-painting ; where one fide of the picture may be faid in fome meafure to balance the other. So univerfally does repetition contribute to our pleafure in the fine arts, that beauty itfelf has been defined by fome writers to confift in a due combination of uniformity and variety. See Sect. XVI. 6.

III. 1. Man is termed by Ariftotle an imitative animal ; this propenfity to imitation not only appears in the actions of children, but in all the cuftoms and fafhions of the world ; many thoufands tread in the beaten paths of others, for one who traverfes regions of his own difcovery. The origin of this propenfity to imitation has not, that I recollect, been deduced from any known principle ; when any action prefents itfelf to the view of a child, as of whetting a knife, or threading a needle, the parts of this action in refpect of time, motion, figure, is imitated by a part of the retina of his eye; to perform this action therefore with his hands is eafier to him than to invent any new action, becaufe it confifts in repeating with another fet of fibres, viz. with the moving mufcles, what he had juft performed by fome parts of the retina; juft as in dancing we transfer the times of motion from the actions of the auditory nerves to the mufcles of the limbs. Imitation therefore confifts of repetition, which we have fhewn above to be the eafieft kind of animal action, and which we

perpetually

perpetually fall into, when we poffefs an accumulation of fenforial power, which is not otherwife called into exertion.

It has been fhewn, that our ideas are configurations of the organs of fenfe, produced originally in confequence of the ftimulus of external bodies. And that thefe ideas, or configurations of the organs of fenfe, refemble in fome property a correfpondent property of external matter; as the parts of the fenfes of fight and of touch, which are excited into action, refemble in figure the figure of the ftimulating body; and probably alfo the colour, and the quantity of denfity, which they perceive. As explained in Sect. XIV. 2. 2. Hence it appears, that our perceptions themfelves are copies, that is, imitations of fome properties of external matter; and the propenfity to imitation is thus interwoven with our exiftence, as it is produced by the ftimuli of external bodies, and is afterwards repeated by our volitions and fenfations, and thus conftitutes all the operations of our minds.

2. Imitations refolve themfelves into four kinds, voluntary, fenfitive, irritative, and affociate. The voluntary imitations are, when we imitate deliberately the actions of others, either by mimicry, as in acting a play, or in delineating a flower; or in the common actions of our lives, as in our drefs, cookery, language, manners, and even in our habits of thinking.

Not only the greateft part of mankind learn all the common arts of life by imitating others, but brute animals feem capable of acquiring knowledge with greater facility by imitating each other, than by any methods by which we can teach them; as dogs and cats, when they are fick, learn of each other to eat grafs; and I fuppofe, that by making an artificial dog perform certain tricks, as in dancing on his hinder legs, a living dog might be eafily induced to imitate them; and that the readieft way of inftructing dumb animals is by practifing them with others of the fame fpecies, which have already learned the arts we wifh to teach them. The important ufe of imitation in ac-

5   quiring

quiring natural language is mentioned in Section XVI. 7. and 8. on Inftinct.

3. The fenfitive imitations are the immediate confequences of pleafure or pain, and thefe are often produced even contrary to the efforts of the will.   Thus many young men on feeing cruel furgical operations become fick, and fome even feel pain in the parts of their own bodies, which they fee tortured or wounded in others; that is, they in fome meafure imitate by the exertions of their own fibres the violent actions, which they witneffed in thofe of others.   In this cafe a double imitation takes place, firft the obferver imitates with the extremities of the optic nerve the mangled limbs, which are prefent before his eyes; then by a fecond imitation he excites fo violent action of the fibres of his own limbs as to produce pain in thofe parts of his own body, which he faw wounded in another.   In thefe pains produced by imitation the effect has fome fimilarity to the caufe, which diftinguifhes them from thofe produced by affociation; as the pains of the teeth, called tooth-edge, which are produced by affociation with difagreeable founds, as explained in Sect. XVI. 10.

The effect of this powerful agent, imitation, in the moral world, is mentioned in Sect. XVI. 7. as it is the foundation of all our intellectual fympathies with the pains and pleafures of others, and is in confequence the fource of all our virtues.   For in what confifts our fympathy with the miferies, or with the joys, of our fellow creatures, but in an involuntary excitation of ideas in fome meafure fimilar or imitative of thofe, which we believe to exift in the minds of the perfons, whom we commiferate or congratulate?

There are certain concurrent or fucceffive actions of fome of the glands, or other parts of the body, which are poffeffed of fenfation, which become intelligible from this propenfity to imitation.   Of thefe are the production of matter by the membranes of the fauces, or by the fkin, in confequence of the venereal difeafe previoufly affecting

the

the parts of generation. Since as no fever is excited, and as neither the blood of such patients, nor even the matter from ulcers of the throat, or from cutaneous ulcers, will by inoculation produce the venereal disease in others, as observed by Mr. Hunter, there is reason to conclude, that no contagious matter is conveyed thither by the blood-vessels, but that a milder matter is formed by the actions of the fine vessels in those membranes imitating each other. See Section XXXIII. 2. 9. In this disease the actions of these vessels producing ulcers on the throat and skin are imperfect imitations of those producing chanker, or gonorrhœa; since the matter produced by them is not infectious, while the imitative actions in the hydrophobia appear to be perfect resemblances, as they produce a material equally infectious with the original one, which induced them.

The contagion from the bite of a mad dog differs from other contagious materials, from its being communicable from other animals to mankind, and from many animals to each other; the phenomena attending the hydrophobia are in some degree explicable on the foregoing theory. The infectious matter does not appear to enter the circulation, as it cannot be traced along the course of the lymphatics from the wound, nor is there any swelling of the lymphatic glands, nor does any fever attend, as occurs in the small-pox, and in many other contagious diseases; yet by some unknown process the disease is communicated from the wound to the throat, and that many months after the injury, so as to produce pain and hydrophobia, with a secretion of infectious saliva of the same kind, as that of the mad dog, which inflicted the wound.

This subject is very intricate.—It would appear, that by certain morbid actions of the salivary glands of the mad dog, a peculiar kind of saliva is produced; which being instilled into a wound of another animal stimulates the cutaneous or mucous glands into morbid actions, but which are ineffectual in respect to the production of a similar contagious

tagious material; but the falivary glands by irritative fympathy are thrown into fimilar action, and produce an infectious faliva fimilar to that inftilled into the wound.

Though in many contagious fevers a material fimilar to that which produced the difeafe, is thus generated by imitation; yet there are other infectious materials, which do not thus propagate themfelves, but which feem to act like flow poifons. Of this kind was the contagious matter, which produced the jail-fever at the affizes at Oxford about a century ago. Which, though fatal to fo many, was not communicated to their nurfes or attendants. In thefe cafes, the imitations of the fine veffels, as above defcribed, appear to be imperfect, and do not therefore produce a matter fimilar to that, which ftimulates them; in this circumftance refembling the venereal matter in ulcers of the throat or fkin, according to the curious difcovery of Mr. Hunter above related, who found, by repeated inoculations, that it would not infect. Hunter on Venereal Difeafe, Part vi. ch. 1.

Another example of morbid imitation is in the production of a great quantity of contagious matter, as in the inoculated fmall-pox, from a fmall quantity of it inferted into the arm, and probably diffufed in the blood. Thefe particles of contagious matter ftimulate the extremities of the fine arteries of the fkin, and caufe them to imitate fome properties of thofe particles of contagious matter, fo as to produce a thoufandfold of a fimilar material. See Sect. XXXIII. 2. 6. Other inftances are mentioned in the Section on Generation, which fhew the probability that the extremities of the feminal glands may imitate certain ideas of the mind, or actions of the organs of fenfe, and thus occafion the male or female fex of the embryon. See Sect. XXXIX. 6.

4. We come now to thofe imitations, which are not attended with fenfation. Of thefe are all the irritative ideas already explained, as when the retina of the eye imitates by its action or configuration the tree or the bench, which I fhun in walking paft without attending to them. Other examples of thefe irritative imitations are daily ob-

L l

fervable

fervable in common life : thus one yawning perfon fhall fet a whole company a yawning ; and fome have acquired winking of the eyes or impediments of fpeech by imitating their companions without being confcious of it.

5. Befides the three fpecies of imitations above defcribed there may be fome affociate motions, which may imitate each other in the kind as well as in the quantity of their action; but it is difficult to diftinguifh them from the affociations of motions treated of in Section XXXV. Where the actions of other perfons are imitated there can be no doubt, or where we imitate a preconceived idea by exertion of our locomotive mufcles, as in painting a dragon ; all thefe imitations may aptly be referred to the fources above defcribed of the propenfity to activity, and the facility of repetition ; at the fame time I do not affirm, that all thofe other apparent fenfitive and irritative imitations may not be refolvable into affociations of a peculiar kind, in which certain diftant parts of fimilar irritability or fenfibility, and which have habitually acted together, may affect each other exactly with the fame kinds of motion; as many parts are known to fympathife in the quantity of their motions. And that therefore they may be ultimately refolvable into affociations of action, as defcribed in Sect. XXXV.

SECT.

SECT. XXIII.

OF THE CIRCULATORY SYSTEM.

I. *The heart and arteries have no antagonist muscles. Veins absorb the blood, propel it forwards, and distend the heart; contraction of the heart distends the arteries. Vena portarum.   II. Glands which take their fluids from the blood. With long necks, with short necks.   III. Absorbent system.   IV. Heat given out from glandular secretions. Blood changes colour in the lungs and in the glands and capillaries.   V. Blood is absorbed by veins, as chyle by lacteal vessels, otherwise they could not join their streams.   VI. Two kinds of stimulus, agreeable and disagreeable. Glandular appetency. Glands originally possessed sensation.*

I. WE now step forwards to illustrate some of the phænomena of diseases, and to trace out their most efficacious methods of cure; and shall commence this subject with a short description of the circulatory system.

As the nerves, whose extremities form our various organs of sense and muscles, are all joined, or communicate, by means of the brain, for the convenience perhaps of the distribution of a subtile ethereal fluid for the purpose of motion; so all those vessels of the body, which carry the grosser fluids for the purposes of nutrition, communicate with each other by the heart.

The heart and arteries are hollow muscles, and are therefore indued with power of contraction in consequence of stimulus, like all other muscular fibres; but, as they have no antagonist muscles, the cavities of the vessels, which they form, would remain for ever

L l 2

closed,

clofed, after they have contracted themfelves, unlefs fome extraneous power be applied to again diftend them. This extraneous power in refpect to the heart is the current of blood, which is perpetually abforbed by the veins from the various glands and capillaries, and pufhed into the heart by a power probably very fimilar to that, which raifes the fap in vegetables in the fpring, which, according to Dr. Hale's experiment on the ftump of a vine, exerted a force equal to a column of water above twenty feet high. This force of the current of blood in the veins is partly produced by their abforbent power, exerted at the beginning of every fine ramification; which may be conceived to be a mouth abforbing blood, as the mouths of the lacteals and lymphatics abforb chyle and lymph. And partly by their intermitted compreffion by the pulfations of their generally concomitant arteries; by which the blood is perpetually propelled towards the heart, as the valves in many veins, and the abforbent mouths in them all, will not fuffer it to return.

The blood, thus forcibly injected into the chambers of the heart, diftends this combination of hollow mufcles; till by the ftimulus of diftention they contract themfelves; and, pufhing forwards the blood into the arteries, exert fufficient force to overcome in lefs than a fecond of time the vis inertiæ, and perhaps fome elafticity, of the very extenfive ramifications of the two great fyftems of the aortal and pulmonary arteries. The power neceffary to do this in fo fhort a time muft be confiderable, and has been varioufly eftimated by different phyfiologifts.

The mufcular coats of the arterial fyftem are then brought into action by the ftimulus of diftention, and propel the blood to the mouths, or through the convolutions, which precede the fecretory apertures of the various glands and capillaries.

In the veffels of the liver there is no intervention of the heart; but the vena portarum, which does the office of an artery, is diftended by the blood poured into it from the mefenteric veins, and is by this

diftention

diftention ftimulated to contract itfelf, and propel the blood to the mouths of the numerous glands, which compofe that vifcus.

II. The glandular fyftem of veffels may be divided into thofe, which take fome fluid from the circulation ; and thofe, which give fomething to it.   Thofe, which take their fluid from the circulation, are the various glands, by which the tears, bile, urine, perfpiration, and many other fecretions are produced; thefe glands probably confift of a mouth to felect, a belly to digeft, and an excretory aperture to emit their appropriated fluids ; the blood is conveyed by the power of the heart and arteries to the mouths of thefe glands, it is there taken up by the living power of the gland, and carried forwards to its belly, and excretory aperture, where a part is feparated, and the remainder abforbed by the veins for further purpofes.

Some of thefe glands are furnifhed with long convoluted necks or tubes, as the feminal ones, which are curioufly feen when injected with quickfilver.   Others feem to confift of fhorter tubes, as that great congeries of glands, which conftitute the liver, and thofe of the kidneys.   Some have their excretory apertures opening into refervoirs, as the urinary and gall-bladders.   And others on the external body, as thofe which fecret the tears, and perfpirable matter.

Another great fyftem of glands, which have very fhort necks, are the capillary veffels; by which the infenfible perfpiration is fecreted on the fkin ; and the mucus of various confiftences, which lubricates the interftices of the cellular membrane, of the mufcular fibres, and of all the larger cavities of the body.   From the want of a long convolution of veffels fome have doubted, whether thefe capillaries fhould be confidered as glands, and have been led to conclude, that the perfpirable matter rather exuded than was fecreted.   But the fluid of perfpiration is not fimple water, though that part of it which exhales into the air may be fuch ; for there is another part of it, which in a ftate of health is abforbed again ; but which, when the abforbents are

difeafed,

diſeaſed, remains on the ſurface of the ſkin, in the form of ſcurf, or indurated mucus. Another thing, which ſhews their ſimilitude to other glands, is their ſenſibility to certain affections of the mind; as is ſeen in the deeper colour of the ſkin in the bluſh of ſhame, or the greater paleneſs of it from fear.

III. Another ſeries of glandular veſſels is called the abſorbent ſyſtem; theſe open their mouths into all the cavities, and upon all thoſe ſurfaces of the body, where the excretory apertures of the other glands pour out their fluids. The mouths of the abſorbent ſyſtem drink up a part or the whole of theſe fluids, and carry them forwards by their living power to their reſpective glands, which are called conglobate glands. There theſe fluids undergo ſome change, before they paſs on into the circulation; but if they are very acrid, the conglobate gland ſwells, and ſometimes ſuppurates, as in inoculation of the ſmall-pox, in the plague, and in venereal abſorptions; at other times the fluid may perhaps continue there, till it undergoes ſome chemical change, that renders it leſs noxious; or, what is more likely, till it is regurgitated by the retrograde motion of the gland in ſpontaneous ſweats or diarrhœas, as diſagreeing food is vomited from the ſtomach.

IV. As all the fluids, that paſs through theſe glands, and capillary veſſels, undergo a chemical change, acquiring new combinations, the matter of heat is at the ſame time given out; this is apparent, ſince whatever increaſes inſenſible perſpiration, increaſes the heat of the ſkin; and when the action of theſe veſſels is much increaſed but for a moment, as in bluſhing, a vivid heat on the ſkin is the immediate conſequence. So when great bilious ſecretions, or thoſe of any other gland, are produced, heat is generated in the part in proportion to the quantity of the ſecretion.

The heat produced on the ſkin by bluſhing may be thought by ſome too ſudden to be pronounced a chemical effect, as the fermenta-
tions

tions or new combinations taking place in a fluid is in general a flower procefs. Yet are there many chemical mixtures in which heat is given out as inftantaneoufly; as in folutions of metals in acids, or in mixtures of effential oils and acids, as of oil of cloves and acid of nitre. So the bruifed parts of an unripe apple become almoft inftantaneoufly fweet; and if the chemico-animal procefs of digeftion be ftopped for but a moment, as by fear, or even by voluntary eructation, a great quantity of air is generated, by the fermentation, which inftantly fucceeds the ftop of digeftion. By the experiments of Dr. Hales it appears, that an apple during fermentation gave up above fix hundred times its bulk of air; and the materials in the ftomach are fuch, and in fuch a fituation, as immediately to run into fermentation, when digeftion is impeded.

As the blood paffes through the fmall veffels of the lungs, which connect the pulmonary artery and vein, it undergoes a change of colour from a dark to a light red; which may be termed a chemical change, as it is known to be effected by an admixture of oxygene, or vital air; which, according to a difcovery of Dr. Prieftley, paffes through the moift membranes, which conftitute the fides of thefe veffels. As the blood paffes through the capillary veffels, and glands, which connect the aorta and its various branches with their correfpondent veins in the extremities of the body, it again lofes the bright red colour, and undergoes fome new combinations in the glands or capillaries, in which the matter of heat is given out from the fecreted fluids. This procefs therefore, as well as the procefs of refpiration, has fome analogy to combuftion, as the vital air or oxygene feems to become united to fome inflammable bafe, and the matter of heat efcapes from the new acid, which is thus produced.

V. After the blood has paffed thefe glands and capillaries, and parted with whatever they chofe to take from it, the remainder is received by the veins, which are a fet of blood-abforbing veffels in general

neral

neral corresponding with the ramifications of the arterial system. At the extremity of the fine convolutions of the glands the arterial force ceases; this in respect to the capillary vessels, which unite the extremities of the arteries with the commencement of the veins, is evident to the eye, on viewing the tail of a tadpole by means of a solar, or even by a common microscope, for globules of blood are seen to endeavour to pass, and to return again and again, before they become absorbed by the mouths of the veins; which returning of these globules evinces, that the arterial force behind them has ceased. The veins are furnished with valves like the lymphatic absorbents; and the great trunks of the veins, and of the lacteals and lymphatic, join together before the ingress of their fluids into the left chamber of the heart; both which evince, that the blood in the veins, and the lymph and chyle in the lacteals and lymphatics, are carried on by a similar force; otherwise the stream, which was propelled with a less power, could not enter the vessels, which contained the stream propelled with a greater power. From whence it appears, that the veins are a system of vessels absorbing blood, as the lacteals and lymphatics are a system of vessels absorbing chyle and lymph. See Sect. XXVII. 1.

VI. The movements of their adapted fluids in the various vessels of the body are carried forwards by the actions of those vessels in consequence of two kinds of stimulus, one of which may be compared to a pleasureable sensation or desire inducing the vessel to seize, and, as it were, to swallow the particles thus selected from the blood; as is done by the mouths of the various glands, veins, and other absorbents, which may be called glandular appetency. The other kind of stimulus may be compared to disagreeable sensation, or aversion, as when the heart has received the blood, and is stimulated by it to push it forwards into the arteries; the same again stimulates the arteries to contract, and carry forwards the blood to their extremities, the glands and capillaries. Thus the mesenteric veins absorb the blood from the

intestines

inteftines by glandular appetency, and carry it forward to the vena por-tarum; which acting as an artery contracts itfelf by difagreeable fti-mulus, and pufhes it to its ramified extremities, the various glands, which conftitute the liver.

It feems probable, that at the beginning of the formation of thefe veffels in the embryon, an agreeable fenfation was in reality felt by the glands during fecretion, as is now felt in the act of fwallowing palatable food; and that a difagreeable fenfation was originally felt by the heart from the diftention occafioned by the blood, or by its che-mical ftimulus; but that by habit thefe are all become irritative mo-tions; that is, fuch motions as do not affect the whole fyftem, except when the veffels are difeafed by inflammation.

M m

SECT.

SECT. XXIV.

OF THE SECRETIONS OF SALIVA, AND OF TEARS, AND OF THE LACRYMAL SACK.

I. *Secretion of saliva increased by mercury in the blood.* 1. *By the food in the mouth. Dryness of the mouth not from a deficiency of saliva.* 2. *By sensitive ideas.* 3. *By volition.* 4. *By distasteful substances. It is secreted in a dilute and saline state. It then becomes more viscid.* 5. *By ideas of distasteful substances.* 6. *By nausea.* 7. *By aversion.* 8. *By catenation with stimulating substances in the ear.* II. 1. *Secretion of tears less in sleep. From stimulation of their excretory duct.* 2. *Lacrymal sack is a gland.* 3. *Its uses.* 4. *Tears are secreted, when the nasal duct is stimulated.* 5. *Or when it is excited by sensation.* 6. *Or by volition.* 7. *The lacrymal sack can regurgitate its contents into the eye.* 8. *More tears are secreted by association with the irritation of the nasal duct of the lacrymal sack, than the puncta lacrymalia can imbibe. Of the gout in the liver and stomach.*

I. THE falival glands drink up a certain fluid from the circumfluent blood, and pour it into the mouth. They are fometimes ftimulated into action by the blood, that furrounds their origin, or by fome part of that heterogeneous fluid : for when mercurial falts, or oxydes, are mixed with the blood, they ftimulate thefe glands into unnatural exertions; and then an unufual quantity of faliva is feparated.

As the faliva fecreted by thefe glands is moft wanted during the maftication of our food, it happens, when the terminations of their ducts in the mouth are ftimulated into action, the falival glands themfelves

3. felves

felves are brought into increafed action at the fame time by affociation, and feparate a greater quantity of their juices from the blood ; in the fame manner as tears are produced in greater abundance during the ftimulus of the vapour of onions, or of any other acrid material in the eye.

The faliva is thus naturally poured into the mouth only during the ftimulus of our food in maftication ; for when there is too great an exhalation of the mucilaginous fecretion from the membranes, which line the mouth, or too great an abforption of it, the mouth becomes dry, though there is no deficiency in the quantity of faliva ; as in thofe who fleep with their mouths open, and in fome fevers.

2. Though during the maftication of our natural food the falival glands are excited into action by the ftimulus on their excretory ducts, and a due quantity of faliva is feparated from the blood, and poured into the mouth ; yet as this maftication of our food is always attended with a degree of pleafure ; and that pleafureable fenfation is alfo connected with our ideas of certain kinds of aliment ; it follows, that when thefe ideas are reproduced, the pleafureable fenfation arifes along with them, and the falival glands are excited into action, and fill the mouth with faliva from this fenfitive affociation, as is frequently feen in dogs, who flaver at the fight of food.

3. We have alfo a voluntary power over the action of thefe falival glands, for we can at any time produce a flow of faliva into our mouth, and fpit out, or fwallow it at will.

4. If any very acrid material be held in the mouth, as the root of pyrethrum, or the leaves of tobacco, the falival glands are ftimulated into ftronger action than is natural, and thence fecrete a much larger quantity of faliva; which is at the fame time more vifcid than in its natural ftate; becaufe the lymphatics, that open their mouths into the ducts of the falival glands, and on the membranes, which line the mouth, are likewife ftimulated into ftronger action, and abforb the

more

more liquid parts of the faliva with greater avidity; and the remainder is left both in greater quantity and more vifcid.

The increafed abforption in the mouth by fome ftimulating fub-ftances, which are called aftringents, as crab juice, is evident from the inftant drynefs produced in the mouth by a fmall quantity of them.

As the extremities of the glands are of exquifite tenuity, as appears by their difficulty of injection, it was neceffary for them to fecrete their fluids in a very dilute ftate; and, probably for the purpofe of ftimulating them into action, a quantity of neutral falt is likewife fe-creted or formed by the gland. This aqueous and faline part of all fecreted fluids is again reabforbed into the habit. More than half of fome fecreted fluids is thus imbibed from the refervoirs, into which they are poured; as in the urinary bladder much more than half of what is fecreted by the kidneys becomes reabforbed by the lymphatics, which are thickly difperfed around the neck of the bladder. This feems to be the purpofe of the urinary bladders of fifh, as otherwife fuch a receptacle for the urine could have been of no ufe to an animal immerfed in water.

5. The idea of fubftances difagreeably acrid will alfo produce a quantity of faliva in the mouth; as when we fmell very putrid va-pours, we are induced to fpit out our faliva, as if fomething difagree-able was actually upon our palates.

6. When difagreeable food in the ftomach produces naufea, a flow of faliva is excited in the mouth by affociation; as efforts to vomit are frequently produced by difagreeable drugs in the mouth by the fame kind of affociation.

7. A preternatural flow of faliva is likewife fometimes occafioned by a difeafe of the voluntary power; for if we think about our faliva, and determine not to fwallow it, or not to fpit it out, an exertion is produced by the will, and more faliva is fecreted againft our wifh;

that

that is, by our averfion, which bears the fame analogy to defire, as pain does to pleafure; as they are only modifications of the fame difpofition of the fenforium. See Clafs IV. 3. 2. 1.

8. The quantity of faliva may alfo be increafed beyond what is natural, by the catenation of the motions of thefe glands with other motions, or fenfations, as by an extraneous body in the ear; of which I have known an inftance; or by the application of ftizolobium, filiqua hirfuta, cowhage, to the feat of the parotis, as fome writers have affirmed.

II. 1. The lacrymal gland drinks up a certain fluid from the circumfluent blood, and pours it on the ball of the eye, on the upper part of the external corner of the eyelids. Though it may perhaps be ftimulated into the performance of its natural action by the blood, which furrounds its origin, or by fome part of that heterogeneous fluid; yet as the tears fecreted by this gland are more wanted at fome times than at others, its fecretion is variable, like that of the faliva above mentioned, and is chiefly produced when its excretory duct is ftimulated; for in our common fleep there feems to be little or no fecretion of tears; though they are occafionally produced by our fenfations in dreams.

Thus when any extraneous material on the eye-ball, or the drynefs of the external covering of it, or the coldnefs of the air, or the acrimony of fome vapours, as of onions, ftimulates the excretory duct of the lacrymal gland, it difcharges its contents upon the ball; a quicker fecretion takes place in the gland, and abundant tears fucceed, to moiften, clean, and lubricate the eye. Thefe by frequent nictitation are diffufed over the whole ball, and as the external angle of the eye in winking is clofed fooner than the internal angle, the tears are gradually driven forwards, and downwards from the lacrymal gland to the puncta lacrymalia.

2. The lacrymal fack, with its puncta lacrymalia, and its nafal duct, is a complete gland; and is fingular in this refpect, that it neither

ther

ther derives its fluid from, nor difgorges it into the circulation. The fimplicity of the ftructure of this gland, and both the extremities of it being on the furface of the body, makes it well worthy our minuter obfervation; as the actions of more intricate and concealed glands may be better underftood from their analogy to this.

3. This fimple gland confifts of two abforbing mouths, a belly, and an excretory duct. As the tears are brought to the internal angle of the eye, thefe two mouths drink them up, being ftimulated into action by this fluid, which they abforb. The belly of the gland, or lacrymal fack, is thus filled, in which the faline part of the tears is abforbed, and when the other end of the gland, or nafal duct, is ftimulated by the drynefs, or pained by the coldnefs of the air, or affected by any acrimonious duft or vapour in the noftrils, it is excited into action together with the fack, and the tears are difgorged upon the membrane, which lines the noftrils; where they ferve a fecond purpofe to moiften, clean, and lubricate, the organ of fmell.

4. When the nafal duct of this gland is ftimulated by any very acrid material, as the powder of tobacco, or volatile fpirits, it not only difgorges the contents of its belly or receptacle (the lacrymal fack), and abforbs haftily all the fluid, that is ready for it in the corner of the eye; but by the affociation of its motions with thofe of the lacrymal gland, it excites that alfo into increafed action, and a large flow of tears is poured into the eye.

5. This nafal duct is likewife excited into ftrong action by fenfitive ideas, as in grief, or joy, and then alfo by its affociations with the lacrymal gland it produces a great flow of tears without any external ftimulus; as is more fully explained in Sect. XVI. 8. on Inftinct.

6. There are fome, famous in the arts of exciting compaffion, who are faid to have acquired a voluntary power of producing a flow of tears in the eye; which, from what has been faid in the fection on Inftinct above-mentioned, I fhould fufpect, is performed by acquiring a voluntary power over the action of this nafal duct.

7. There

7. There is another circumſtance well worthy our attention, that when by any accident this naſal duct is obſtructed, the lacrymal ſack, which is the belly or receptacle of this gland, by ſlight preſſure of the finger is enabled to diſgorge its contents again into the eye; perhaps the bile in the ſame manner, when the biliary ducts are obſtructed, is returned into the blood by the veſſels which ſecrete it ?

8. A very important though minute occurrence muſt here be obſerved, that though the lacrymal gland is only excited into action, when we weep at a diſtreſsful tale, by its aſſociation with this naſal duct, as is more fully explained in Sect. XVI. 8; yet the quantity of tears ſecreted at once is more than the puncta lacrymalia can readily abſorb; which ſhews *that the motions occaſioned by aſſociations are fre-quently more energetic than the original motions, by which they were oc-caſioned.* Which we ſhall have occaſion to mention hereafter, to il-luſtrate, why pains frequently exiſt in a part diſtant from the cauſe of them, as in the other end of the urethra, when a ſtone ſtimulates the neck of the bladder. And why inflammations frequently ariſe in parts diſtant from their cauſe, as the gutta roſea of drinking people, from an inflamed liver.

The inflammation of a part is generally preceded by a torpor or quieſcence of it; if this exiſts in any large congeries of glands, as in the liver, or any membranous part, as the ſtomach, pain is produced, and chillineſs in conſequence of the torpor of the veſſels. In this ſitu-ation ſometimes an inflammation of the parts ſucceeds the torpor; at other times a diſtant more ſenſible part becomes inflamed; whoſe ac-tions have previouſly been aſſociated with it; and the torpor of the firſt part ceaſes. This I apprehend happens, when the gout of the foot ſucceeds a pain of the biliary duct, or of the ſtomach. Laſtly, it ſome-times happens, that the pain of torpor exiſts without any conſequent inflammation of the affected part, or of any diſtant part aſſociated with it, as in the membranes about the temple and eye-brows in hemicra-nia, and in thoſe pains, which occaſion convulſions; if this happens

to

to gouty people, when it affects the liver, I suppose epileptic fits are produced; and, when it affects the stomach, death is the consequence. In these cases the pulse is weak, and the extremities cold, and such medicines as stimulate the quiescent parts into action, or which induce inflammation in them, or in any distant part, which is associated with them, cures the present pain of torpor, and saves the patient.

I have twice seen a gouty inflammation of the liver, attended with jaundice; the patients after a few days were both of them affected with cold fits, like ague-fits, and their feet became affected with gout, and the inflammation of their livers ceased. It is probable, that the uneasy sensations about the stomach, and indigestion, which precedes gouty paroxysms, are generally owing to torpor or slight inflammation of the liver, and biliary ducts; but where great pain with continued sickness, with feeble pulse, and sensation of cold, affect the stomach in patients debilitated by the gout, that it is a torpor of the stomach itself, and destroys the patient from the great connexion of that viscus with the vital organs.   See Sect. XXV. 17.

SECT.

SECT. XXV.

OF THE STOMACH AND INTESTINES.

1. *Of swallowing our food.   Ruminating animals.   2. Action of the stomach.   3. Action of the intestines.   Irritative motions connected with these.   4. Effects of repletion.   5. Stronger action of the stomach and intestines from more stimulating food.   6. Their action inverted by still greater stimuli.   Or by disgustful ideas. Or by volition.   7. Other glands strengthen or invert their motions by sympathy. 8. Vomiting performed by intervals.   9. Inversion of the cutaneous absorbents. 10. Increased secretion of bile and pancreatic juice.   11. Inversion of the lacteals. 12. And of the bile-ducts.   13. Case of a cholera.   14. Further account of the inversion of lacteals.   15. Iliac passion.   Valve of the colon.   16. Cure of the iliac passion.   17. Pain of gall-stone distinguished from pain of the stomach. Gout of the stomach from torpor, from inflammation.   Intermitting pulse owing to indigestion.   To overdose of foxglove.   Weak pulse from emetics.   Death from a blow on the stomach.   From gout of the stomach.*

1. THE throat, stomach and inteſtines, may be conſidered as one great gland; which, like the lacrymal ſack above mentioned, neither begins nor ends in the circulation.   Though the act of maſticating our aliment belongs to the ſenſitive claſs of motions, for the pleaſure of its taſte induces the muſcles of the jaw into action;   yet the deglutition of it when maſticated is generally, if not always, an irritative motion, occaſioned by the application of the food already maſticated to the origin of the pharix;   in the ſame manner as we often ſwallow our ſpittle without attending to it.

N n

The

The ruminating clafs of animals have the power to invert the motion of their gullet, and of their firft ftomach, from the ftimulus of this aliment, when it is a little further prepared; as is their daily practice in chewing the cud; and appears to the eye of any one, who attends to them, whilft they are employed in this fecond maftication of their food.

2. When our natural aliment arrives into the ftomach, this organ is ftimulated into its proper vermicular action; which beginning at the upper orifice of it, and terminating at the lower one, gradually mixes together and pufhes forwards the digefting materials into the inteftine beneath it.

At the fame time the glands, that fupply the gaftric juices, which are neceffary to promote the chemical part of the procefs of digeftion, are ftimulated to difcharge their contained fluids, and to feparate a further fupply from the blood-veffels: and the lacteals or lymphatics, which open their mouths into the ftomach, are ftimulated into action, and take up fome part of the digefting materials.

3. The remainder of thefe digefting materials is carried forwards into the upper inteftines, and ftimulates them into their periftaltic motion fimilar to that of the ftomach; which continues gradually to mix the changing materials, and pafs them along through the valve of the colon to the excretory end of this great gland, the fphincter ani.

The digefting materials produce a flow of bile, and of pancreatic juice, as they pafs along the duodenum, by ftimulating the excretory ducts of the liver and pancreas, which terminate in that inteftine: and other branches of the abforbent or lymphatic fyftem, called lacteals, are excited to drink up, as it paffes, thofe parts of the digefting meterials, that are proper for their purpofe, by its ftimulus on their mouths.

4. When the ftomach and inteftines are thus filled with their proper food, not only the motions of the gaftric glands, the pancreas,

liver,

liver, and lacteal veffels, are excited into action; but at the fame time the whole tribe of irritative motions are exerted with greater energy, a greater degree of warmth, colour, plumpnefs, and moifture, is given to the fkin from the increafed action of thofe glands called capillary veffels; pleafureable fenfation is excited, the voluntary motions are lefs eafily exerted, and at length fufpended; and fleep fucceeds, unlefs it be prevented by the ftimulus of furrounding objects, or by voluntary exertion, or by an acquired habit, which was originally produced by one or other of thefe circumftances, as is explained in Sect. XXI. on Drunkennefs.

At this time alfo, as the blood-veffels become replete with chyle, more urine is feparated into the bladder, and lefs of it is reabforbed; more mucus poured into the cellular membranes, and lefs of it reabforbed; the pulfe becomes fuller, and fofter, and in general quicker. The reafon why lefs urine and cellular mucus is abforbed after a full meal with fufficient drink is owing to the blood-veffels being fuller: hence one means to promote abforption is to decreafe the refiftance by emptying the veffels by venefection. From this decreafed abforption the urine becomes pale as well as copious, and the fkin appears plump as.well as florid.

By daily repetition of thefe movements they all become connected together, and make a diurnal circle of irritative action, and if one of this chain be difturbed, the whole is liable to be put into diforder. See Sect. XX. on Vertigo.

5. When the ftomach and inteftines receive a quantity of food, whofe ftimulus is greater than ufual, all their motions, and thofe of the glands and lymphatics, are ftimulated into ftronger action than ufual, and perform their offices with greater vigour and in lefs time: fuch are the effects of certain quantities of fpice or of vinous fpirit.

6. But if the quantity or duration of thefe ftimuli are ftill further increafed, the ftomach and throat are ftimulated into a motion, whofe direction is contrary to the natural one above defcribed; and they re-

gurgitate

gurgitate the materials, which they contain, inftead of carrying them forwards. This retrograde motion of the ftomach may be compared to the ftretchings of wearied limbs the contrary way, and is well elucidated by the following experiment. Look earneftly for a minute or two on an area an inch fquare of pink filk, placed in a ftrong light, the eye becomes fatigued, the colour becomes faint, and at length vanifhes, for the fatigued eye can no longer be ftimulated into direct motions; then on clofing the eye a green fpectrum will appear in it, which is a colour directly contrary to pink, and which will appear and difappear repeatedly, like the efforts in vomiting. See Section XXIX. 11.

Hence all thofe drugs, which by their bitter or aftringent ftimulus increafe the action of the ftomach, as camomile and white vitriol, if their quantity is increafed above a certain dofe become emetics.

Thefe inverted motions of the ftomach and throat are generally produced from the ftimulus of unnatural food, and are attended with the fenfation of naufea or ficknefs: but as this fenfation is again connected with an idea of the diftafteful food, which induced it; fo an idea of naufeous food will alfo fometimes excite the action of naufea; and that give rife by affociation to the inverfion of the motions of the ftomach and throat. As fome, who have had horfe-flefh or dogs-flefh given them for beef or mutton, are faid to have vomited many hours afterwards, when they have been told of the impofition.

I have been told of a perfon, who had gained a voluntary command over thefe inverted motions of the ftomach and throat, and fupported himfelf by exhibiting this curiofity to the public. At thefe exhibitions he fwallowed a pint of red rough goofeberries, and a pint of white fmooth ones, brought them up in fmall parcels into his mouth, and reftored them feparately to the fpectators, who called for red or white as they pleafed, till the whole were redelivered.

7. At the fame time that thefe motions of the ftomach and throat are ftimulated into inverfion, fome of the other irritative motions,

that

that had acquired more immediate connexions with the ftomach, as thofe of the gaftric glands, are excited into ftronger action by this affociation; and fome other of thefe motions, which are more eafily excited, as thofe of the gaftric lymphatics, are inverted by their affociation with the retrograde motions of the ftomach, and regurgitate their contents, and thus a greater quantity of mucus, and of lymph, or chyle, is poured into the ftomach, and thrown up along with its contents.

8. Thefe inverfions of the motion of the ftomach in vomiting are performed by intervals, for the fame reafon that many other motions are reciprocally exerted and relaxed; for during the time of exertion the ftimulus, or fenfation, which caufed this exertion, is not perceived; but begins to be perceived again, as foon as the exertion ceafes, and is fome time in again producing its effect. As explained in Sect. XXXIV. on Volition, where it is fhewn, that the contractions of the fibres, and the fenfation of pain, which occafioned that exertion, cannot exift at the fame time. The exertion ceafes from another caufe alfo, which is the exhauftion of the fenforial power of the part, and thefe two caufes frequently operate together.

9. At the times of thefe inverted efforts of the ftomach not only the lymphatics, which open their mouths into the ftomach, but thofe of the fkin alfo, are for a time inverted; for fweats are fometimes pufhed out during the efforts of vomiting without an increafe of heat.

10. But if by a greater ftimulus the motions of the ftomach are inverted ftill more violently or more permanently, the duodenum has its periftaltic motions inverted at the fame time by their affociation with thofe of the ftomach; and the bile and pancreatic juice, which it contains, are by the inverted motions brought up into the ftomach, and difcharged along with its contents; while a greater quantity of bile and pancreatic juice is poured into this inteftine; as the glands,

glands, that fecrete them, are by their affociation with the motions of the inteſtine excited into ſtronger action than uſual.

11. The other inteſtines are by affociation excited into more power-ful action, while the lymphatics, that open their mouths into them, ſuffer an inverſion of their motions correſponding with the lymphatics of the ſtomach, and duodenum; which with a part of the abundant ſecretion of bile is carried downwards, and contributes both to ſtimu-late the bowels, and to increaſe the quantity of the evacuations. This inverſion of the motion of the lymphatics appears from the quantity of chyle, which comes away by ſtools; which is otherwiſe abſorbed as ſoon as produced, and by the immenſe quantity of thin fluid, which is evacuated along with it.

12. But if the ſtimulus, which inverts the ſtomach, be ſtill more powerful, or more permanent, it ſometimes happens, that the motions of the biliary glands, and of their excretory ducts, are at the ſame time inverted, and regurgitate their contained bile into the blood-veſ-fels, as appears by the yellow colour of the ſkin, and of the urine; and it is probable the pancreatic ſecretion may ſuffer an inverſion at the ſame time, though we have yet no mark by which this can be aſcertained.

13. Mr. ——— eat two putrid pigeons out of a cold pigeon-pye, and drank about a pint of beer and ale along with them, and immedi-ately rode about five miles. He was then ſeized with vomiting, which was after a few periods ſucceeded by purging; theſe con-tinued alternately for two hours; and the purging continued by intervals for ſix or eight hours longer. During this time he could not force himſelf to drink more than one pint in the whole, this great inability to drink was owing to the nauſea, or inverted motions of the ſtomach, which the voluntary exertion of ſwallowing could ſeldom and with difficulty overcome; yet he diſcharged in the whole at leaſt ſix quarts; whence came this

quantity

quantity of liquid ? Firſt, the contents of the ſtomach were emitted, then of the duodenum, gall-bladder, and pancreas, by vomiting. After this the contents of the lower bowels, then the chyle, that was in the lacteal veſſels, and in the receptacle of chyle, was regurgitated into the inteſtines by a retrograde motion of theſe veſſels. And afterwards the mucus depoſited in the cellular membrane, and on the ſurface of all the other membranes, ſeems to have been abſorbed ; and with the fluid abſorbed from the air to have been carried up their reſpective lymphatic branches by the increaſed energy of their natural motions, and down the viſceral lymphatics, or lacteals, by the inverſion of their motions.

14. It may be difficult to invent experiments to demonſtrate the truth of this inverſion of ſome branches of the abſorbent ſyſtem, and increaſed abſorption of others, but the analogy of theſe veſſels to the inteſtinal canal, and the ſymptoms of many diſeaſes, render this opinion more probable than many other received opinions of the animal œconomy.

In the above inſtance, after the yellow excrement was voided, the fluid ceaſed to have any ſmell, and appeared like curdled milk, and then a thinner fluid, and ſome mucus, were evacuated : did not theſe ſeem to partake of the chyle, of the mucus fluid from all the cells of the body, and laſtly, of the atmoſpheric moiſture ? All theſe facts may be eaſily obſerved by any one, who takes a briſk purge.

15. Where the ſtimulus on the ſtomach, or on ſome other part of the inteſtinal canal, is ſtill more permanent, not only the lacteal veſſels, but the whole canal itſelf, becomes inverted from its aſſociations : this is the iliac paſſion, in which all the fluids mentioned above are thrown up by the mouth. At this time the valve in the colon, from the inverted motions of that bowel, and the inverted action of this living valve, does not prevent the regurgitation of its contents.

The

The ftructure of this valve may be reprefented by a flexile leathern pipe ftanding up from the bottom of a veffel of water : its fides collapfe by the preffure of the ambient fluid, as a fmall part of that fluid paffes through it ; but if it has a living power, and by its inverted action keeps itfelf open, it becomes like a rigid pipe, and will admit the whole liquid to pafs. See Sect. XXIX. 2. 5.

In this cafe the patient is averfe to drink, from the conftant inverfion of the motions of the ftomach, and yet many quarts are daily ejected from the ftomach, which at length fmell of excrement, and at laft feem to be only a thin mucilaginous or aqueous liquor.

From whence is it poffible, that this great quantity of fluid for many fucceffive days can be fupplied, after the cells of the body have given up their fluids, but from the atmofphere ? When the cutaneous branch of abforbents acts with unnatural ftrength, it is probable the inteftinal branch has its motions inverted, and thus a fluid is fupplied without entering the arterial fyftem. Could oiling or painting the fkin give a check to this difeafe ?

So when the ftomach has its motions inverted, the lymphatics of the ftomach, which are moft ftrictly affociated with it, invert their motions at the fame time. But the more diftant branches of lymphatics, which are lefs ftrictly affociated with it, act with increafed energy; as the cutaneous lymphatics in the cholera, or iliac paffion, above defcribed. And other irritative motions become decreafed, as the pulfations of the arteries, from the extra-derivation or exhauftion of the fenforial power.

Sometimes when ftronger vomiting takes place the more diftant branches of the lymphatic fyftem invert their motions with thofe of the ftomach, and loofe ftools are produced, and cold fweats.

So when the lacteals have their motions inverted, as during the operation of ftrong purges, the urinary and cutaneous abforbents have their motions increafed to fupply the want of fluid in the blood, as in great thirft; but after a meal with fufficient potation the urine is pale,

that

that is, the urinary abforbents act weakly, no fupply of water being wanted for the blood. And when the inteftinal abforbents act too violently, as when too great quantities of fluid have been drank, the urinary abforbents invert their motions to carry off the fuperfluity, which is a new circumftance of affociation, and a temporary diabetes fupervenes.

16. I have had the opportunity of feeing four patients in the iliac paffion, where the ejected material fmelled and looked like excrement. Two of thefe were fo exhaufted at the time I faw them, that more blood could not be taken from them, and as their pain had ceafed, and they continued to vomit up every thing which they drank, I fufpected that a mortification of the bowel had already taken place, and as they were both women advanced in life, and a mortification is produced with lefs preceding pain in old and weak people, thefe both died. The other two, who were both young men, had ftill pain and ftrength fufficient for further venefection, and they neither of them had any appearance of hernia, both recovered by repeated bleeding, and a fcruple of calomel given to one, and half a dram to the other, in very fmall pills: the ufual means of clyfters, and purges joined with opiates, had been in vain attempted. I have thought an ounce or two of crude mercury in lefs violent difeafes of this kind has been of ufe, by contributing to reftore its natural motion to fome part of the inteftinal canal, either by its weight or ftimulus; and that hence the whole tube recovered its ufual affociations of progreffive periftaltic motion. I have in three cafes feen crude mercury given in fmall dofes, as one or two ounces twice a day, have great effect in ftopping pertinaceous vomitings.

17. Befides the affections above defcribed, the ftomach is liable, like many other membranes of the body, to torpor without confequent inflammation: as happens to the membranes about the head in fome cafes of hemicrania, or in general head-ach. This torpor of the fto-

mach

mach is attended with indigeftion, and confequent flatulency, and with pain, which is ufually called the cramp of the ftomach, and is relievable by aromatics, effential oils, alcohol, or opium.

The intrufion of a gall-ftone into the common bile-duct from the gall-bladder is fometimes miftaken for a pain of the ftomach, as neither of them are attended with fever; but in the paffage of a gall-ftone, the pain is confined to a lefs fpace, which is exactly where the common bile-duct enters the duodenum, as explained in Section XXX. 3. Whereas in this gaftrodynia the pain is diffufed over the whole ftomach; and, like other difeafes from torpor, the pulfe is weaker, and the extremities colder, and the general debility greater, than in the paffage of a gall-ftone; for in the former the debility is the confequence of the pain, in the latter it is the caufe of it.

Though the firft fits of the gout, I believe, commence with a torpor of the liver; and the ball of the toe becomes inflamed inftead of the membranes of the liver in confequence of this torpor, as a coryza or catarrh frequently fucceeds a long expofure of the feet to cold, as in fnow, or on a moift brick-floor; yet in old or exhaufted conftitutions, which have been long habituated to its attacks, it fometimes commences with a torpor of the ftomach, and is transferable to every membrane of the body. When the gout begins with torpor of the ftomach, a painful fenfation of cold occurs, which the patient compares to ice, with weak pulfe, cold extremities, and ficknefs; this in its flighter degree is relievable by fpice, wine, or opium; in its greater degree it is fucceeded by fudden death, which is owing to the fympathy of the ftomach with the heart, as explained below.

If the ftomach becomes inflamed in confequence of this gouty torpor of it, or in confequence of its fympathy with fome other part, the danger is lefs. A ficknefs and vomiting continues many

3　　　　　　　　　days,

days, or even weeks, the ftomach rejecting every thing ftimulant, even opium or alcohol, together with much vifcid mucus; till the inflammation at length ceafes, as happens when other membranes, as thofe of the joints, are the feat of gouty inflammation; as obferved in Sect. XXIV. 2. 8.

The fympathy, or affociation of motions, between thofe of the ftomach and thofe of the heart, are evinced in many difeafes. Firft, many people are occafionally affected with an intermiffion of their pulfe for a few days, which then ceafes again. In this cafe there is a ftop of the motion of the heart, and at the fame time a tendency to eructation from the ftomach. As foon as the patient feels a tendency to the intermiffion of the motion of his heart, if he voluntarily brings up wind from his ftomach, the ftop of the heart does not occur. From hence I conclude that the ftop of digeftion is the primary difeafe; and that air is inftantly generated from the aliment, which begins to ferment, if the digeftive pro-cefs is impeded for a moment, (fee Sect. XXIII. 4.); and that the ftop of the heart is in confequence of the affociation of the motions of thefe vifcera, as explained in Sect. XXXV. 1. 4.; but if the little air, which is inftantly generated during the temporary torpor of the ftomach, be evacuated, the digeftion recommences, and the tempo-rary torpor of the heart does not follow. One patient, whom I lately faw, and who had been five or fix days much troubled with this in-termiffion of a pulfation of his heart, and who had hemicrania with fome fever, was immediately relieved from them all by lofing ten ounces of blood, which had what is termed an inflammatory cruft on it.

Another inftance of this affociation between the motions of the ftomach and heart is evinced by the exhibition of an over dofe of foxglove, which induces an inceffant vomiting, which is attended with very flow, and fometimes intermitting pulfe.—Which continues

in

in ſpite of the exhibition of wine and opium for two or three days. To the ſame aſſociation muſt be aſcribed the weak pulſe, which conſtantly attends the exhibition of emetics during their operation. And alſo the ſudden deaths, which have been occaſioned in boxing by a blow on the ſtomach; and laſtly, the ſudden death of thoſe, who have been long debilitated by the gout, from the torpor of the ſtomach. See Sect. XXV. 1. 4.

SECT.

SECT. XXVI.

OF THE CAPILLARY GLANDS AND MEMBRANES.

I. 1. *The capillary veffels are glands.* 2. *Their excretory ducts. Experiments on the mucus of the inteftines, abdomen, cellular membrane, and on the humours of the eye.* 3. *Scurf on the head, cough, catarrh, diarrhœa, gonorrhœa.* 4. *Rheumatifm. Gout. Leprofy.* II. 1. *The moft minute membranes are unorganized.* 2. *Larger membranes are compofed of the ducts of the capillaries, and the mouths of the abforbents.* 3. *Mucilaginous fluid is fecreted on their furfaces.* III. *Three kinds of rheumatifm.*

I. 1. THE capillary veffels are like all the other glands except the abforbent fyftem, inafmuch as they receive blood from the arteries, feparate a fluid from it, and return the remainder by the veins.

2. This feries of glands is of the moft extenfive ufe, as their excretory ducts open on the whole external fkin forming its perfpirative pores, and on the internal furfaces of every cavity of the body. Their fecretion on the fkin is termed infenfible perfpiration, which in health is in part reabforbed by the mouths of the lymphatics, and in part evaporated in the air; the fecretion on the membranes, which line the larger cavities of the body, which have external openings, as the mouth and inteftinal canal, is termed mucus, but is not however coagulable by heat; and the fecretion on the membranes of thofe cavities of the body, which have no external openings, is called lymph or
water,

water, as in the cavities of the cellular membrane, and of the abdomen; this lymph however is coagulable by the heat of boiling water.   Some mucus nearly as vifcid as the white of egg, which was
difcharged by ftool, did not coagulate, though I evaporated it to one
fourth of the quantity, nor did the aqueous and vitreous humours of a
fheep's eye coagulate by the like experiment: but the ferofity from
an anafarcous leg, and that from the abdomen of a dropfical perfon,
and the cryftalline humour of a fheep's eye, coagulated in the fame
heat.

3. When any of thefe capillary glands are ftimulated into greater
irritative actions, than is natural, they fecrete a more copious material; and as the mouths of the abforbent fyftem, which open in their
vicinity, are at the fame time ftimulated into greater action, the
thinner and more faline part of the fecreted fluid is taken up again;
and the remainder is not only more copious but alfo more vifcid than
natural.   This is more or lefs troublefome or noxious according to the
importance of the functions of the part affected: on the fkin and
bronchiæ, where this fecretion ought naturally to evaporate, it becomes fo vifcid as to adhere to the membrane; on the tongue it forms
a pellicle, which can with difficulty be fcraped off; produces the fcurf
on the heads of many people; and the mucus, which is fpit up by
others in coughing.   On the noftrils and fauces, when the fecretion
of thefe capillary glands is increafed, it is termed fimple catarrh;
when in the inteftines, a mucous diarrhœa; and in the urethra, or
vagina, it has the name of gonorrhœa, or fluor albus.

4. When thefe capillary glands become inflamed, a ftill more vifcid
or even cretaceous humour is produced upon the furfaces of the membranes, which is the caufe or the effect of rheumatifm, gout, leprofy,
and of hard tumours of the legs, which are generally termed fcorbutic;
all which will be treated of hereafter.

II.   1. The whole furface of the body, with all its cavities and contents,

tents, are covered with membrane.   It lines every veffel, forms every cell, and binds together all the mufcular and perhaps the offeous fibres of the body; and is itfelf therefore probably a fimpler fubftance than thofe fibres.   And as the containing veffels of the body from the largeft to the leaft are thus lined and connected with membranes, it follows that thefe membranes themfelves confifted of unorganized materials.

For however fmall we may conceive the diameters of the minuteft veffels of the body, which efcape our eyes and glaffes, yet thefe veffels muft confift of coats or fides, which are made up of an unorganized material, and which are probably produced from a gluten, which hardens after its production, like the filk or web of caterpillars and fpiders.   Of this material confift the membranes, which line the fhells of eggs, and the fhell itfelf, both which are unorganized, and are formed from mucus, which hardens after it is formed, either by the abforption of its more fluid part, or by its uniting with fome part of the atmofphere.   Such is alfo the production of the fhells of fnails, and of fhell-fifh, and I fuppofe of the enamel of the teeth.

2. But though the membranes, that compofe the fides of the moft minute veffels, are in truth unorganized materials, yet the larger membranes, which are perceptible to the eye, feem to be compofed of an intertexture of the mouths of the abforbent fyftem, and of the excretory ducts of the capillaries, with their concomitant arteries, veins, and nerves : and from this conftruction it is evident, that thefe membranes muft poffefs great irritability to peculiar ftimuli, though they are incapable of any motions, that are vifible to the naked eye : and daily experience fhews us, that in their inflamed ftate they have the greateft fenfibility to pain, as in the pleurify and paronychia.

3. On all thefe membranes a mucilaginous or aqueous fluid is fecreted, which moiftens and lubricates their furfaces, as was ex-

plained

plained in Section XXIII. 2. Some have doubted, whether this mucus is feparated from the blood by an appropriated fet of glands, or exudes through the membranes, or is an abrafion or deftruction of the furface of the membrane itfelf, which is continually repaired on the other fide of it, but the great analogy between the capillary veffels, and the other glands, countenances the former opinion; and evinces, that thefe capillaries are the glands, that fecrete it; to which we muft add, that the blood in paffing thefe capillary veffels undergoes a change in its colour from florid to purple, and gives out a quantity of heat; from whence, as in other glands, we muft conclude that fomething is fecreted from it.

III. The feat of rheumatifm is in the membranes, or upon them; but there are three very diftinct difeafes, which commonly are confounded under this name. Firft, when a membrane becomes affected with torpor, or inactivity of the veffels which compofe it, pain and coldnefs fucceed, as in the hemicrania, and other headachs, which are generally termed nervous rheumatifm; they exift whether the part be at reft or in motion, and are generally attended with other marks of debility.

Another rheumatifm is faid to exift, when inflammation and fwelling, as well as pain, affect fome of the membranes of the joints, as of the ancles, wrifts, knees, elbows, and fometimes of the ribs. This is accompanied with fever, is analogous to pleurify, and other inflammations, and is termed the acute rheumatifm.

A third difeafe is called chronic rheumatifm, which is diftinguifhed from that firft mentioned, as in this the pain only affects the patient during the motion of the part, and from the fecond kind of rheumatifm above defcribed, as it is not attended with quick pulfe or inflammation. It is generally believed to fucceed the acute rheu-

matifm

matifm of the fame part, and that fome coagulable lymph, or cretaceous, or calculous material, has been left on the membrane; which
gives pain, when the mufcles move over it, as fome extraneous body
would do, which was too infoluble to be abforbed.   Hence there is
an analogy between this chronic rheumatifm and the difeafes which
produce gravel or gout-ftones; and it may perhaps receive relief from
the fame remedies, fuch as aerated fal foda.

SECT. XXVII.

OF HÆMORRHAGES.

I. *The veins are abforbent veffels.* 1. *Hæmorrhages from inflammation. Cafe of hæmorrhage from the kidney cured by cold bathing. Cafe of hæmorrhage from the nofe cured by cold immerfion.* II. *Hæmorrhage from venous paralyfis. Of Piles. Black ftools. Petechiæ. Confumption. Scurvy of the lungs. Blacknefs of the face and eyes in epileptic fits. Cure of hæmorrhages from venous inability.*

I. AS the imbibing mouths of the abforbent fyftem already defcribed open on the furface, and into the larger cavities of the body, fo there is another fyftem of abforbent veffels, which are not commonly efteemed fuch, I mean the veins, which take up the blood from the various glands and capillaries, after their proper fluids or fecretions have been feparated from it.

The veins refemble the other abforbent veffels; as the progreffion of their contents is carried on in the fame manner in both, they alike abforb their appropriated fluids, and have valves to prevent its regurgitation by the accidents of mechanical violence. This appears firft, becaufe there is no pulfation in the very beginnings of the veins, as is feen by microfcopes; which muft happen, if the blood was carried into them by the action of the arteries. For though the concurrence of various venous ftreams of blood from different diftances muft prevent any pulfation in the larger branches, yet in the very beginnings of all thefe branches a pulfation muft unavoidably exift, if the circulation

lation in them was owing to the intermitted force of the arteries. Secondly, the venous abforption of blood from the penis, and from the teats of female animals after their erection, is ftill more fimilar to the lymphatic abforption, as it is previoufly poured into cells, where all arterial impulfe muft ceafe.

There is an experiment, which feems to evince this venous abforption, which confifts in the external application of a ftimulus to the lips, as of vinegar, by which they become inftantly pale ; that is, the bibulous mouths of the veins by this ftimulus are excited to abforb the blood fafter, than it can be fupplied by the ufual arterial exertion. See Sect. XXIII. 5.

There are two kinds of hæmorrhages frequent in difeafes, one is where the glandular or capillary action is too powerfully exerted, and propels the blood forwards more haftily, than the veins can abforb it ; and the other is, where the abforbent power of the veins is diminifhed, or a branch of them is become totally paralytic.

The former of thefe cafes is known by the heat of the part, and the general fever or inflammation that accompanies the hæmorrhage. An hæmorrhage from the nofe or from the lungs is fometimes a crifis of inflammatory difeafes, as of the hepatitis and gout, and generally ceafes fpontaneoufly, when the veffels are confiderably emptied. Sometimes the hæmorrhage recurs by daily periods accompanying the hot fits of fever, and ceafing in the cold fits, or in the intermiffions ; this is to be cured by removing the febrile paroxyfms, which will be treated of in their place.   Otherwife it is cured by venefection, by the internal or external preparations of lead, or by the application of cold, with an abftemious diet, and diluting liquids, like other inflammations.   Which by inducing a quiefcence on thofe glandular parts, that are affected, prevents a greater quantity of blood from being protruded forwards, than the veins are capable of abforbing.

Mr. B——— had an hæmorrhage from his kidney, and parted with not lefs than a pint of blood a day (by conjecture) along with his

urine

urine for above a fortnight: venefections, mucilages, balfams, preparations of lead, the bark, alum, and dragon's blood, opiates, with a
large blifter on his loins, were feparately tried, in large dofes, to no
purpofe.   He was then directed to bathe in a cold fpring up to the
middle of his body only, the upper part being covered, and the
hæmorrhage diminifhed at the firft, and ceafed at the fecond immerfion.

In this cafe the external capillaries were rendered quiefcent by the
coldnefs of the water, and thence a lefs quantity of blood was circulated through them; and the internal capillaries, or other glands,
became quiefcent from their irritative affociations with the external
ones; and the hæmorrhage was ftopped a fufficient time for the ruptured veffels to contract their apertures, or for the blood in thofe apertures to coagulate.

Mrs. K——— had a continued hæmorrhage from her nofe for
fome days; the ruptured veffel was not to be reached by plugs up the
noftrils, and the fenfibility of her fauces was fuch that nothing
could be borne behind the uvula.   After repeated venefection, and
other common applications, fhe was directed to immerfe her whole
head into a pail of water, which was made colder by the addition of feveral handfuls of falt, and the hæmorrhage immediately
ceafed, and returned no more; but her pulfe continued hard,
and fhe was neceffitated to lofe blood from the arm on the fucceeding day.

Query, might not the cold bath inftantly ftop hæmorrhages from
the lungs in inflammatory cafes?—for the fhortnefs of breath of
thofe, who go fuddenly into cold water, is not owing to the
accumulation of blood in the lungs, but to the quiefcence of the
pulmonary capillaries from affociation, as explained in Section XXXII.
3. 2.

II. The other kind of hæmorrhage is known from its being attended with a weak pulfe, and other fymptoms of general debility,
                                                            and

and very frequently occurs in thofe, who have difeafed livers, owing to intemperance in the ufe of fermented liquors. Thefe conftitutions are fhewn to be liable to paralyfis of the lymphatic abforbents, producing the various kinds of dropfies in Section XXIX. 5. Now if any branch of the venous fyftem lofes its power of abforption, the part fwells, and at length burfts and difcharges the blood, which the capillaries or other glands circulate through them.

It fometimes happens that the large external veins of the legs burft, and effufe their blood; but this occurs moft frequently in the veins of the inteftines, as the vena portarum is liable to fuffer from a fchirrus of the liver oppofing the progreffion of the blood, which is abforbed from the inteftines. Hence the piles are a fymptom of hepatic obftruction, and hence the copious difcharges downwards or upwards of a black material, which has been called melancholia, or black bile; but is no other than the blood, which is probably difcharged from the veins of the inteftines.

J. F. Meckel, in his Experimenta de Finibus Vaforum, publifhed at Berlin, 1772, mentions his difcovery of a communication of a lymphatic veffel with the gaftric branch of the vena portarum. It is poffible, that when the motion of the lymphatic becomes retrograde in fome difeafes, that blood may obtain a paffage into it, where it anaftomofes with the vein, and thus be poured into the inteftines. A difcharge of blood with the urine fometimes attends diabetes, and may have its fource in the fame manner.

Mr. A————, who had been a hard drinker, and had the gutta rofacea on his face and breaft, after a ftroke of the palfy voided near a quart of a black vifcid material by ftool: on diluting it with water it did not become yellow, as it muft have done if it had been infpiffated bile, but continued black like the grounds of coffee.

But any other part of the venous fyftem may become quiefcent or totally paralytic as well as the veins of the inteftines: all which

5

occur

occur more frequently in thofe who have difeafed livers, than in any others. Hence troublefome bleedings of the nofe, or from the lungs with a weak pulfe; hence hæmorrhages from the kidneys, too great menftruation; and hence the oozing of blood from every part of the body, and the petechiæ in thofe fevers, which are termed putrid, and which is erroneoufly afcribed to the thinnefs of the blood: for the blood in inflammatory difeafes is equally fluid before it coagulates in the cold air.

Is not that hereditary confumption, which occurs chiefly in dark-eyed people about the age of twenty, and commences with flight pulmonary hæmorrhages without fever, a difeafe of this kind?—Thefe hæmorrhages frequently begin during fleep, when the irritability of the lungs is not fufficient in thefe patients to carry on the circulation without the affiftance of volition; for in our waking hours, the motions of the lungs are in part voluntary, efpecially if any difficulty of breathing renders the efforts of volition neceffary. See Clafs I. 2. 1. 2. and Clafs III. 2. 1. 10. Another fpecies of pulmonary confumption which feems more certainly of fcrophulous origin is defcribed in the next Section, No. 2.

I have feen two cafes of women, of about forty years of age, both of whom were feized with quick weak pulfe, with difficult refpiration, and who fpit up by coughing much vifcid mucus mixed with dark coloured blood. They had both large vibices on their limbs, and petechiæ; in one the feet were in danger of mortification, in the other the legs were œdematous. To relieve the difficult refpiration, about fix ounces of blood were taken from one of them, which to my furprife was fizy, like inflamed blood: they had both palpitations or unequal pulfations of the heart. They continued four or five weeks with pale and bloated countenances, and did not ceafe fpitting phlegm mixed with black blood, and the pulfe feldom flower than 130 or 135 in a minute. This blood, from its

dark

dark colour, and from the many vibices and petechiæ, feems to have been venous blood; the quicknefs of the pulfe, and the irregularity of the motion of the heart, are to be afcribed to debility of that part of the fyftem; as the extravafation of blood originated from the defect of venous abforption. The approximation of thefe two cafes to fea-fcurvy is peculiar, and may allow them to be called fcorbutus pulmonalis. Had thefe been younger fubjects, and the paralyfis of the veins had only affected the lungs, it is probable the difeafe would have been a pulmonary confumption.

Laft week I faw a gentleman of Birmingham, who had for ten days laboured under great palpitation of his heart, which was fo diftinctly felt by the hand, as to difcountenance the idea of there being a fluid in the pericardium. He frequently fpit up mucus ftained with dark coloured blood, his pulfe very unequal and very weak, with cold hands and nofe. He could not lie down at all, and for about ten days paft could not fleep a minute together, but waked perpetually with great uneafinefs. Could thofe fymptoms be owing to very extenfive adhefions of the lungs? or is this a fcorbutus pulmonalis? After a few days he fuddenly got fo much better as to be able to fleep many hours at a time by the ufe of one grain of powder of foxglove twice a day, and a grain of opium at night. After a few days longer, the bark was exhibited, and the opium continued with fome wine; and the palpitations of his heart became much relieved, and he recovered his ufual degree of health.

In epileptic fits the patients frequently become black in the face, from the temporary paralyfis of the venous fyftem of this part. I have known two inftances where the blacknefs has continued many days. M. P———, who had drank intemperately, was feized with the epilepfy when he was in his fortieth year; in one of thefe fits the white part of his eyes was left totally black with effufed blood; which was attended with no pain or heat, and was in a few weeks gradually abforbed, changing colour as is ufual with vibices from bruifes.

The

The hæmorrhages produced from the inability of the veins to abforb the refluent blood, is cured by opium, the preparations of fteel, lead, the bark, vitriolic acid, and blifters; but thefe have the effect with much more certainty, if a venefection to a few ounces, and a moderate cathartic with four or fix grains of calomel be premifed, where the patient is not already too much debilitated; as one great means of promoting the abforption of any fluid confifts in previoufly emptying the veffels, which are to receive it.

SECT.

SECT. XXVIII.

OF THE PARALYSIS OF THE ABSORBENT SYSTEM.

I. Paralyſis of the laǎeals, atrophy.   Diſtaſte to animal food.   II. Cauſe of dropſy.   Cauſe of herpes.   Meſenteric conſumption.   Pulmonary conſumption. Why ulcers in the lungs are ſo difficult to heal.

THE term paralyſis has generally been uſed to expreſs the loſs of voluntary motion, as in the hemiplagia, but may with equal propriety be applied to expreſs the diſobediency of the muſcular fibres to the other kinds of ſtimulus; as to thoſe of irritation or ſenſation.

I. There is a ſpecies of atrophy, which has not been well underſtood; when the abſorbent veſſels of the ſtomach and inteſtines have been long inured to the ſtimulus of too much ſpirituous liquor, they at length, either by the too ſudden omiſſion of fermented or ſpirituous potation, or from the gradual decay of nature, become in a certain degree paralytic; now it is obſerved in the larger muſcles of the body, when one ſide is paralytic, the other is more frequently in motion, owing to the leſs expenditure of ſenſorial power in the paralytic limbs; ſo in this caſe the other part of the abſorbent ſyſtem acts with greater force, or with greater perſeverance, in conſequence of the paralyſis of the laǎeals; and the body becomes greatly emaciated in a ſmall time.

Q q

I have

I have feen feveral patients in this difeafe, of which the following are the circumftances.   1. They were men about fifty years of age, and had lived freely in refpect to fermented liquors.   2. They loft their appetite to animal food.   3. They became fuddenly emaciated to a great degree.   4. Their fkins were dry and rough.   5. They coughed and expectorated with difficulty a vifcid phlegm.   6. The membrane of the tongue was dry and red, and liable to become ulcerous.

The inability to digeft animal food, and the confequent diftafte to it, generally precedes the dropfy, and other difeafes, which originate from fpirituous potation.   I fuppofe when the ftomach becomes in-irritable, that there is at the fame time a deficiency of gaftric acid; hence milk feldom agrees with thefe patients, unlefs it be previoufly curdled, as they have not fufficient gaftric acid to curdle it; and hence vegetable food, which is itfelf acefcent, will agree with their ftomachs longer than animal food, which requires more of the gaftric acid for its digeftion.

In this difeafe the fkin is dry from the increafed abforption of the cutaneous lymphatics, the fat is abforbed from the increafed abforption of the cellular lymphatics, the mucus of the lungs is too vifcid to be eafily fpit up by the increafed abforption of the thinner parts of it, the membrana fneideriana becomes dry, covered with hardened mucus, and at length becomes inflamed and full of apthæ, and either thefe floughs, or pulmonary ulcers, terminate the fcene.

II. The immediate caufe of dropfy is the paralyfis of fome other branches of the abforbent fyftem, which are called lymphatics, and which open into the larger cavities of the body, or into the cells of the cellular membrane; whence thofe cavities or cells become diftended with the fluid, which is hourly fecreted into them for the purpofe of lubricating their furfaces.   As is more fully explained in No. 5. of the next Section.

As

As thofe lymphatic veffels confift generally of a long neck or mouth, which drinks up its appropriated fluid, and of a conglobate gland, in which this fluid undergoes fome change, it happens, that fometimes the mouth of the lymphatic, and fometimes the belly or glandular part of it, becomes totally or partially paralytic.   In the former cafe, where the mouths of the cutaneous lymphatics become torpid or quiefcent, the fluid fecreted on the fkin ceafes to be abforbed, and erodes the fkin by its faline acrimony, and produces eruptions termed herpes, the difcharge from which is as falt, as the tears, which are fecreted too faft to be reabforbed, as in grief, or when the puncta lacrymalia are obftructed, and which running down the cheek redden and inflame the fkin.

When the mouths of the lymphatics, which open on the mucous membrane of the noftrils, become torpid, as on walking into the air in a frofty morning; the mucus, which continues to be fecreted, has not its aqueous and faline part reabforbed, which running over the upper lip inflames it, and has a falt tafte, if it falls on the tongue.

When the belly, or glandular part of thefe lymphatics, becomes torpid, the fluid abforbed by its mouth ftagnates, and forms a tumour in the gland.   This difeafe is called the fcrophula.   If thefe glands fuppurate externally, they gradually heal, as thofe of the neck; if they fuppurate without an opening on the external habit, as the mefenteric glands, a hectic fever enfues, which deftroys the patient; if they fuppurate in the lungs, a pulmonary confumption enfues, which is believed thus to differ from that defcribed in the preceding Section, in refpect to its feat or proximate caufe.

It is remarkable, that matter produced by fuppuration will lie concealed in the body many weeks, or even months, without producing hectic fever; but as foon as the wound is opened, fo as to

Q q 2

admit

admit air to the furface of the ulcer, a hectic fever fupervenes, even in very few hours, which is probably owing to the azotic part of the atmofphere rather than to the oxygene; becaufe thofe medicines, which contain much oxygene, as the calces or oxydes of metals, externally applied, greatly contribute to heal ulcers, of thefe are the folutions of lead and mercury, and copper in acids, or their precipitates.

Hence when ulcers are to be healed by the firft intention, as it is called, it is neceffary carefully to exclude the air from them. Hence we have one caufe, which prevents pulmonary ulcers from healing; which is their being perpetually expofed to the air.

Both the dark-eyed patients, which are affected with pulmonary ulcers from deficient venous abforption, as defcribed in Section XXVII. 2. and the light-eyed patients from deficient lymphatic abforption, which we are now treating of, have generally large apertures of the iris; thefe large pupils of the eyes are a common mark of want of irritability; and it generally happens, that an increafe of fenfibility, that is, of motions in confequence of fenfation, attends thefe conftitutions. See Sect. XXXI. 2. Whence inflammations may occur in thefe from ftagnated fluids more frequently than in thofe conftitutions, which poffefs more irritability and lefs fenfibility.

Great expectations in refpect to the cure of confumptions, as well as of many other difeafes, are produced by the very ingenious exertions of Dr. BEDDOES; who has eftablifhed an apparatus for breathing various mixtures of airs or gaffes, at the hot-wells near Briftol, which well deferves the attention of the public.

Dr. BEDDOES very ingenioufly concludes, from the florid colour of the blood of confumptive patients, that it abounds in oxygene; and that the rednefs of their tongues, and lips, and the fine blufh of their cheeks fhew the prefence of the fame principle; like flefh reddened

7

by

by nitre.   And adds, that the circumſtance of the conſumptions of pregnant women being ſtopped in their progreſs during pregnancy, at which time their blood may be ſuppoſed to be in part deprived of its oxygene, by oxygenating the blood of the fœtus, is a forceable argument in favour of this theory; which muſt ſoon be confirmed or confuted by his experiments.   Sèe Eſſay on Scurvy, Conſumption, &c. by Dr. Beddoes. Murray. London.   Alſo Letter to Dr. Darwin, by the ſame.   Murray.   London.

SECT. XXIX.

ON THE RETROGRADE MOTIONS OF THE ABSORBENT SYSTEM.

I. *Account of the abforbent fyftem.* II. *The valves of the abforbent veffels may fuffer their fluids to regurgitate in fome difeafes.* III. *Communication from the alimentary canal to the bladder by means of the abforbent veffels.* IV. *The phænomena of diabetes explained.* V. 1. *The phænomena of dropfies explained.* 2. *Cafes of the ufe of foxglove.* VI. *Of cold fweats.* VII. *Tranflations of matter, of chyle, of milk, of urine, operation of purging drugs applied externally.* VIII. *Circumftances by which the fluids, that are effufed by the retrograde motions of the abforbent veffels, are diftinguifhed.* IX. *Retrograde motions of vegetable juices.* X. *Objections anfwered.* XI. *The caufes, which induce the retrograde motions of animal veffels, and the medicines by which the natural motions are reftored.*

N. B. *The following Section is a tranflation of a part of a Latin thefis written by the late Mr. Charles Darwin, which was printed with his prize-differtation on a criterion between matter and mucus in 1780. Sold by Cadell, London.*

I. *Account of the Abforbent Syftem.*

1. THE abforbent fyftem of veffels in animal bodies confifts of feveral branches, differing in refpect to their fituations, and to the fluids, which they abforb.

The inteftinal abforbents open their mouths on the internal furfaces of the inteftines; their office is to drink up the chyle and the

3   other

other fluids from the alimentary canal; and they are termed lacteals, to diftinguifh them from the other abforbent veffels, which have been termed lymphatics.

Thofe, whofe mouths are difperfed on the external fkin, imbibe a great quantity of water from the atmofphere, and a part of the perfpirable matter, which does not evaporate, and are termed cutaneous abforbents.

Thofe, which arife from the internal furface of the bronchia, and which imbibe moifture from the atmofphere, and a part of the bronchial mucus, are called pulmonary abforbents.

Thofe, which open their innumerable mouths into the cells of the whole cellular membrane; and whofe ufe is to take up the fluid, which is poured into thofe cells, after it has done its office there; may be called cellular abforbents.

Thofe, which arife from the internal furfaces of the membranes, which line the larger cavities of the body, as the thorax, abdomen, fcrotum, pericardium, take up the mucus poured into thofe cavities; and are diftinguifhed by the names of their refpective cavities.

Whilft thofe, which arife from the internal furfaces of the urinary bladder, gall-bladder, falivary ducts, or other receptacles of fecreted fluids, may take their names from thofe fluids; the thinner parts of which it is their office to abforb: as urinary, bilious, or falivary abforbents.

2. Many of thefe abforbent veffels, both lacteals and lymphatics, like fome of the veins, are replete with valves: which feem defigned to affift the progrefs of their fluids, or at leaft to prevent their regurgitation; where they are fubjected to the intermitted preffure of the mufcular, or arterial actions in their neighbourhood.

Thefe valves do not however appear to be neceffary to all the abforbents, any more than to all the veins; fince they are not found to exift in the abforbent fyftem of fifh; according to the difcoveries of the

ingenious,

ingenious, and much lamented Mr. Hewfon. Philof. Tranf. v. 59, Enquiries into the Lymph. Syft. p. 94.

3. Thefe abforbent veffels are alfo furnifhed with glands, which are called conglobate glands; whofe ufe is not at prefent fufficiently inveftigated; but it is probable that they refemble the conglomerate glands both in ftructure and in ufe, except that their abforbent mouths are for the conveniency of fituation placed at a greater dif-tance from the body of the gland. The conglomerate glands open their mouths immediately into the fanguiferous veffels, which bring the blood, from whence they abforb their refpective fluids, quite up to the gland: but thefe conglobate glands collect their adapted fluids from very diftant membranes, or cyfts, by means of mouths furnifh-ed with long necks for this purpofe; and which are called lacteals, or lymphatics.

4. The fluids, thus collected from various parts of the body, pafs by means of the thoracic duct into the left fubclavian near the ju-gular vein; except indeed that thofe collected from the right fide of the head and neck, and from the right arm, are carried into the right fubclavian vein: and fometimes even the lymphatics from the right fide of the lungs are inferted into the right fubclavian vein; whilft thofe of the left fide of the head open but juft into the fummit of the thoracic duct.

5. In the abforbent fyftem there are many anaftomofes of the vef-fels, which feem of great confequence to the prefervation of health. Thefe anaftomofes are difcovered by diffection to be very frequent between the inteftinal and urinary lymphatics, as mentioned by Mr. Hewfon, (Phil. Tranf. v. 58).

6. Nor do all the inteftinal abforbents feem to terminate in the thoracic duct, as appears from fome curious experiments of D. Monro, who gave madder to fome animals, having previoufly put a ligature on the thoracic duct, and found their bones, and the ferum of their blood, coloured red.

II. *The*

II. *The Valves of the Abforbent Syftem may fuffer their Fluids to regurgitate in fome Difeafes.*

1. The many valves, which occur in the progrefs of the lymphatic and lacteal veffels, would feem infuperable obftacles to the regurgitation of their contents. But as thefe valves are placed in veffels, which are indued with life, and are themfelves indued with life alfo; and are very irritable into thofe natural motions, which abforb, or propel the fluids they contain; it is poffible, in fome difeafes, where thefe valves or veffels are ftimulated into unnatural exertions, or are become paralytic, that during the diaftole of the part of the veffel to which the valve is attached, the valve may not fo completely clofe, as to prevent the relapfe of the lymph or chyle. This is rendered more probable, by the experiments of injecting mercury, or water, or fuet, or by blowing air down thefe veffels; all which pafs the valves very eafily, contrary to the natural courfe of their fluids, when the veffels are thus a little forcibly dilated, as mentioned by Dr. Haller, Elem. Phyfiol. t. iii. f. 4.

" The valves of the thoracic duct are few, fome affert they are not more than twelve, and that they do not very accurately perform their office, as they do not clofe the whole area of the duct, and thence may permit chyle to repafs them downwards. In living animals, however, though not always, yet more frequently than in the dead, they prevent the chyle from returning. The principal of thefe valves is that, which prefides over the infertion of the thoracic duct, into the fubclavian vein; many have believed this alfo to perform the office of a valve, both to admit the chyle into the vein, and to preclude the blood from entering the duct; but in my opinion it is fcarcely fufficient for this purpofe." Haller, Elem. Phyf. t. vii. p. 226.

R r

2. The

2. The mouths of the lymphatics feem to admit water to pafs through them after death, the inverted way, eafier than the natural one; fince an inverted bladder readily lets out the water with which it is filled; whence it may be inferred, that there is no obftacle at the mouths of thefe veffels to prevent the regurgitation of their contained fluids.

I was induced to repeat this experiment, and having accurately tied the ureters and neck of a frefh ox's bladder, I made an opening at the fundus of it; and then, having turned it infide outwards, filled it half full with water, and was furprifed to fee it empty itfelf fo haftily. I thought the experiment more appofite to my purpofe by fufpending the bladder with its neck downwards, as the lymphatics are chiefly fpread upon this part of it; as fhewn by Dr. Watfon, Philof. Tranf. v. 59. p. 392.

3. In fome difeafes, as in the diabetes and fcrophula, it is probable the valves themfelves are difeafed, and are thence incapable of preventing the return of the fluids they fhould fupport. Thus the valves of the aorta itfelf have frequently been found fchirrous, according to the diffections of Monf. Lieutaud, and have given rife to an interrupted pulfe, and laborious palpitations, by fuffering a return of part of the blood into the heart. Nor are any parts of the body fo liable to fchirrofity as the lymphatic glands and veffels, infomuch that their fchirrofities have acquired a diftinct name, and been termed fchrophula.

4. There are valves in other parts of the body, analogous to thofe of the abforbent fyftem, and which are liable, when difeafed, to regurgitate their contents: thus the upper and lower orifices of the ftomach are clofed by valves, which, when too great quantities of warm water have been drank with a defign to promote vomiting, have fometimes refifted the utmoft efforts of the abdominal mufcles, and diaphragm: yet, at other times, the upper valve, or cardia, eafily permits the evacuation of the contents of the ftomach; whilft

the

the inferior valve, or pylorus, permits the bile, and other contents of the duodenum, to regurgitate into the ftomach.

5. The valve of the colon is well adapted to prevent the retrograde motion of the excrements; yet, as this valve is poffeffed of a living power, in the iliac paffion, either from fpafm, or other unnatural exertions, it keeps itfelf open, and either fuffers or promotes the retrograde movements of the contents of the inteftines below; as in ruminating animals the mouth of the firft ftomach feems to be fo conftructed, as to facilitate or affift the regurgitation of the food; the rings of the œfophagus afterwards contracting themfelves in inverted order.   De Haen, by means of a fyringe, forced fo much water into the rectum inteftinum of a dog, that he vomited it in a full ftream from his mouth; and in the iliac paffion above mentioned, excrements and clyfter are often evacuated by the mouth.   See Section XXV. 15.

6. The puncta lacrymalia, with the lacrymal fack and nafal duct, compofe a complete gland, and much refemble the inteftinal canal the puncta lacrymalia are abforbent mouths, that take up the tears from the eye, when they have done their office there, and convey them into the noftrils; but when the nafal duct is obftructed, and the lacrymal fack diftended with its fluid, on preffure with the finger the mouths of this gland (puncta lacrymalia) will readily difgorge the fluid, they had previoufly abforbed, back into the eye.

7. As the capillary veffels receive blood from the arteries, and feparating the mucus, or perfpirable matter from it, convey the remainder back by the veins; thefe capillary veffels are a fet of glands, in every refpect fimilar to the fecretory veffels of the liver, or other large congeries of glands.   The beginnings of thefe capillary veffels have frequent anaftomofes into each other, in which circumftance they are refembled by the lacteals; and like the mouths or beginnings of other glands, they are a fet of abforbent veffels, which drink up the blood which is brought to them by the arteries, as the chyle is drank

R r 2

up

up by the lacteals: for the circulation of the blood through the capillaries is proved to be independent of arterial impulfe; fince in the blufh of fhame, and in partial inflammations, their action is increafed, without any increafe of the motion of the heart.

8. Yet not only the mouths, or beginnings of thefe anaftomofing capillaries are frequently feen by microfcopes, to regurgitate fome particles of blood, during the ftruggles of the animal; but retrograde motion of the blood, in the veins of thofe animals, from the very heart to the extremity of the limbs, is obfervable, by intervals, during the diftreffes of the dying creature. Haller, Elem. Phyfiol. t. i. p. 216.   Now, as the veins have perhaps all of them a valve fomewhere between their extremities and the heart, here is ocular demonftration of the fluids in this difeafed condition of the animal, repaffing through venous valves: and it is hence highly probable, from the ftricteft analogy, that if the courfe of the fluids, in the lymphatic veffels, could be fubjected to microfcopic obfervation, they would alfo, in the difeafed ftate of the animal, be feen to repafs the valves, and the mouths of thofe veffels, which had previoufly abforbed them, or promoted their progreffion.

III. *Communication from the Alimentary Canal to the Bladder, by means of the Abforbent Veffels.*

MANY medical philofophers, both ancient and modern, have fufpected that there was a nearer communication between the ftomach and the urinary bladder, than that of the circulation: they were led into this opinion from the great expedition with which cold water, when drank to excefs, paffes off by the bladder; and from the fimilarity of the urine, when produced in this hafty manner, with the material that was drank.

The

The former of thefe circumftances happens perpetually to thofe who drink abundance of cold water, when they are much heated by exercife, and to many at the beginning of intoxication.

Of the latter, many inftances are recorded by Etmuller, t. xi. p. 716. where fimple water, wine, and wine with fugar, and emulfions, were returned by urine unchanged.

There are other experiments, that feem to demonftrate the exiftence of another paffage to the bladder, befides that through the kidneys. Thus Dr. Kratzenftein put ligatures on the ureters of a dog, and then emptied the bladder by a catheter; yet in a little time the dog drank greedily, and made a quantity of water, (Difputat. Morbor. Halleri. t. iv. p. 63.) A fimilar experiment is related in the Philofophical Tranfactions, with the fame event, (No. 65, 67, for the year 1670.)

Add to this, that in fome morbid cafes the urine has continued to pafs, after the fuppuration or total deftruction of the kidneys; of which many inftances are referred to in the Elem. Phyfiol. t. vii. p. 379. of Dr. Haller.

From all which it muft be concluded, that fome fluids have paffed from the ftomach or abdomen, without having gone through the fanguiferous circulation: and as the bladder is fupplied with many lymphatics, as defcribed by Dr. Watfon, in the Philof. Tranf. v. 59. p. 392. and as no other veffels open into it befides thefe and the ureters, it feems evident, that the unnatural urine, produced as above defcribed, when the ureters were tied, or the kidneys obliterated, was carried into the bladder by the retrograde motions of the urinary branch of the lymphatic fyftem.

The more certainly to afcertain the exiftence of another communication between the ftomach and bladder, befides that of the circulation, the following experiment was made, to which I muft beg your patient attention:—A friend of mine (June 14, 1772) on drinking repeatedly of cold fmall punch, till he began to be intoxicated, made a

quantity

quantity of colourlefs urine. He then drank about two drams of nitre diffolved in fome of the punch, and eat about twenty ftalks of boiled afparagus: on continuing to drink more of the punch, the next urine that he made was quite clear, and without fmell; but in a little time another quantity was made, which was not quite fo colourlefs, and had a ftrong fmell of the afparagus: he then loft about four ounces of blood from the arm.

The fmell of afparagus was not at all perceptible in the blood, neither when frefh taken, nor the next morning, as myfelf and two others accurately attended to; yet this fmell was ftrongly perceived in the urine, which was made juft before the blood was taken from his arm.

Some bibulous paper, moiftened in the ferum of this blood, and fuffered to dry, fhewed no figns of nitre by its manner of burning. But fome of the fame paper, moiftened in the urine, and dried, on being ignited, evidently fhewed the prefence of nitre. This blood and the urine ftood fome days expofed to the fun in the open air, till they were evaporated to about a fourth of their original quantity, and began to ftink: the paper, which was then moiftened with the concentrated urine, fhewed the prefence of much nitre by its manner of burning; whilft that moiftened with the blood fhewed no fuch appearance at all.

Hence it appears, that certain fluids at the beginning of intoxication, find another paffage to the bladder befides the long courfe of the arterial circulation; and as the inteftinal abforbents are joined with the urinary lymphatics by frequent anaftomofes, as Hewfon has demonftrated; and as there is no other road, we may juftly conclude, that thefe fluids pafs into the bladder by the urinary branch of the lymphatics, which has its motions inverted during the difeafed ftate of the animal.

A gentleman, who had been fome weeks affected with jaundice, and whofe urine was in confequence of a very deep yellow, took fome
cold

cold fmall punch, in which was diffolved about a dram of nitre; he then took repeated draughts of the punch, and kept himfelf in a cool room, till on the approach of flight intoxication he made a large quantity of water; this water had a flight yellow tinge, as might be expected from a fmall admixture of bile fecreted from the kidneys; but if the whole of it had paffed through the fanguiferous veffels, which were now replete with bile (his whole fkin being as yellow as gold) would not this urine alfo, as well as that he had made for weeks before, have been of a deep yellow? Paper dipped in this water, and dryed, and ignited, fhewed evident marks of the prefence of nitre, when the flame was blown out.

IV. *The Phænomena of the Diabetes explained, and of fome Diarrhœas.*

THE phænomena of many difeafes are only explicable from the retrograde motions of fome of the branches of the lymphatic fyftem; as the great and immediate flow of pale urine in the beginning of drunkennefs; in hyfteric paroxyfms; from being expofed to cold air; or to the influence of fear or anxiety.

Before we endeavour to illuftrate this doctrine, by defcribing the phænomena of thefe difeafes, we muft premife one circumftance; that all the branches of the lymphatic fyftem have a certain fympathy with each other, infomuch that when one branch is ftimulated into unufual kinds or quantities of motion, fome other branch has its motions either increafed, or decreafed, or inverted at the fame time. This kind of fympathy can only be proved by the concurrent teftimony of numerous facts, which will be related in the courfe of the work. I fhall only add here, that it is probable, that this fympathy does not depend on any communication of nervous filaments, but on

habit;

habit; owing to the various branches of this fyſtem having frequently been ſtimulated into action at the ſame time.

There are a thouſand inſtances of involuntary motions aſſociated in this manner; as in the act of vomiting, while the motions of the ſtomach and œſophagus are inverted, the pulſations of the arterial ſyſtem by a certain ſympathy become weaker; and when the bowels or kidneys are ſtimulated by poiſon, a ſtone, or inflammation, into more violent action; the ſtomach and œſophagus by ſympathy invert their motions.

1. When any one drinks a moderate quantity of vinous ſpirit, the whole ſyſtem acts with more energy by conſent with the ſtomach and inteſtines, as is ſeen from the glow on the ſkin, and the increaſe of ſtrength and activity; but when a greater quantity of this inebriating material is drank, at the ſame time that the lacteals are excited into greater action to abſorb it; it frequently happens, that the urinary branch of abſorbents, which is connected with the lacteals by many anaſtomoſes, inverts its motions, and a great quantity of pale unanimalized urine is diſcharged. By this wiſe contrivance too much of an unneceſſary fluid is prevented from entering the circulation—This may be called the drunken diabetes, to diſtinguiſh it from the other temporary diabetes, which occur in hyſteric diſeaſes, and from continued fear or anxiety.

2. If this idle ingurgitation of too much vinous ſpirit be daily practiſed, the urinary branch of abſorbents at length gains an habit of inverting its motions, whenever the lacteals are much ſtimulated; and the whole or a great part of the chyle is thus daily carried to the bladder without entering the circulation, and the body becomes emaciated. This is one kind of chronic diabetes, and may be diſtinguiſhed from the others by the taſte and appearance of the urine; which is ſweet, and the colour of whey, and may be termed the chyliferous diabetes.

3. Many

3. Many children have a fimilar depofition of chyle in their urine, from the irritation of worms in their inteftines, which ftimulating the mouths of the lacteals into unnatural action, the urinary branch of the abforbents becomes inverted, and carries part of the chyle to the bladder: part of the chyle alfo has been carried to the iliac and lumbar glands, of which inftances are recorded by Haller, t. vii. 225. and which can be explained on no other theory: but the diffections of the lymphatic fyftem of the human body, which have yet been publifhed, are not fufficiently extenfive for our purpofe; yet if we may reafon from comparative anatomy, this tranflation of chyle to the bladder is much illuftrated by the account given of this fyftem of veffels in a turtle, by Mr. Hewfon, who obferved, " That the lacteals near the root of the mefentery anaftomofe, fo as to form a net-work, from which feveral large branches go into fome confiderable lymphatics lying near the fpine; and which can be traced almoft to the anus, and particularly to the kidneys. Philof. Tranf. v. 59. p. 199—Enquiries, p. 74.

4. At the fame time that the urinary branch of abforbents, in the beginning of diabetes, is excited into inverted action, the cellular branch is excited by the fympathy above mentioned, into more energetic action; and the fat, that was before depofited, is reabforbed and thrown into the blood veffels; where it floats, and was miftaken for chyle, till the late experiments of the ingenious Mr. Hewfon demonftrated it to be fat.

This appearance of what was miftaken for chyle in the blood, which was drawn from thefe patients, and the obftructed liver, which very frequently accompanies this difeafe, feems to have led Dr. Mead to fufpect the diabetes was owing to a defect of fanguification; and that the fchirrofity of the liver was the original caufe of it: but as the fchirrhus of the liver is moft frequently owing to the fame caufes, that produce the diabetes and dropfies; namely, the great ufe

S f

of

of fermented liquors; there is no wonder they should exist together, without being the consequence of each other.

5. If the cutaneous branch of absorbents gains a habit of being excited into stronger action, and imbibes greater quantities of moisture from the atmosphere, at the same time that the urinary branch has its motions inverted, another kind of diabetes is formed, which may be termed the aqueous diabetes. In this diabetes the cutaneous absorbents frequently imbibe an amazing quantity of atmospheric moisture; insomuch that there are authentic histories, where many gallons a day, for many weeks together, above the quantity that has been drank, have been discharged by urine.

Dr. Keil, in his Medicina Statica, found that he gained eighteen ounces from the moist air of one night; and Dr. Percival affirms, that one of his hands imbibed, after being well chafed, near an ounce and half of water, in a quarter of an hour. (Transact. of the College, London, vol. ii. p. 102). Home's Medic. Facts, p. 2. sect. 3.

The pale urine in hysterical women, or which is produced by fear or anxiety, is a temporary complaint of this kind; and it would in reality be the same disease, if it was confirmed by habit.

6. The purging stools, and pale urine, occasioned by exposing the naked body to cold air, or sprinkling it with cold water, originate from a similar cause; for the mouths of the cutaneous lymphatics being suddenly exposed to cold become torpid, and cease, or nearly cease, to act; whilst, by the sympathy above described, not only the lymphatics of the bladder and intestines cease also to absorb the more aqueous and saline part of the fluids secreted into them; but it is probable that these lymphatics invert their motions, and return the fluids, which were previously absorbed, into the intestines and bladder. At the very instant that the body is exposed naked to the cold air, an unusual movement is felt in the bowels; as is experienced by boys going into the cold bath: this could not occur from an obstruction of

the

the perfpirable matter, fince there is not time for that to be returned to the bowels by the courfe of the circulation.

There is alfo a chronic aqueous diarrhœa, in which the atmofpheric moifture, drank up by the cutaneous and pulmonary lymphatics, is poured into the inteftines, by the retrograde motions of the lacteals. This difeafe is moft fimilar to the aqueous diabetes, and is frequently exchanged for it: a diftinct inftance of this is recorded by Benningerus, Cent. v. Obf. 98. in which an aqueous diarrhœa fucceeded an aqueous diabetes, and deftroyed the patient.   There is a curious example of this, defcribed by Sympfon (De Re Medica)—" A young man (fays he) was feized with a fever, upon which a diarrhœa came on, with great ftupor; and he refufed to drink any thing, though he was parched up with exceffive heat: the better to fupply him with moifture, I directed his feet to be immerfed in cold water; immediately I obferved a wonderful decreafe of water in the veffel, and then an impetuous ftream of a fluid, fcarcely coloured, was difcharged by ftool, like a cataract."

7. There is another kind of diarrhœa, which has been called cæliaca; in this difeafe the chyle, drank up by the lacteals of the fmall inteftines, is probably poured into the large inteftines, by the retrograde motions of their lacteals: as in the chyliferous diabetes, the chyle is poured into the bladder, by the retrograde motions of the urinary branch of abforbents.

The chyliferous diabetes, like this chyliferous diarrhœa, produces fudden atrophy; fince the nourifhment, which ought to fupply the hourly wafte of the body, is expelled by the bladder, or rectum: whilft the aqueous diabetes, and the aqueous diarrhœa produce exceffive thirft; becaufe the moifture, which is obtained from the atmofphere, is not conveyed to the thoracic receptacle, as it ought to be, but to the bladder, or lower inteftines; whence the chyle, blood, and whole fyftem of glands, are robbed of their proportion of humidity.

8. There

8. There is a third ſpecies of diabetes, in which the urine is mucilaginous, and appears ropy in pouring it from one veſſel into another; and will ſometimes coagulate over the fire. This diſeaſe appears by intervals, and ceaſes again, and ſeems to be occaſioned by a previous dropſy in ſome part of the body. When ſuch a collection is reabſorbed, it is not always returned into the circulation; but the ſame irritation that ſtimulates one lymphatic branch to reabſorb the depoſited fluid, inverts the urinary branch, and pours it into the bladder. Hence this mucilaginous diabetes is a cure, or the conſequence of a cure, of a worſe diſeaſe, rather than a diſeaſe itſelf.

Dr. Cotunnius gave half an ounce of cream of tartar, every morning, to a patient, who had the anaſarca; and he voided a great quantity of urine; a part of which, put over the fire, coagulated, on the evaporation of half of it, ſo as to look like the white of an egg. De Iſchiade Nervos.

This kind of diabetes frequently precedes a dropſy; and has this remarkable circumſtance attending it, that it generally happens in the night; as during the recumbent ſtate of the body, the fluid, that was accumulated in the cellular membrane, or in the lungs, is more readily abſorbed, as it is leſs impeded by its gravity. I have ſeen more than one inſtance of this diſeaſe. Mr. D. a man in the decline of life, who had long accuſtomed himſelf to ſpirituous liquor, had ſwelled legs, and other ſymptoms of approaching anaſarca; about once in a week, or ten days, for ſeveral months, he was ſeized, on going to bed, with great general uneaſineſs, which his attendants reſembled to an hyſteric fit; and which terminated in a great diſcharge of viſcid urine; his legs became leſs ſwelled, and he continued in better health for ſome days afterwards. I had not the opportunity to try if this urine would coagulate over the fire, when part of it was evaporated, which I imagine would be the criterion of this kind of diabetes; as the mucilaginous fluid depoſited in the cells and cyſts of the body, which have no communication with the external air, ſeems to acquire, by ſtagna-

tion,

tion, this property of coagulation by heat, which the fecreted mucus of the inteftines and bladder do not appear to poffefs; as I have found by experiment : and if any one fhould fuppofe this coagulable urine was feparated from the blood by the kidneys, he may recollect, that in the moft inflammatory difeafes, in which the blood is moft replete, or moft ready to part with the coagulable lymph, none of this appears in the urine.

9. Different kinds of diabetes require different methods of cure. For the firft kind, or chyliferous diabetes, after clearing the ftomach and inteftines, by ipecacuanha and rhubarb, to evacuate any acid material, which may too powerfully ftimulate the mouths of the lacteals, repeated and large dofes of tincture of cantharides have been much recommended. The fpecific ftimulus of this medicine, on the neck of the bladder, is likely to excite the numerous abforbent veffels, which are fpread on that part, into ftronger natural actions, and by that means prevent their retrograde ones ; till, by perfifting in the ufe of the medicine, their natural habits of motions might again be eftablifh-ed. Another indication of cure, requires fuch medicines, as by lin-ing the inteftines with mucilaginous fubftances, or with fuch as con-fift of fmooth particles, or which chemically deftroy the acrimony of their contents, may prevent the too great action of the inteftinal ab-forbents. For this purpofe, I have found the earth precipitated from a folution of alum, by means of fixed alcali, given in the dofe of half a dram every fix hours, of great advantage, with a few grains of rhu-barb, fo as to procure a daily evacuation.

The food fhould confift of materials that have the leaft ftimulus, with calcareous water, as of Briftol and Matlock ; that the mouths of the lacteals may be as little ftimulated as is neceffary for their proper abforption ; left with their greater exertions, fhould be connected by fympathy, the inverted motions of the urinary lymphatics.

The fame method may be employed with equal advantage in the aqueous diabetes, fo great is the fympathy between the fkin and the

ftomach.

ftomach.   To which, however, fome application to the fkin might
be ufefully added; as rubbing the patient all over with oil, to prevent
the too great action of the cutaneous abforbents.   I knew an experiment of this kind made upon one patient with apparent advantage.

The mucilaginous diabetes will require the fame treatment, which
is moft efficacious in the dropfy, and will be defcribed below.   I muft
add, that the diet and medicines above mentioned, are ftrongly recommended by various authors, as by Morgan, Willis, Harris, and
Etmuller; but more hiftories of the fuccefsful treatment of thefe
difeafes are wanting to fully afcertain the moft efficacious methods of
cure.

In a letter from Mr. Charles Darwin, dated April 24, 1778, Edinburgh, is the fubfequent paffage:—" A man who had long laboured
under a diabetes died yefterday in the clinical ward.   He had for fome
time drank four, and paffed twelve pounds of fluid daily; each pound
of urine contained an ounce of fugar.   He took, without confiderable
relief, gum kino, fanguis draconis melted with alum, tincture of cantharides, ifinglafs, gum arabic, crabs eyes, fpirit of hartfhorn, and eat
ten or fifteen oyfters thrice a day.   Dr. Home, having read my thefis,
bled him, and found that neither the frefh blood nor the ferum tafted
fweet.   His body was opened this morning—every vifcus appeared in
a found and natural ftate, except that the left kidney had a very fmall
pelvis, and that there was a confiderable enlargement of moft of the
mefenteric lymphatic glands.   I intend to infert this in my thefis, as
it coincides with the experiment, where fome afparagus was eaten at
the beginning of intoxication, and its fmell perceived in the urine,
though not in the blood."

The following cafe of chyliferous diabetes is extracted from fome
letters of Mr. Hughs, to whofe unremitted care the infirmary at Stafford for many years was much indebted.   Dated October 10,
1778.

                                                    Richard

Richard Davis, aged 33, a whitefmith by trade, had drank hard by intervals; was much troubled with fweating of his hands, which incommoded him in his occupation, but which ceafed on his frequently dipping them in lime.   About feven months ago he began to make large quantities of water; his legs are œdematous, his belly tenfe, and he complains of a rifing in his throat, like the globus hyftericus: he eats twice as much as other people, drinks about fourteen pints of fmall beer a day, befides a pint of ale, fome milk-porridge, and a bafon of broth, and he makes about eighteen pints of water a day.

He tried alum, dragon's blood, fteel, blue vitriol, and cantharides in large quantities, and duly repeated, under the care of Dr. Underhill, but without any effect; except that on the day after he omitted the cantharides, he made but twelve pints of water, but on the next day this good effect ceafed again.

November 21.—He made eighteen pints of water, and he now, at Dr. Darwin's requeft, took a grain of opium every four hours, and five grains of aloes at night; and had a flannel fhirt given him.

22.—Made fixteen pints.   23.—Thirteen pints: drinks lefs.

24.—Increafed the opium to a grain and quarter every four hours: he made twelve pints.

25.—Increafed the opium to a grain and half: he now makes ten pints; and drinks eight pints in a day.

The opium was gradually increafed during the next fortnight, till he took three grains every four hours, but without any further dimunition of his water.   During the ufe of the opium he fweat much in the nights, fo as to have large drops ftand on his face and all over him. The quantity of opium was then gradually decreafed, but not totally omitted, as he continued to take about a grain morning and evening.

January 17.—He makes fourteen pints of water a day.   Dr. Underhill now directed him two fcruples of common rofin triturated

with

with as much fugar, every fix hours; and three grains of opium every night.

19.—Makes fifteen pints of water: fweats at night.

21.—Makes feventeen pints of water; has twitchings of his limbs in a morning, and pains of his legs: he now takes a dram of rofin for a dofe, and continues the opium.

23.—Water more coloured, and reduced to fixteen pints, and he thinks has a brackifh tafte.

26.—Water reduced to fourteen pints.

28.—Water thireen pints: he continues the opium, and takes four fcruples of the rofin for a dofe.

February 1.—Water twelve pints.

4.—Water eleven pints: twitchings lefs: takes five fcruples for a dofe.

8.—Water ten pints: has had many ftools.

12.—Appetite lefs: purges very much.

After this the rofin either purged him, or would not ftay on his ftomach; and he gradually relapfed nearly to his former condition, and in a few months funk under the difeafe.

October 3, Mr. Hughs evaporated two quarts of the water, and obtained from it four ounces and half of a hard and brittle faccharine mafs, like treacle which had been fome time boiled. Four ounces of blood, which he took from his arm with defign to examine it, had the common appearances, except that the ferum refembled cheefe-whey; and that on the evidence of four perfons, two of whom did not know what it was they tafted, *the ferum had a faltifh tafte.*

From hence it appears, that the faccharine matter, with which the urine of thefe patients fo much abounds, does not enter the blood-veffels like the nitre and afparagus mentioned above; but that the procefs of digeftion refembles the procefs of the germination of ve-getables, or of making barley into malt; as the vaft quantity of fugar

3   found

found in the urine muft be made from the food which he took (which was double that taken by others), and from the fourteen pints of fmall beer which he drank. And, fecondly, as the ferum of the blood was not fweet, the chyle appears to have been conveyed to the bladder without entering the circulation of the blood, fince fo large a quantity of fugar, as was found in the urine, namely, twenty ounces a day, could not have previoufly exifted in the blood without being perceptible to the tafte.

November 1. Mr. Hughes diffolved two drams of nitre in a pint of a decoction of the roots of afparagus, and added to it two ounces of tincture of rhubarb: the patient took a fourth part of this mixture every five minutes, till he had taken the whole.—In about half an hour he made eighteen ounces of water, which was very manifeftly tinged with the rhubarb; the fmell of afparagus was doubtful.

He then loft four ounces of blood, the ferum of which was not fo opake as that drawn before, but of a yellowifh caft, as the ferum of the blood ufually appears.

Paper, dipped three or four times in the tinged urine and dried again, did not fcintillate when it was fet on fire; but when the flame was blown out, the fire ran along the paper for half an inch; which, when the fame paper was unimpregnated, it would not do; nor when the fame paper was dipped in urine made before he took the nitre, and dried in the fame manner.

Paper, dipped in the ferum of the blood and dried in the fame manner as in the urine, did not fcintillate when the flame was blown out, but burnt exactly in the fame manner as the fame paper dipped in the ferum of blood drawn from another perfon.

This experiment, which is copied from a letter of Mr. Hughes, as well as the former, feems to evince the exiftence of another paffage from the inteftines to the bladder, in this difeafe, befides that of the fanguiferous fyftem; and coincides with the curious experiment related in fection the third, except that the fmell of the afparagus was

T t   not

not here perceived, owing perhaps to the roots having been made use of inſtead of the heads.

The riſing in the throat of this patient, and the twitchings of his limbs, ſeem to indicate ſome ſimilarity between the diabetes and the hyſteric diſeaſe, beſides the great flow of pale urine, which is common to them both.

Perhaps if the meſenteric glands were nicely inſpected in the diſ-ſections of theſe patients; and if the thoracic duct, and the larger branches of the lacteals, and if the lymphatics, which ariſe from the bladder, were well examined by injection, or by the knife, the cauſe of diabetes might be more certainly underſtood.

The opium alone, and the opium with the roſin, ſeem much to have ſerved this patient, and might probably have effected a cure, if the diſeaſe had been ſlighter, or the medicine had been exhibited, be-fore it had been confirmed by habit during the ſeven months it had continued. The increaſe of the quantity of water on beginning the large doſes of roſin was probably owing to his omitting the morning doſes of opium.

V. *The Phænomena of Dropſies explained.*

I. Some inebriates have their paroxyſms of inebriety terminated by much pale urine, or profuſe ſweats, or vomiting, or ſtools; others have their paroxyſms terminated by ſtupor, or ſleep, without the above evacuations.

The former kind of theſe inebriates have been obſerved to be more liable to diabetes and dropſy; and the latter to gout, gravel, and le-proſy. Evoe! attend ye bacchanalians! ſtart at this dark train of evils, and, amid your immodeſt jeſts, and idiot laughter, recollect,

Quem Deus vult perdere, prius dementat.

In

In thofe who are fubject to diabetes and dropfy, the abforbent vef-fels are naturally more irritable than in the latter; and by being fre-quently difturbed or inverted by violent ftimulus, and by their too great fympathy with each other, they become at length either entire-ly paralytic, or are only fufceptible of motion from the ftimulus of very acrid materials; as every part of the body, after having been ufed to great irritations, becomes lefs affected by fmaller ones.   Thus we cannot diftinguifh objects in the night, for fome time after we come out of a ftrong light, though the iris is prefently dilated; and the air of a fummer evening appears cold, after we have been expofed to the heat of the day.

There are no cells in the body, where dropfy may not be produced, if the lymphatics ceafe to abforb that mucilaginous fluid, which is perpetually depofited in them, for the purpofe of lubricating their furfaces.

If the lymphatic branch, which opens into the cellular membrane, either does its office imperfectly, or not at all; thefe cells become re-plete with a mucilaginous fluid, which, after it has ftagnated fome time in the cells, will coagulate over the fire; and is erroneoufly called water.   Wherever the feat of this difeafe is, (unlefs in the lungs or other pendent vifcera) the mucilaginous liquid above mentioned will fubfide to the moft depending parts of the body, as the feet and legs, when thofe are lower than the head and trunk; for all thefe cells have communications with each other.

When the cellular abforbents are become infenfible to their ufual irritations, it moft frequently happens, but not always, that the cuta-neous branch of abforbents, which is ftrictly affociated with them, fuffers the like inability.   And then, as no water is abforbed from the atmofphere, the urine is not only lefs diluted at the time of its fecre-tion, and confequently in lefs quantity and higher coloured: but great thirft is at the fame time induced, for as no water is abforbed from the atmofphere to dilute the chyle and blood, the lacteals and other ab-

T t 2   forbent

forbent veffels, which have not loft their powers, are excited into more conftant or more violent action, to fupply this deficiency; whence the urine becomes ftill lefs in quantity, and of a deeper colour, and turbid like the yolk of an egg, owing to a greater abforption of its thinner parts.   From this ftronger action of thofe abforbents, which ftill retain their irritability, the fat is alfo abforbed, and the whole body becomes emaciated.   This increafed exertion of fome branches of the lymphatics, while others are totally or partially paralytic, is refembled by what conftantly occurs in the hemiplagia ; when the patient has loft the ufe of the limbs on one fide, he is inceffantly moving thofe of the other ; for the moving power, not having accefs to the paralytic limbs, becomes redundant in thofe which are not difeafed.

The paucity of urine and thirft cannot be explained from a greater quantity of mucilaginous fluid being depofited in the cellular membrane : for though thefe fymptoms have continued many weeks, or even months, this collection frequently does not amount to more than very few pints.   Hence alfo the difficulty of promoting copious fweats in anafarca is accounted for, as well as the great thirft, paucity of urine, and lofs of fat ; fince, when the cutaneous branch of abforbents is paralytic, or nearly fo, there is already too fmall a quantity of aqueous fluid in the blood : nor can thefe torpid cutaneous lymphatics be readily excited into retrograde motions.

Hence likewife we underftand, why in the afcites, and fome other dropfies, there is often no thirft, and no paucity of urine ; in thefe cafes the cutaneous abforbents continue to do their office.

Some have believed, that dropfies were occafioned by the inability of the kidneys, from having only obferved the paucity of urine ; and have thence laboured much to obtain diuretic medicines ; but it is daily obfervable, that thofe who die of a total inability to make water, do not become dropfical in confequence of it : Fernelius mentions one, who laboured under a perfect fuppreffion of urine during twenty days

before

before his death, and yet had no fymptoms of dropfy.   Pathol. l. vi.
c. 8.   From the fame idea many phyficians have reftrained their patients from drinking, though their thirft has been very urgent; and
fome cafes have been publifhed, where this cruel regimen has been
thought advantageous: but others of nicer obfervation are of opinion,
that it has always aggravated the diftreffes of the patient; and though
it has abated his fwellings, yet by inducing a fever it has haftened his
diffolution.   See Tranfactions of the College, London, vol. ii. p. 235.
Cafes of Dropfy by Dr. G. Baker.

The cure of anafarca, fo far as refpects the evacuation of the accumulated fluid, coincides with the idea of the retrograde action of the
lymphatic fyftem.   It is well known that vomits, and other drugs,
which induce ficknefs or naufea; at the fame time that they evacuate
the ftomach, produce a great abforption of the lymph accumulated in
the cellular membrane.   In the operation of a vomit, not only the
motions of the ftomach and duodenum become inverted, but alfo thofe
of the lymphatics and lacteals, which belong to them; whence a
great quantity of chyle and lymph is perpetually poured into the ftomach and inteftines, during the operation, and evacuated by the
mouth.   Now at the fame time, other branches of the lymphatic
fyftem, viz. thofe which open on the cellular membrane, are brought
into more energetic action, by the fympathy above mentioned, and an
increafe of their abforption is produced.

Hence repeated vomits, and cupreous falts, and fmall dofes of fquill
or foxglove, are fo efficacious in this difeafe.   And as draftic purges
act alfo by inverting the motions of the lacteals; and thence the other
branches of lymphatics are induced into more powerful natural action,
by fympathy, and drink up the fluids from all the cells of the body;
and by their anaftomofes, pour them into the lacteal branches; which,
by their inverted actions, return them into the inteftines; and they are
thus evacuated from the body:—thefe purges alfo are ufed with fuccefs
in difcharging the accumulated fluid in anafarca.

II. The

II. The following cafes are related with defign to afcertain the particular kinds of dropfy in which the digitalis purpurea, or common foxglove, is preferable to fquill, or other evacuants, and were firft publifhed in 1780, in a pamphlet entitled Experiments on mucilaginous and purulent Matter, &c. Cadell. London. Other cafes of dropfy, treated with digitalis, were afterwards publifhed by Dr. Darwin in the Medical Tranfactions, vol. iii. in which there is a miftake in refpect to the dofe of the powder of foxglove, which fhould have been from five grains to one, inftead of from five grains to ten.

Anafarca of the Lungs.

1. A lady, between forty and fifty years of age, had been indifpofed fome time, was then feized with cough and fever, and afterwards expectorated much digefted mucus. This expectoration fuddenly ceafed, and a confiderable difficulty of breathing fupervened, with a pulfe very irregular both in velocity and ftrength; fhe was much diftreffed at firft lying down, and at firft rifing; but after a minute or two bore either of thofe attitudes with eafe. She had no pain or numbnefs in her arms; fhe had no hectic fever, nor any cold fhiverings, and the urine was in due quantity, and of the natural colour.

The difficulty of breathing was twice confiderably relieved by fmall dofes of ipecacuanha, which operated upwards and downwards, but recurred in a few days: fhe was then directed a decoction of foxglove, (digitalis purpurea) prepared by boiling four ounces of the frefh leaves from two pints of water to one pint; to which was added two ounces of vinous fpirit: fhe took three large fpoonfuls of this mixture every two hours, till fhe had taken it four times; a continued ficknefs fupervened, with frequent vomiting, and a copious flow of urine: thefe evacuations continued at intervals for two or three days, and

relieved

relieved the difficulty of breathing—She had some relapses afterwards, which were again relieved by the repetition of the decoction of foxglove.

2. A gentleman, about sixty years of age, who had been addicted to an immoderate use of fermented liquors, and had been very corpulent, gradually lost his strength and flesh, had great difficulty of breathing, with legs somewhat swelled, and a very irregular pulse. He was very much distressed at first lying down, and at first rising from his bed, yet in a minute or two was easy in both those attitudes. He made straw-coloured urine in due quantity, and had no pain or numbness of his arms.

He took a large spoonful of the decoction of foxglove, as above, every hour, for ten or twelve successive hours, had incessant sickness for about two days, and passed a large quantity of urine; upon which his breath became quite easy, and the swelling of his legs subsided; but as his whole constitution was already sinking from the previous intemperance of his life, he did not survive more than three or four months.

Hydrops Pericardii.

3. A gentleman of temperate life and sedulous application to business, between thirty and forty years of age, had long been subject, at intervals, to an irregular pulse: a few months ago he became weak, with difficulty of breathing, and dry cough. In this situation a physician of eminence directed him to abstain from all animal food and fermented liquor, during which regimen all his complaints increased; he now became emaciated, and totally lost his appetite; his pulse very irregular both in velocity and strength; with great difficulty of breathing, and some swelling of his legs; yet he could lie down horizontally in his bed, though he got little sleep, and passed a due

8

quantity

quantity of urine, and of the natural colour: no fullneſs or hardneſs could be perceived about the region of the liver; and he had no pain or numbneſs in his arms.

One night he had a moſt profuſe ſweat all over his body and limbs, which quite deluged his bed, and for a day or two ſomewhat relieved his difficulty of breathing, and his pulſe became leſs irregular: this copious ſweat recurred three or four times at the intervals of five or ſix days, and repeatedly alleviated his ſymptoms.

He was directed one large ſpoonful of the above decoction of fox-glove every hour, till it procured ſome conſiderable evacuation: after he had taken it eleven ſucceſſive hours he had a few liquid ſtools, at-tended with a great flow of urine, which laſt had a dark tinge, as if mixed with a few drops of blood: he continued ſick at intervals for two days, but his breath became quite eaſy, and his pulſe quite regu-lar, the ſwelling of his legs diſappeared, and his appetite and ſleep returned.

He then took three grains of white vitriol twice a day, with ſome bitter medicines, and a grain of opium with five grains of rhubarb every night; was adviſed to eat fleſh meat, and ſpice, as his ſtomach would bear it, with ſmall beer, and a few glaſſes of wine; and had iſſues made in his thighs; and has ſuffered no relapſe.

4. A lady, about fifty years of age, had for ſome weeks great dif-ficulty of breathing, with very irregular pulſe, and conſiderable ge-neral debility: ſhe could lie down in bed, and the urine was in due quantity and of the natural colour, and ſhe had no pain or numbneſs of her arms.

She took one large ſpoonful of the above decoction of foxglove every hour, for ten or twelve ſucceſſive hours; was ſick, and made a quantity of pale urine for about two days, and was quite relieved both of the difficulty of breathing, and the irregularity of her pulſe. She then took a grain of opium, and five grains of rhubarb, every

night,

night, for many weeks; with fome flight chalybeate and bitter medicines, and has fuffered no relapfe.

Hydrops Thoracis.

5. A tradefman, about fifty years of age, became weak and fhort of breath, efpecially on increafe of motion, with pain in one arm, about the infertion of the biceps mufcle. He obferved he fometimes in the night made an unufual quantity of pale water. He took calomel, alum, and peruvian bark, and all his fymptoms increafed: his legs began to fwell confiderably; his breath became more difficult, and he could not lie down in bed; but all this time he made a due quantity of ftraw-coloured water.

The decoction of foxglove was given as in the preceding cafes, which operated chiefly by purging, and feemed to relieve his breath for a day or two; but alfo feemed to contribute to weaken him.—He became after fome weeks univerfally dropfical, and died comatous.

6. A young lady of delicate conftitution, with light eyes and hair, and who had perhaps lived too abftemioufly both in refpect to the quantity and quality of what fhe eat and drank, was feized with great difficulty of breathing, fo as to threaten immediate death. Her extremities were quite cold, and her breath felt cold to the back of one's hand. She had no fweat, nor could lie down for a fingle moment; and had previoufly, and at prefent, complained of great weaknefs and pain and numbnefs of both her arms; had no fwelling of her legs, no thirft, water in due quantity and colour. Her fifter, about a year before, was afflicted with fimilar fymptoms, was repeatedly blooded, and died univerfally dropfical.

A grain of opium was given immediately, and repeated every fix hours with evident and amazing advantage; afterwards a blifter, with chalybeates, bitters, and effential oils, were exhibited, but nothing

U u

had

had such eminent effect in relieving the difficulty of breathing and coldness of her extremities as opium, by the use of which in a few weeks she perfectly regained her health, and has suffered no relapse.

Ascites.

7. A young lady of delicate constitution having been exposed to great fear, cold, and fatigue, by the overturn of a chaise in the night, began with pain and tumour in the right hypochondrium: in a few months a fluctuation was felt throughout the whole abdomen, more distinctly perceptible indeed about the region of the stomach; since the integuments of the lower part of the abdomen generally become thickened in this disease by a degree of anasarca. Her legs were not swelled, no thirst, water in due quantity and colour.—She took the foxglove so as to induce sickness and stools, but without abating the swelling, and was obliged at length to submit to the operation of tapping.

8. A man about sixty-seven, who had long been accustomed to spirituous potation, had some time laboured under ascites; his legs somewhat swelled; his breath easy in all attitudes; no appetite; great thirst; urine in exceedingly small quantity, very deep coloured, and turbid; pulse equal. He took the foxglove in such quantity as vomited him, and induced sickness for two days; but procured no flow of urine, or diminution of his swelling; but was thought to leave him considerably weaker.

9. A corpulent man, accustomed to large potation of fermented liquors, had vehement cough, difficult breathing, anasarca of his legs, thighs, and hands, and considerable tumour, with evident fluctuation of his abdomen; his pulse was equal; his urine in small quantity, of deep colour, and turbid. These swellings had been twice consider-

ably

ably abated by draftic cathartics. He took three ounces of a decoction of foxglove (made by boiling one ounce of the fresh leaves in a pint of water) every three hours, for two whole days; it then began to vomit and purge him violently, and promoted a great flow of urine; he was by these evacuations completely emptied in twelve hours. After two or three months all these symptoms returned, and were again relieved by the use of the foxglove; and thus in the space of about three years he was about ten times evacuated, and continued all that time his usual potations: excepting at first, the medicine operated only by urine, and did not appear considerably to weaken him—The last time he took it, it had no effect; and a few weeks afterwards he vomited a great quantity of blood, and expired.

QUERIES.

1. As the first six of these patients had a due discharge of urine, and of the natural colour, was not the seat of the disease confined to some part of the thorax, and the swelling of the legs rather a symptom of the obstructed circulation of the blood, than of a paralysis of the cellular lymphatics of those parts?

2. When the original disease is a general anasarca, do not the cutaneous lymphatics always become paralytic at the same time with the cellular ones, by their greater sympathy with each other? and hence the paucity of urine, and the great thirst, distinguish this kind of dropsy?

3. In the anasarca of the lungs, when the disease is not very great, though the patients have considerable difficulty of breathing at their first lying down, yet after a minute or two their breath becomes easy again; and the same occurs at their first rising. Is not this owing to the time necessary for the fluid in the cells of the lungs to change

its

its place, fo as the leaft to incommode refpiration in the new attitude?

4. In the dropfy of the pericardium does not the patient bear the horizontal or perpendicular attitude with equal eafe? Does this circumftance diftinguifh the dropfy of the pericardium from that of the lungs and of the thorax?

5. Do the univerfal fweats diftinguifh the dropfy of the pericardium, or of the thorax? and thofe, which cover the upper parts of the body only, the anafarca of the lungs?

6. When in the dropfy of the thorax, the patient endeavours to lie down, does not the extravafated fluid comprefs the upper parts of the bronchia, and totally preclude the accefs of air to every part of the lungs; whilft in the perpendicular attitude the inferior parts of the lungs only are compreffed? Does not fomething fimilar to this occur in the anafarca of the lungs, when the difeafe is very great, and thus prevent thofe patients alfo from lying down?

7. As a principal branch of the fourth cervical nerve of the left fide, after having joined a branch of the third and of the fecond cervical nerves, defcending between the fubclavian vein and artery, is received in a groove formed for it in the pericardium, and is obliged to make a confiderable turn outwards to go over the prominent part of it, where the point of the head is lodged, in its courfe to the diaphragm; and as the other phrenic nerve of the right fide has a ftraight courfe to the diaphragm; and as many other confiderable branches of this fourth pair of cervical nerves are fpread on the arms; does not a pain in the left arm diftinguifh a difeafe of the pericardium, as in the angina pectoris, or in the dropfy of the pericardium? and does not a pain or weaknefs in both arms diftinguifh the dropfy of the thorax?

8. Do not the dropfies of the thorax and pericardium frequently exift together, and thus add to the uncertainty and fatality of the difeafe?

9. Might

9. Might not the foxglove be ferviceable in hydrocephalus internus, in hydrocele, and in white fwellings of the joints?

VI. *Of cold Sweats.*

THERE have been hiftories given of chronical immoderate fweatings, which bear fome analogy to the diabetes. Dr. Willis mentions a lady then living, whofe fweats were for many years fo profufe, that all her bed-clothes were not only moiftened, but deluged with them every night; and that many ounces, and fometimes pints, of this fweat, were received in veffels properly placed, as it trickled down her body. He adds, that fhe had great thirft, had taken many medicines, and fubmitted to various rules of life, and changes of climate, but ftill continued to have thefe immoderate fweats. Pharmac. ration. de fudore anglico.

Dr. Willis has alfo obferved, that the fudor anglicanus which appeared in England, in 1483, and continued till 1551, was in fome refpects fimilar to the diabetes; and as Dr. Caius, who faw this difeafe, mentions the vifcidity, as well as the quantity of thefe fweats, and adds, that the extremities were often cold, when the internal parts were burnt up with heat and thirft, with great and fpeedy emaciation and debility: there is great reafon to believe, that the fluids were abforbed from the cells of the body by the cellular and cyftic branches of the lymphatics, and poured on the fkin by the retrograde motions of the cutaneous ones.

Sydenham has recorded, in the ftationary fever of the year 1685, the vifcid fweats flowing from the head, which were probably from the fame fource as thofe in the fweating plague above mentioned.

It is very common in dropfies of the cheft or lungs to have the difficulty of breathing relieved by copious fweats, flowing from the head and neck. Mr. P. about 50 years of age, had for many weeks

been

been afflicted with anasarca of his legs and thighs, attended with difficulty of breathing; and had repeatedly been relieved by squill, other bitters, and chalybeates.—One night the difficulty of breathing became so great, that it was thought he must have expired; but so copious a sweat came out of his head and neck, that in a few hours some pints, by estimation, were wiped off from those parts, and his breath was for a time relieved.   This dyspnœa and these sweats recurred at intervals, and after some weeks he ceased to exist. The skin of his head and neck felt cold to the hand, and appeared pale at the time these sweats flowed so abundantly; which is a proof, that they were produced by an inverted motion of the absorbents of those parts: for sweats, which are the consequence of an increased action of the sanguiferous system, are always attended with a warmth of the skin, greater than is natural, and a more florid colour; as the sweats from exercise, or those that succeed the cold fits of agues. Can any one explain how these partial sweats should relieve the difficulty of breathing in anasarca, but by supposing that the pulmonary branch of absorbents drank up the fluid in the cavity of the thorax, or in the cells of the lungs, and threw it on the skin, by the retrograde motions of the cutaneous branch? for, if we could suppose, that the increased action of the cutaneous glands or capillaries poured upon the skin this fluid, previously absorbed from the lungs; why is not the whole surface of the body covered with sweat? why is not the skin warm? Add to this, that the sweats above mentioned were clammy or glutinous, which the condensed perspirable matter is not; whence it would seem to have been a different fluid from that of common perspiration.

Dr. Dobson, of Liverpool, has given a very ingenious explanation of the acid sweats, which he observed in a diabetic patient—he thinks part of the chyle is secreted by the skin, and afterwards undergoes an acetous fermentation.—Can the chyle get thither, but by an inverted motion of the cutaneous lymphatics? in the same manner as it is car-

ried

ried to the bladder, by the inverted motions of the urinary lymphatics. Medic. Obfervat. and Enq. London, vol. v.

Are not the cold fweats in fome fainting fits, and in dying people, owing to an inverted motion of the cutaneous lymphatics? for in thefe there can be no increafed arterial or glandular action.

Is the difficulty of breathing, arifing from anarfaca of the lungs, relieved by fweats from the head and neck; whilft that difficulty of breathing, which arifes from a dropfy of the thorax, or pericardium, is never attended with thefe fweats of the head? and thence can thefe difeafes be diftinguifhed from each other? Do the periodic returns of nocturnal afthma rife from a temporary dropfy of the lungs, collected during their more torpid ftate in found fleep, and then re-abforbed by the vehement efforts of the difordered organs of refpiration, and carried off by the copious fweats about the head and neck?

More extenfive and accurate diffections of the lymphatic fyftem are wanting to enable us to unravel thefe knots of fcience.

VII. *Tranflations of Matter, of Chyle, of Milk, of Urine. Operation. of purging Drugs applied externally.*

1. Tʜᴇ tranflations of matter from one part of the body to another, can only receive an explanation from the doctrine of the occafional retrograde motions of fome branches of the lymphatic fyftem: for how can matter, abforbed and mixed with the whole mafs of blood, be fo haftily collected again in any one part? and is it not an immutable law, in animal bodies, that each gland can fecrete no other, but its own proper fluid? which is, in part, fabricated in the very gland by an animal procefs, which it there undergoes: of thefe purulent tranflations innumerable and very remarkable inftances are recorded.

2. The chyle, which is feen among the materials thrown up by violent

lent

lent vomiting, or in purging ftools, can only come thither by its having been poured into the bowels by the inverted motions of the lacteals: for our aliment is not converted into chyle in the ftomach or inteftines by a chemical procefs, but is made in the very mouths of the lacteals; or in the mefenteric glands; in the fame manner as other fecreted fluids are made by an animal procefs in their adapted glands.

Here a curious phænomenon in the exhibition of mercury is worth explaining:—If a moderate dofe of calomel, as fix or ten grains, be fwallowed, and within one or two days a cathartic is given, a falivation is prevented: but after three or four days, a falivation having come on, repeated purges every day, for a week or two, are required to eliminate the mercury from the conftitution. For this acrid metallic preparation, being abforbed by the mouths of the lacteals, continues, for a time arrefted by the mefenteric glands, (as the variolous or venereal poifons fwell the fubaxillar or inguinal glands): which, during the operation of a cathartic, is returned into the inteftines by the inverted action of the lacteals, and thus carried out of the fyftem.

Hence we underftand the ufe of vomits or purges, to thofe who have fwallowed either contagious or poifonous materials, even though exhibited a day or even two days after fuch accidents; namely, that by the retrograde motions of the lacteals and lymphatics, the material ftill arrefted in the mefenteric, or other glands, may be eliminated from the body.

3. Many inftances of milk and chyle found in ulcers are given by Haller, El. Phyfiol. t. vii. p. 12, 23, which admit of no other explanation than by fuppofing, that the chyle, imbibed by one branch of the abforbent fyftem, was carried to the ulcer, by the inverted motions of another branch of the fame fyftem.

4. Mrs. P. on the fecond day after delivery, was feized with a violent purging, in which, though opiates, mucilages, the bark, and teftacea were profufely ufed, continued many days, till at length fhe recovered. During the time of this purging, no milk could be drawn

from

from her breafts ; but the ftools appeared like the curd of milk broken
into fmall pieces.   In this cafe, was not the milk taken up from the
follicles of the pectoral glands, and thrown on the inteftines, by a retrogreffion of the inteftinal abforbents ? for how can we for a moment
fufpect that the mucous glands of the inteftines could feparate pure
milk from the blood ? Doctor Smelly has obferved, that loofe ftools,
mixed with milk, which is curdled in the inteftines, frequently relieves the turgefcency of the breafts of thofe who ftudioufly repel
their milk.   Cafes in Midwifery, 43, No. 2. 1.

    5. J. F. Meckel obferved in a patient, whofe urine was in fmall
quantity and high coloured, that a copious fweat under the arm-pits,
of a perfectly urinous fmell, ftained the linen ; which ceafed again
when the ufual quantity of urine was difcharged by the urethra. Here
we muft believe from analogy, that the urine was firft fecreted in the
kidneys, then re-abforbed by the increafed action of the urinary lymphatics, and laftly carried to the axillæ by the retrograde motions of
the lymphatic branches of thofe parts.   As in the jaundice it is neceffary, that the bile fhould firft be fecreted by the liver, and re-abforbed into the circulation, to produce the yellownefs of the fkin ; as
was formerly demonftrated by the late Dr. Monro, (Edin. Medical
Effays) and if in this patient the urine had been re-abforbed into the
mafs of blood, as the bile in the jaundice, why was it not detected in
other parts of the body, as well as in the arm-pits ?

    6. Cathartic and vermifuge medicines applied externally to the ab-
domen, feem to be taken up by the cutaneous branch of lymphatics,
and poured on the inteftines by the retrograde motions of the lacteals,
without having paffed the circulation.

    For when the draftic purges are taken by the mouth, they excite
the lacteals of the inteftines into retrograde motions, as appears from
the chyle, which is found coagulated among the fæces, as was fhewn
above, (fect 2 and 4.)   And as the cutaneous lymphatics are joined
with the lacteals of the inteftines, by frequent anaftomofes ; it would

                              X x   be

be more extraordinary, when a ſtrong purging drug, abſorbed by the
ſkin, is carried to the anaſtomoſing branches of the lacteals unchanged,
if it ſhould not excite them into retrograde action as efficaciouſly, as
if it was taken by the mouth, and mixed with the food of the ſtomach.

VIII. *Circumſtances by which the Fluids, that are effuſed by the retrograde Motions of the abſorbent Veſſels, are diſtinguiſhed.*

1. WE frequently obſerve an unuſual quantity of mucus or other
fluids in ſome diſeaſes, although the action of the glands, by which
thoſe fluids are ſeparated from the blood, is not unuſually increaſed;
but when the power of abſorption alone is diminiſhed. Thus the ca-
tarrhal humour from the noſtrils of ſome, who ride in froſty weather;
and the tears, which run down the cheeks of thoſe, who have an obſtruction of the puncta lacrymalia; and the ichor of thoſe phagedenic
ulcers, which are not attended with inflammation, are all inſtances of
this circumſtance.

Theſe fluids however are eaſily diſtinguiſhed from others by their
abounding in ammoniacal or muriatic ſalts; whence they inflame the
circumjacent ſkin: thus in the catarrh the upper lip becomes red and
ſwelled from the acrimony of the mucus, and patients complain of
the ſaltneſs of its taſte. The eyes and cheeks are red with the corro-
ſive tears, and the ichor of ſome herpetic eruptions erodes far and wide
the contiguous parts, and is pungently ſalt to the taſte, as ſome patients have informed me.

Whilſt, on the contrary, thoſe fluids, which are effuſed by the retrograde action of the lymphatics, are for the moſt part mild and in-
nocent; as water, chyle, and the natural mucus: or they take their
properties from the materials previouſly abſorbed, as in the coloured

or

or vinous urine, or that fcented with afparagus, defcribed before.

2. Whenever the fecretion of any fluid is increafed, there is at the fame time an increafed heat in the part; for the fecreted fluid, as the bile, did not previoufly exift in the mafs of blood, but a new combination is produced in the gland.   Now as folutions are attended with cold, fo combinations are attended with heat; and it is probable the fum of the heat given out by all the fecreted fluids of animal bodies may be the caufe of their general heat above that of the atmofphere.

Hence the fluids derived from increafed fecretions are readily diftinguifhed from thofe originating from the retrograde motions of the lymphatics: thus an increafe of heat either in the difeafed parts, or diffufed over the whole body, is perceptible, when copious bilious ftools are confequent to an inflamed liver; or a copious mucous falivation from the inflammatory angina.

3. When any fecreted fluid is produced in an unufual quantity, and at the fame time the power of abforption is increafed in equal proportion, not only the heat of the gland becomes more intenfe, but the fecreted fluid becomes thicker and milder, its thinner and faline parts being re-abforbed : and thefe are diftinguifhable both by their greater confiftence, and by their heat, from the fluids, which are effufed by the retrograde motions of the lymphatics ; as is obfervable towards the termination of gonorrhœa, catarrh, chincough, and in thofe ulcers, which are faid to abound with laudable pus.

4. When chyle is obferved in ftools, or among the materials ejected by vomit, we may be confident it muft have been brought thither by the retrograde motions of the lacteals ; for chyle does not previoufly exift amid the contents of the inteftines, but is made in the very mouths of the lacteals, as was before explained.

5. When chyle, milk, or other extraneous fluids are found in the urinary bladder, or in any other excretory receptacle of a gland ; no one can for a moment believe, that thefe have been collected from

X x 2

the

the mafs of blood by a morbid fecretion, as it contradicts all analogy.

> ————Aurea duræ
> Mala ferant quercus? Narcifco floreat alnus?
> Pinguia corticibus fudent electra myricæ?
>
> VIRGIL.

IX. *Retrograde Motions of Vegetable Juices.*

THERE are befides fome motions of the fap in vegetables, which
bear analogy to our prefent fubject; and as the vegetable tribes are by
many philofophers held to be inferior animals, it may be a matter of
curiofity at leaft to obferve, that their abforbent veffels feem evidently,
at times, to be capable of a retrograde motion. Mr. Perault cut off
a forked branch of a tree, with the leaves on; and inverting one of
the forks into a veffel of water, obferved, that the leaves on the other
branch continued green much longer than thofe of a fimilar branch,
cut off from the fame tree; which fhews, that the water from the
veffel was carried up one part of the forked branch, by the retrograde
motion of its veffels. and fupplied nutriment fome time to the other
part of the branch, which was out of the water. And the celebrated
Dr. Hales found, by numerous very accurate experiments, that the
fap of trees rofe upwards during the warmer hours of the day, and in
part defcended again during the cooler ones. Vegetable Statics.

It is well known that the branches of willows, and of many other
trees, will either take root in the earth or engraft on other trees,
fo as to have their natural direction inverted, and yet flourifh with
vigour.

Dr. Hope has alfo made this pleafing experiment, after the manner
of Hales—he has placed a forked branch, cut from one tree, erect between two others; then cutting off a part of the bark from one fork

                         applied

applied it to a fimilar branch of one of the trees in its vicinity; and the fame of the other fork; fo that a tree is feen to grow fufpended in the air, between two other trees; which fupply their fofter friend with due nourifhment.

Miranturque novas frondes, et non fua poma.

All thefe experiments clearly evince, that the juices of vegetables can occafionally pafs either upwards or downwards in their abforbent fyftem of veffels.

X. Objections anfwered.

The following experiment, at firft view, would feem to invalidate this opinion of the retrograde motions of the lymphatic veffels, in fome difeafes.

About a gallon of milk having been given to an hungry fwine, he was fuffered to live about an hour, and was then killed by a ftroke or two on his head with an axe.—On opening his belly the lacteals were well feen filled with chyle; on irritating many of the branches of them with a knife, they did not appear to empty themfelves haftily; but they did however carry forwards their contents in a little time.

I then paffed a ligature round feveral branches of lacteals, and irritated them much with a knife beneath the ligature, but could not make them regurgitate their contained fluid into the bowels.

I am not indeed certain, that the nerve was not at the fame time included in the ligature, and thus the lymphatic rendered unirritable or lifelefs; but this however is certain, that it is not any quantity of any ftimulus, which induces the veffels of animal bodies to revert their motions; but a certain quantity of a certain ftimulus, as appears from wounds in the ftomach, which do not produce vomiting; and wounds of the inteftines, which do not produce the cholera morbus.

At

At Nottingham, a few years ago, two fhoemakers quarrelled, and one of them with a knife, which they ufe in their occupation, ftabbed his companion about the region of the ftomach. On opening the abdomen of the wounded man after his death the food and medicines he had taken were in part found in the cavity of the belly, on the outfide of the bowels; and there was a wound about half an inch long at the bottom of the ftomach; which I fuppofe was diftended with liquor and food at the time of the accident; and thence was more liable to be injured at its bottom: but during the whole time he lived, which was about ten days, he had no efforts to vomit, nor ever even complained of being fick at the ftomach! Other cafes fimilar to this are mentioned in the philofophical tranfactions.

Thus, if you vellicate the throat with a feather, naufea is produced; if you wound it with a penknife, pain is induced, but not ficknefs. So if the foles of the feet of children or their armpits are tickled, convulfive laughter is excited, which ceafes the moment the hand is applied, fo as to rub them more forcibly.

The experiment therefore above related upon the lacteals of a dead pig, which were included in a ftrict ligature, proves nothing; as it is not the quantity, but the kind of ftimulus, which excites the lymphatic veffels into retrograde motion.

XI. *The Caufes which induce the retrograde Motions of animal Veffels; and the Medicines by which the natural Motions are reftored.*

1. Such is the conftruction of animal bodies, that all their parts, which are fubjected to lefs ftimuli than nature defigned, perform their functions with lefs accuracy: thus, when too watery or too acefcent food is taken into the ftomach, indigeftion, and flatulency, and heartburn fucceed.

2. Another law of irritation, connate with our exiftence, is, that
all

all thofe parts of the body, which have previoufly been expofed to too great a quantity of fuch ftimuli, as ftrongly affect them, become for fome time afterwards difobedient to the natural quantity of their adapted ftimuli.—Thus the eye is incapable of feeing objects in an obfcure room, though the iris is quite dilated, after having been expofed to the meridian fun.

3. There is a third law of irritation, that all the parts of our bodies, which have been lately fubjected to lefs ftimulus, than they have been accuftomed to, when they are expofed to their ufual quantity of ftimulus, are excited into more energetic motions : thus when we come from a dufky cavern into the glare of daylight, our eyes are dazzled; and after emerging from the cold bath, the fkin becomes warm and red.

4. There is a fourth law of irritation, that all the parts of our bodies, which are fubjected to ftill ftronger ftimuli for a length of time, become torpid, and refufe to obey even thefe ftronger ftimuli; and thence do their offices very imperfectly.—Thus, if any one looks earneftly for fome minutes on an area, an inch diameter, of red filk, placed on a fheet of white paper, the image of the filk will gradually become pale, and at length totally vanifh.

5. Nor is it the nerves of fenfe alone, as the optic and auditory nerves, that thus become torpid, when the ftimulus is withdrawn or their irritability decreafed; but the motive mufcles, when they are deprived of their natural ftimuli, or of their irritability, become torpid and paralytic; as is feen in the tremulous hand of the drunkard in a morning; and in the awkward ftep of age.

The hollow mufcles alfo, of which the various veffels of the body are conftructed, when they are deprived of their natural ftimuli, or of their due degree of irritability, not only become tremulous, as the arterial pulfations of dying people; but alfo frequently invert their motions, as in vomiting, in hyfteric fuffocations, and diabetes above defcribed.

I muft

I muſt beg your patient attention, for a few moments whilſt I endeavour to explain, how the retrograde actions of our hollow muſcles are the conſequence of their debility; as the tremulous actions of the ſolid muſcles are the conſequence of their debility. When, through fatigue, a muſcle can act no longer; the antagoniſt muſcles, either by their inanimate elaſticity, or by their animal action, draw the limb into a contrary direction: in the ſolid muſcles, as thoſe of locomotion, their actions are aſſociated in tribes, which have been accuſtomed to ſynchronous action only; hence when they are fatigued, only a ſingle contrary effort takes place; which is either tremulous, when the fatigued muſcles are again immediately brought into action; or it is a pandiculation, or ſtretching, where they are not immediately again brought into action.

Now the motions of the hollow muſcles, as they in general propel a fluid along their cavities, are aſſociated in trains, which have been accuſtomed to ſucceſſive actions: hence when one ring of ſuch a muſcle is fatigued from its too great debility, and is brought into retrograde action, the next ring from its aſſociation falls ſucceſſively into retrograde action; and ſo on throughout the whole canal. See Sect. XXV. 6.

6. But as the retrograde motions of the ſtomach, œſophagus, and fauces in vomiting are, as it were, apparent to the eye; we ſhall conſider this operation more minutely, that the ſimilar operations in the more recondite parts of our ſyſtem may be eaſier underſtood.

From certain nauſeous ideas of the mind, from an ungrateful taſte in the mouth, or from fœtid ſmells, vomiting is ſometimes inſtantly excited; or even from a ſtroke on the head, or from the vibratory motions of a ſhip; all which originate from aſſociation, or ſympathy. See Sect. XX. on Vertigo.

But when the ſtomach is ſubjected to a leſs ſtimulus than is natural, according to the firſt law of irritation mentioned above, its motions become diſturbed, as in hunger; firſt pain is produced,

then

then ficknefs, and at length vain efforts to vomit, as many authors inform us.

But when a great quantity of wine, or of opium, is fwallowed, the retrograde motions of the ftomach do not occur till after feveral minutes, or even hours; for when the power of fo ftrong a ftimulus ceafes, according to the fecond law of irritation, mentioned above, the periftaltic motions become tremulous, and at length retrograde; as is well known to the drunkard, who on the next morning has ficknefs and vomitings.

When a ftill greater quantity of wine, or of opium, or when naufeous vegetables, or ftrong bitters, or metallic falts, are taken into the ftomach, they quickly induce vomiting; though all thefe in lefs dofes excite the ftomach into more energetic action, and ftrengthen the digeftion; as the flowers of chamomile, and the vitriol of zinc: for, according to the fourth law of irritation, the ftomach will not long be obedient to a ftimulus fo much greater than is natural; but its action becomes firft tremulous and then retrograde.

7. When the motions of any veffels become retrograde, lefs heat of the body is produced; for in paroxyfms of vomiting, of hyfteric affections, of diabetes, of afthma, the extremities of the body are cold: hence we may conclude, that thefe fymptoms arife from the debility of the parts in action; for an increafe of mufcular action is always attended with increafe of heat.

8. But as animal debility is owing to defect of ftimulus, or to defect of irritability, as fhewn above, the method of cure is eafily deduced: when the vafcular mufcles are not excited into their due action by the natural ftimuli, we fhould exhibit thofe medicines, which poffefs a ftill greater degree of ftimulus; amongft thefe are the fœtids, the volatiles, aromatics, bitters, metallic falts, opiates, wine, which indeed fhould be given in fmall dofes, and frequently repeated. To thefe fhould be added conftant, but moderate exercife, cheerfulnefs of mind, and change of country to a warmer climate; and perhaps occafionally the external ftimulus of blifters.

Y y

It

It is alſo frequently uſeful to diminiſh the quantity of natural ſtimulus for a ſhort time, by which afterwards the irritability of the ſyſtem becomes increaſed; according to the third law of irritation above-mentioned, hence the uſe of baths ſomewhat colder than animal heat, and of equitation in the open air.

The catalogue of diſeaſes owing to the retrograde motions of lymphatics is here omitted, as it will appear in the ſecond volume of this work. The following is the concluſion to this theſis of Mr. CHARLES DARWIN.

THUS have I endeavoured in a conciſe manner to explain the numerous diſeaſes, which deduce their origin from the inverted motions of the hollow muſcles of our bodies: and it is probable, that Saint Vitus's dance, and the ſtammering of ſpeech, originate from a ſimilar inverted order of the aſſociated motions of ſome of the ſolid muſcles; which, as it is foreign to my preſent purpoſe, I ſhall not here diſcuſs.

I beg, illuſtrious profeſſors, and ingenious fellow-ſtudents, that you will recollect how difficult a taſk I have attempted, to evince the retrograde motions of the lymphatic veſſels, when the veſſels themſelves for ſo many ages eſcaped the eyes and glaſſes of philoſophers: and if you are not yet convinced of the truth of this theory, hold, I entreat you, your minds in ſuſpenſe, till ANATOMY draws her ſword with happier omens, cuts aſunder the knots, which entangle PHYSIOLOGY; and, like an augur inſpecting the immolated victim, announces to mankind the wiſdom of HEAVEN.

SECT.

SECT. XXX.

PARALYSIS OF THE LIVER AND KIDNEYS.

1. Bile-ducts less irritable after having been stimulated much. 2. Jaundice from paralysis of the bile ducts cured by electric shocks. 3. From bile-stones. Experiments on bile-stones. Oil vomit. 4. Palsy of the liver, two cases. 5. Schirrosity of the liver. 6. Large livers of geese. II. Paralysis of the kidneys. III. Story of Prometheus.

1. FROM the ingurgitation of spirituous liquors into the stomach and duodenum, the termination of the common bile-duct in that bowel becomes stimulated into unnatural action, and a greater quantity of bile is produced from all the secretory vessels of the liver, by the association of their motions with those of their excretory ducts; as has been explained in Section XXIV. and XXV. but as all parts of the body, that have been affected with stronger stimuli for any length of time, become less susceptible of motion, from their natural weaker stimuli, it follows, that the motions of the secretory vessels, and in consequence the secretion of bile, is less than is natural during the intervals of sobriety. 2. If this ingurgitation of spirituous liquors has been daily continued in considerable quantity, and is then suddenly intermitted, a languor or paralysis of the common bile-duct is induced; the bile is prevented from being poured into the intestines; and as the bilious absorbents are stimulated into stronger action by its accumulation, and by the acrimony or viscidity, which it acquires by

Y y 2

delay,

delay, it is abforbed, and carried to the receptacle of the chyle; or otherwife the fecretory veffels of the liver, by the above-mentioned ftimulus, invert their motions, and regurgitate their contents into the blood, as fometimes happens to the tears in the lachrymal fack, fee Sect. XXIV. 2. 7. and one kind of jaundice is brought on.

There is reafon to believe, that the bile is moft frequently returned into the circulation by the inverted motions of thefe hepatic glands, for the bile does not feem liable to be abforbed by the lymphatics, for it foaks through the gall-ducts, and is frequently found in the cellular membrane. This kind of jaundice is not generally attended with pain, neither at the extremity of the bile-duct, where it enters the duodenum, nor on the region of the gall-bladder.

Mr. S. a gentleman between 40 and 50 years of age, had had the jaundice about fix weeks, without pain, ficknefs, or fever; and had taken emetics, cathartics, mercurials, bitters, chalybeates, effential oil, and ether, without apparent advantage. On a fuppofition that the obftruction of the bile might be owing to the paralyfis, or torpid action of the common bile-duct, and the ftimulants taken into the ftomach feeming to have no effect, I directed half a fcore fmart electric fhocks from a coated bottle, which held about a quart, to be paffed through the liver, and along the courfe of the common gall-duct, as near as could be gueffed, and on that very day the ftools became yellow; he continued the electric fhocks a few days more, and his fkin gradually became clear.

3. The bilious vomiting and purging, that affects fome people by intervals of a few weeks, is a lefs degree of this difeafe; the bile-duct is lefs irritable than natural, and hence the bile becomes accumulated in the gall-bladder, and hepatic ducts, till by its quantity, acrimony or vifcidity, a greater degree of irritation is produced, and it is fuddenly evacuated, or laftly from the abforption of the more liquid parts of the bile, the remainder becomes infpiffated, and chryftallizes into

4

maffes

maffes too large to pafs, and forms another kind of jaundice, where the bile-duct is not quite paralytic, or has regained its irritability.

This difeafe is attended with much pain, which at firft is felt at the pit of the ftomach, exactly in the centre of the body, where the bile-duct enters the duodenum; afterwards, when the fize of the bile-ftones increafe, it is alfo felt on the right fide, where the gall-bladder is fituated. The former pain at the pit of the ftomach recurs by intervals, as the bile-ftone is pufhed againft the neck of the duct; like the paroxyfms of the ftone in the urinary bladder, the other is a more dull and conftant pain.

Where thefe bile-ftones are too large to pafs, and the bile-ducts poffefs their fenfibility, this becomes a very painful and hopelefs difeafe. I made the following experiments with a view to their chemical folution.

Some fragments of the fame bile-ftone were put into the weak fpirit of marine falt, which is fold in the fhops, and into folution of mild alcali; and into a folution of cauftic alcali; and into oil of turpentine; without their being diffolved. All thefe mixtures were after fome time put into a heat of boiling water, and then the oil of turpentine diffolved its fragments of bile-ftone, but no alteration was produced upon thofe in the other liquids except fome change of their colour.

Some fragments of the fame bile-ftone were put into vitriolic æther, and were quickly diffolved without additional heat. Might not æther mixed with yolk of egg or with honey be given advantageoufly in bilious concretions?

I have in two inftances feen from 30 to 50 bile-ftones come away by ftool, about the fize of large peafe, after having given fix grains of calomel in the evening, and four ounces of oil of almonds or olives on the fucceeding morning. I have alfo given half a pint of good olive or almond oil as an emetic during the painful fit, and

repeated

repeated it in half an hour, if the firſt did not operate, with frequent good effect.

4. Another diſeaſe of the liver, which I have ſeveral times obſerved, conſiſts in the inability or paralyſis of the ſecretory veſſels. This diſeaſe has generally the ſame cauſe as the preceding one, the too frequent potation of ſpirituous liquors, or the too ſudden omiſſion of them, after the habit is confined; and is greater or leſs in proportion, as the whole or a part of the liver is affected, and as the inability or paralyſis is more or leſs complete.

This palſy of the liver is known from theſe ſymptoms, the patients have generally paſſed the meridian of life, have drank fermented liquors daily, but perhaps not been opprobrious drunkards; they loſe their appetite, then their fleſh and ſtrength diminiſh in conſequence, there appears no bile in their ſtools, nor in their urine, nor is any hardneſs or ſwelling perceptible on the region of the liver. But what is peculiar to this diſeaſe, and diſtinguiſhes it from all others at the firſt glance of the eye, is the bombycinous colour of the ſkin, which, like that of full-grown ſilkworms, has a degree of tranſparency with a yellow tint not greater than is natural to the ſerum of the blood.

Mr. C. and Mr. B. both very ſtrong men, between 50 and 60 years of age, who had drank ale at their meals inſtead of ſmall beer, but were not reputed hard-drinkers, ſuddenly became weak, loſt their appetite, fleſh, and ſtrength, with all the ſymptoms above enumerated, and died in about two months from the beginning of their malady. Mr. C. became anaſarcous a few days before his death, and Mr. B. had frequent and great hæmorrhages from an iſſue, and ſome parts of his mouth, a few days before his death. In both theſe caſes calomel, bitters and chalybeates were repeatedly uſed without effect.

One of the patients deſcribed above, Mr. C, was by trade a plumber; both of them could digeſt no food, and died apparently for

want

want of blood.   Might not the transfusion of blood be used in thefe
cafes with advantage?

5. When the paralyfis of the hepatic glands is lefs complete, or lefs
univerfal, a fchirrofity of fome part of the liver is induced; for the
fecretory veffels retaining fome of their living power take up a fluid
from the circulation, without being fufficiently irritable to carry it
forwards to their excretory duets; hence the body, or receptacle of
each gland, becomes inflated, and this diftenfion increafes, till by its
very great ftimulus inflammation is produced, or till thofe parts of
the vifcus become totally paralytic.   This difeafe is diftinguifhable
from the foregoing by the palpable hardnefs or largenefs of the liver;
and as the hepatic glands are not totally paralytic, or the whole liver
not affected, fome bilé continues to be made.   The inflammations of
this vifcus, confequent to the fchirrofity of it, belong to the difeafes
of the fenfitive motions, and will be treated of hereafter.

6. The ancients are faid to have poffeffed an art of increafing the
livers of geefe to a fize greater than the remainder of the goofe.
Martial. l. 13. epig. 58.—This is faid to have been done by fat and
figs. Horace, l. 2. fat. 8.—Juvenal fets thefe large livers before an
epicure as a great rarity. Sat. 5. l. 114; and Perfius, fat. 6. l. 71.
Pliny fays thefe large goofe-livers were foaked in mulled milk, that
is, I fuppofe, milk mixed with honey and wine; and adds, " that it
is uncertain whether Scipio Metellus, of confular dignity, or M.
Seftius, a Roman knight, was the great difcoverer of this excellent
difh." A modern traveller, I believe Mr. Brydone, afferts that the
art of enlarging the livers of geefe ftill exifts in Sicily; and it is to
be lamented that he did not import it into his native country, as fome
method of affecting the human liver might perhaps have been collected from it; befides the honour he might have acquired in improving our giblet pies.

Our wifer caupones, I am told, know how to fatten their fowls,
as well as their geefe, for the London markets, by mixing gin inftead

of

of figs and fat with their food; by which they are said to become sleepy, and to fatten apace, and probably acquire enlarged livers; as the swine are asserted to do, which are fed on the sediments of barrels in the distilleries; and which so frequently obtains in those, who ingurgitate much ale, or wine, or drams.

II. The irritative diseases of the kidneys, pancreas, spleen, and other glands, are analogous to those of the liver above described, differing only in the consequences attending their inability to action. For instance, when the secretory vessels of the kidneys become disobedient to the stimulus of the passing current of blood, no urine is separated or produced by them; their excretory mouths become filled with concreted mucus, or calculus matter, and in eight or ten days stupor and death supervenes in consequence of the retention of the feculent part of the blood.

This disease in a slighter degree, or when only a part of the kidney is affected, is succeeded by partial inflammation of the kidney in consequence of previous torpor. In that case greater actions of the secretory vessels occur, and the nucleus of gravel is formed by the inflamed mucous membranes of the tubuli uriniferi, as farther explained in its place.

This torpor, or paralysis of the secretory vessels of the kidneys, like that of the liver, owes its origin to their being previously habituated to too great stimulus; which in this country is generally owing to the alcohol contained in ale or wine; and hence must be registered amongst the diseases owing to inebriety; though it may be caused by whatever occasionally inflames the kidney; as too violent riding on horseback, or the cold from a damp bed, or by sleeping on the cold ground; or perhaps by drinking in general too little aqueous fluids.

III. I shall conclude this section on the diseases of the liver induced by spirituous liquors, with the well-known story of Prometheus, which seems indeed to have been invented by physicians in those

ancient

ancient times, when all things were clothed in hieroglyphic, or in fable. Prometheus was painted as ftealing fire from heaven, which might well reprefent the inflammable fpirit produced by fermentation; which may be faid to animate or enliven the man of clay: whence the conquefts of Bacchus, as well as the temporary mirth and noife of his devotees. But the after punifhment of thofe, who fteal this ac-curfed fire, is a vulture gnawing the liver; and well allegorifes the poor inebriate lingering for years under painful hepatic dif-eafes. When the expediency of laying a further tax on the diftilla-tion of fpirituous liquors from grain was canvaffed before the Houfe of Commons fome years ago, it was faid of the diftillers, with great truth, " *They take the bread from the people, and convert it into poifon!*" Yet is this manufactory of difeafe permitted to continue, as appears by its paying into the treafury above 9ᱍ0,000*l.* near a million of money annually. And thus, under the names of rum, brandy, gin, whifky, ufquebaugh, wine, cyder, beer, and porter, alcohol is become the bane of the Chriftian world, as opium of the Mahometan.

> Evoe ! parce, liber ?
> Parce, gravi metuende thirfo !
>
> Hor.

S E C T.   XXXI.

OF TEMPERAMENTS.

I. *The temperament of decreased irritability known by weak pulse, large pupils of the eyes, cold extremities. Are generally supposed to be too irritable. Bear pain better than labour. Natives of North-America contrasted with those upon the coast of Africa. Narrow and broad-shouldered people. Irritable constitutions bear labour better than pain.* II. *Temperament of increased sensibility. Liable to intoxication, to inflammation, hæmontoe, gutta serena, enthusiasm, delirium, reverie. These constitutions are indolent to voluntary exertions, and dull to irritations. The natives of South-America, and brute animals of this temperament.* III. *Of increased voluntarity; these are subject to locked jaw, convulsions, epilepsy, mania. Are very active, bear cold, hunger, fatigue. Are suited to great exertions. This temperament distinguishes mankind from other animals.* IV. *Of increased association. These have great memories, are liable to quartan agues, and stronger sympathics of parts with each other.* V. *Change of temperaments into one another.*

ANTIENT writers have spoken much of temperaments, but without sufficient precision. By temperament of the system should be meant a permanent predisposition to certain classes of diseases: without this definition a temporary predisposition to every distinct malady might be termed a temperament. There are four kinds of constitution, which permanently deviate from good health, and are perhaps sufficiently marked to be distinguished from each other, and constitute the temperaments or predispositions to the irritative, sensitive, voluntary, and associate classes of diseases.

I. *The*

I. *The Temperament of decreafed Irritability.*

Tʜᴇ difeafes, which are caufed by irritation, moft frequently originate from the defect of it ; for thofe, which are immediately owing
to the excefs of it, as the hot fits of fever, are generally occafioned
by an accumulation of fenforial power in confequence of a previous
defect of irritation, as in the preceding cold fits of fever. Whereas
the difeafes, which are caufed by fenfation and volition, moft frequently originate from the excefs of thofe fenforial powers, as will
be explained below.

The temperament of decreafed irritability appears from the following circumftances, which fhew that the mufcular fibres or organs of
fenfe are liable to become torpid or quiefcent from lefs defect of ftimulation than is productive of torpor or quiefcence in other conftitutions.

1. The firft is the weak pulfe, which in fome conftitutions is at
the fame time quick. 2. The next moft marked criterion of this temperament is the largenefs of the aperture of the iris, or pupil of the
eye, which has been reckoned by fome a beautiful feature in the female countenance, as an indication of delicacy, but to an experienced
obferver it is an indication of debility, and is therefore a defect, not
an excellence. The third moft marked circumftance in this conftitution is, that the extremities, as the hands and feet, or nofe and
ears, are liable to become cold and pale in fituations in refpect to
warmth, where thofe of greater ftrength are not affected. Thofe
of this temperament are fubject to hyfteric affections, nervous fevers,
hydrocephalus, fcrophula, and confumption, and to all other difeafes
of debility.

Thofe, who poffefs this kind of conftitution, are popularly fuppofed to be more irritable than is natural, but are in reality lefs fo.

Z z 2   This

This miftake has arifen from their generally having a greater quick-nefs of pulfe, as explained in Sect. XII. 1. 4. XII. 3. 3.; but this frequency of pulfe is not neceffary to the temperament, like the debility of it.

Perfons of this temperament are frequently found amongft the fofter fex, and amongft narrow-fhouldered men; who are faid to bear labour worfe, and pain better than others. This laft circumftance is fuppofed to have prevented the natives of North America from having been made flaves of by the Europeans. They are a narrow-fhouldered race of people, and will rather expire under the lafh, than be made to labour. Some nations of Afia have fmall hands, as may be feen by the handles of their fcymetars; which with their narrow fhoulders fhew, that they have not been accuftomed to fo great labour with their hands and arms, as the European nations in agriculture, and thofe on the coafts of Africa in fwimming and rowing. Dr. Maningham, a popular accoucheur in the beginning of this century, obferves in his aphorifms, that broad-fhouldered men procreate broad-fhouldered children. Now as labour ftrengthens the mufcles employed, and increafes their bulk, it would feem that a few generations of labour or of indolence may in this refpect change the form and temperament of the body.

On the contrary, thofe who are happily poffeffed of a great degree of irritability, bear labour better than pain; and are ftrong, active, and ingenious. But there is not properly a temperament of increafed irritability tending to difeafe, becaufe an increafed quantity of irritative motions generally induces an increafe of pleafure or pain, as in intoxication, or inflammation; and then the new motions are the immediate confequences of increafed fenfation, not of increafed irritation; which have hence been fo perpetually confounded with each other.

II. *Temperament*

II. *Temperament of Senfibility.*

THERE is not properly a temperament, or predifpofition to difeafe, from decreafed fenfibility, fince irritability and not fenfibility is immediately neceffary to bodily health.   Hence it is the excefs of fenfation alone, as it is the defect of irritation, that moft frequently produces difeafe.   This temperament of increafed fenfibility is known from the increafed activity of all thofe motions of the organs of fenfe and mufcles, which are exerted in confequence of pleafure or pain, as in the beginning of drunkennefs, and in inflammatory fever.   Hence thofe of this conftitution are liable to inflammatory difeafes, as hepatitis; and to that kind of confumption which is hereditary, and commences with flight repeated hœmoptoe.   They have high-coloured lips, frequently dark hair and dark eyes with large pupils, and are in that cafe fubject to gutta ferena.   They are liable to enthufiafm, delirium, and reverie.   In this laft circumftance they are liable to ftart at the clapping of a door; becaufe the more intent any one is on the paffing current of his ideas, the greater furprife he experiences on their being diffevered by fome external violence, as explained in Sect. XIX. on reverie.

As in thefe conftitutions more than the natural quantities of fenfitive motions are produced by the increafed quantity of fenfation exifting in the habit, it follows, that the irritative motions will be performed in fome degree with lefs energy, owing to the great expenditure of fenforial power on the fenfitive ones.   Hence thofe of this temperament do not attend to flight ftimulations, as explained in Sect. XIX.   But when a ftimulus is fo great as to excite fenfation, it produces greater fenfitive actions of the fyftem than in others; fuch as delirium or inflammation.   Hence they are liable to be abfent in company; fit or lie long in one pofture; and in winter have the fkin

of

of their legs burnt into various colours by the fire.   Hence alſo they are fearful of pain; covet muſic and ſleep; and delight in poetry and romance.

As the motions in conſequence of ſenſation are more than natural, it alſo happens from the greater expenditure of ſenſorial power on them, that the voluntary motions are leſs eaſily exerted.   Hence the ſubjects of this temperament are indolent in reſpect to all voluntary exertions, whether of mind or body.

A race of people of this deſcription ſeems to have been found by the Spaniards in the iſlands of America, where they firſt landed, ten of whom are ſaid not to have conſumed more food than one Spaniard, nor to have been capable of more than one tenth of the exertion of a Spaniard.   Robertſon's Hiſtory.—In a ſtate ſimilar to this the greateſt part of the animal world paſs their lives, between ſleep or inactive reverie, except when they are excited by the call of hunger.

III. *The Temperament of increaſed Voluntarity.*

THOSE of this conſtitution differ from both the laſt mentioned in this, that the pain, which gradually ſubſides in the firſt, and is productive of inflammation or delirium in the ſecond, is in this ſucceded by the exertion of the muſcles or ideas, which are moſt frequently connected with volition; and they are thence ſubject to locked jaw, convulſions, epilepſy, and mania, as explained in Sect. XXXIV.   Thoſe of this temperament attend to the ſlighteſt irritations or ſenſations, and immediately exert themſelves to obtain or avoid the objects of them; they can at the ſame time bear cold and hunger better than others, of which Charles the Twelfth of Sweden was an inſtance. They are ſuited and generally prompted to all great exertions of genius or labour, as their deſires are more extenſive and more vehement, and their powers of attention and of labour greater.   It is this facility

8   of

of voluntary exertion, which diſtinguiſhes men from brutes, and which has made them lords of the creation.

IV. *The Temperament of increaſed Aſſociation.*

THIS conſtitution conſiſts in the too great facility, with which the fibrous motions acquire habits of aſſociation, and by which theſe aſ-ſociations become proportionably ſtronger than in thoſe of the other temperaments.   Thoſe of this temperament are ſlow in voluntary exertions, or in thoſe dependent on ſenſation, or on irritation.   Hence great memories have been ſaid to be attended with leſs ſenſe and leſs imagination from Ariſtotle down to the preſent time ; for by the word memory theſe writers only underſtood the unmeaning repetition of words or numbers in the order they were received, without any vo-luntary efforts of the mind.

In this temperament thoſe aſſociations of motions, which are com-monly termed ſympathies, act with greater certainty and energy, as thoſe between diſturbed viſion and the inverſion of the motion of the ſtomach, as in ſea-ſickneſs ; and the pains in the ſhoulder from hepatic inflammation.   Add to this, that the catenated circles of actions are of greater extent than in the other conſtitutions.   Thus if a ſtrong vo-mit or cathartic be exhibited in this temperament, a ſmaller quantity will produce as great an effect, if it be given ſome weeks afterwards ; whereas in other temperaments this is only to be expected, if it be exhibited in a few days after the firſt doſe.   Hence quartan agues are formed in thoſe of this temperament, as explained in Section XXXII. on diſeaſes from irritation, and other intermittents are liable to recur from ſlight cauſes many weeks after they have been cured by the bark.

V. The

V. The firſt of theſe temperaments differs from the ſtandard of health from defect, and the others from exceſs of ſenſorial power; but it ſometimes happens that the ſame individual, from the changes introduced into his habit by the different ſeaſons of the year, modes or periods of life, or by accidental diſeaſes, paſſes from one of theſe temperaments to another. Thus a long uſe of too much fermented liquor produces the temperament of increaſed ſenſibility; great indolence and ſolitude that of decreaſed irritability; and want of the neceſſaries of life that of increaſed voluntarity.

SECT.

SECT. XXXII.

DISEASES OF IRRITATION.

I. *Irritative fevers with strong pulse. With weak pulse. Symptoms of fever. Their source.* II. 1. *Quick pulse is owing to decreased irritability.* 2. *Not in sleep or in apoplexy.* 3. *From inanition. Owing to deficiency of sensorial power.* III. 1. *Causes of fever. From defect of heat. Heat from secretions. Pain of cold in the loins and forehead.* 2. *Great expense of sensorial power in the vital motions. Immersion in cold water. Succeeding glow of heat. Difficult respiration in cold bathing explained. Why the cold bath invigorates. Bracing and relaxation are mechanical terms.* 3. *Uses of cold bathing. Uses of cold air in fevers.* 4. *Ague fits from cold air. Whence their periodical returns.* IV. *Defect of distention a cause of fever. Deficiency of blood. Transfusion of blood.* V. 1. *Defect of momentum of the blood from mechanic stimuli.* 2. *Air injected into the blood-vessels.* 3. *Exercise increases the momentum of the blood.* 4. *Sometimes bleeding increases the momentum of it.* VI. *Influence of the sun and moon on diseases. The chemical stimulus of the blood. Menstruation obeys the lunations. Queries.* VII. *Quiescence of large glands a cause of fever. Swelling of the præcordia.* VIII. *Other causes of quiescence, as hunger, bad air, fear, anxiety.* IX. 1. *Symptoms of the cold fit.* 2. *Of the hot fit.* 3. *Second cold fit why.* 4. *Inflammation introduced, or delirium, or stupor.* X. *Recapitulation. Fever not an effort of nature to relieve herself. Doctrine of spasm.*

I. WHEN the contractile sides of the heart and arteries perform a greater number of pulsations in a given time, and move through a greater area at each pulsation, whether these motions are occasioned by the stimulus of the acrimony or quantity of the blood, or by their association with other irritative motions, or by the increased irritability

3 A   of

of the arterial fyftem, that is, by an increafed quantity of fenforial power, one kind of fever is produced; which may be called Synocha irritativa, or Febris irritativa pulfu forti, or irritative fever with ftrong pulfe.

When the contractile fides of the heart and arteries perform a greater number of pulfations in a given time, but move through a much lefs area at each pulfation, whether thefe motions are occafioned by defect of their natural ftimuli, or by the defect of other irritative motions with which they are affociated, or from the inirritability of the arterial fyftem, that is, from a decreafed quantity of fenforial power, another kind of fever arifes; which may be termed, Typhus irritativus, or Febris irritativa pulfu debili, or irritative fever with weak pulfe. The former of thefe fevers is the fynocha of nofologifts, and the latter the typhus mitior, or nervous fever. In the former there appears to be an increafe of fenforial power, in the latter a deficiency of it; which is fhewn to be the immediate caufe of ftrength and weaknefs, as defined in Sect. XII. 1. 3.

It fhould be added, that a temporary quantity of ftrength or debility may be induced by the defect or excefs of ftimulus above what is natural; and that in the fame fever *debility always exifts during the cold fit, though ftrength does not always exift during the hot fit.*

Thefe fevers are always connected with, and generally induced by, the difordered irritative motions of the organs of fenfe, or of the inteftinal canal, or of the glandular fyftem, or of the abforbent fyftem; and hence are always complicated with fome or many of thefe difordered motions, which are termed the fymptoms of the fever, and which compofe the great variety in thefe difeafes.

The irritative fevers both with ftrong and with weak pulfe, as well as the fenfitive fevers with ftrong and with weak pulfe, which are to be defcribed in the next fection, are liable to periodical remiffions, and then they take the name of intermittent fevers, and are diftinguifhed by the periodical times of their accefs.

II. For

II. For the better illuftration of the phenomena of irritative fevers we muft refer the reader to the circumftances of irritation explained in Sect. XII. and fhall commence this intricate fubject by fpeaking of the quick pulfe, and proceed by confidering many of the caufes, which either feparately or in combination moft frequently produce the cold fits of fevers.

1. If the arteries are dilated but to half their ufual diameters, though they contract twice as frequently in a given time, they will circulate only half their ufual quantity of blood; for as they are cylinders, the blood which they contain muft be as the fquares of their diameters. Hence when the pulfe becomes quicker and fmaller in the fame proportion, the heart and arteries act with lefs energy than in their natural ftate. See Sect. XII. 1. 4.

That this quick fmall pulfe is owing to want of irritability, appears, firft, becaufe it attends other fymptoms of want of irritability; and, fecondly, becaufe on the application of a ftimulus greater than ufual, it becomes flower and larger. Thus in cold fits of agues, in hyfteric palpitations of the heart, and when the body is much exhaufted by hæmorrhages, or by fatigue, as well as in nervous fevers, the pulfe becomes quick and fmall; and fecondly, in all thofe cafes if an increafe of ftimulus be added, by giving a little wine or opium; the quick fmall pulfe becomes flower and larger, as any one may eafily experience on himfelf, by counting his pulfe after drinking one or two glaffes of wine, when he is faint from hunger or fatigue.

Now nothing can fo ftrongly evince that this quick fmall pulfe is owing to defect of irritability, than that an additional ftimulus, above what is natural, makes it become flower and larger immediately: for what is meant by a defect of irritability, but that the arteries and heart are not excited into their ufual exertions by their ufual quantity of ftimulus? but if you increafe the quantity of ftimulus, and they immediately act with their ufual energy, this proves their previous want of their natural degree of irritability. Thus the trembling

3 A 2

hands

hands of drunkards in a morning become steady, and acquire strength to perform their usual offices, by the accustomed stimulus of a glass or two of brandy.

2. In sleep and in apoplexy the pulse becomes slower, which is not owing to defect of irritability, for it is at the same time larger; and thence the quantity of the circulation is rather increased than diminished. In these cases the organs of sense are closed, and the voluntary power is suspended, while the motions dependent on internal irritations, as those of digestion and secretion, are carried on with more than their usual vigour; which has led superficial observers to confound these cases with those arising from want of irritability. Thus if you lift up the eyelid of an apoplectic patient, who is not actually dying, the iris will, as usual, contract itself, as this motion is associated with the stimulus of light; but it is not so in the last stages of nervous fevers, where the pupil of the eye continues expanded in the broad day-light: in the former case there is a want of voluntary power, in the latter a want of irritability.

Hence also those constitutions which are deficient in quantity of irritability, and which possess too great sensibility, as during the pain of hunger, of hysteric spasms, or nervous headachs, are generally supposed to have too much irritability; and opium, which in its due dose is a most powerful stimulant, is erroneously called a sedative; because by increasing the irritative motions it decreases the pains arising from defect of them.

Why the pulse should become quicker both from an increase of irritation, as in the synocha irritativa, or irritative fever with strong pulse; and from the decrease of it, as in the typhus irritativus, or irritative fever with weak pulse; seems paradoxical. The former circumstance needs no illustration; since if the stimulus of the blood, or the irritability of the sanguiferous system be increased, and the strength of the patient not diminished, it is plain that the motions must be performed quicker and stronger.

In

In the latter circumftance the weaknefs of the mufcular power of the heart is foon over-balanced by the elafticity of the coats of the arteries, which they poffefs befides a mufcular power of contraction; and hence the arteries are diftended to lefs than their ufual diameters. The heart being thus ftopped when it is but half emptied, begins fooner to dilate again; and the arteries being dilated to lefs than their ufual diameters, begin fo much fooner to contract themfelves; infomuch, that in the laft ftages of fevers with weaknefs the frequency of pulfation of the heart and arteries becomes doubled; which, however, is never the cafe in fevers with ftrength, in which they feldom exceed 118 or 120 pulfations in a minute. It muft be added, that in thefe cafes, while the pulfe is very fmall and very quick, the heart often feels large, and labouring to one's hand; which coincides with the above explanation, fhewing that it does not completely empty itfelf.

3. In cafes however of debility from paucity of blood, as in animals which are bleeding to death in the flaughter-houfe, the quick pulfations of the heart and arteries may be owing to their not being diftended to more than half their ufual diaftole; and in confequence they muft contract fooner, or more frequently, in a given time. As weak people are liable to a deficient quantity of blood, this caufe may occafionally contribute to quicken the pulfe in fevers with debility, which may be known by applying one's hand upon the heart as above; but the principal caufe I fuppofe to confift in the diminution of fenforial power. When a mufcle contains, or is fupplied with but little fenforial power, its contraction foon ceafes, and in confequence may foon recur, as is feen in the trembling hands of people weakened by age or by drunkennefs. See Sect. XII. 1. 4. XII. 3. 4.

It may neverthelefs frequently happen, that both the deficiency of ftimulus, as where the quantity of blood is leffened (as defcribed in No. 4. of this fection), and the deficiency of fenforial power, as in

5   thofe

thofe of the temperament of inirritability, defcribed in Sect. XXXI.
occur at the fame time ; which will thus add to the quicknefs of the
pulfe and to the danger of the difeafe.

III.   1. A certain degree of heat is neceffary to mufcular motion,
and is, in confequence, effential to life.   This is obferved in thofe animals and infects which pafs the cold feafon in a torpid ftate, and which
revive on being warmed by the fire.   This neceffary ftimulus of heat
has two fources ; one from the fluid atmofphere of heat, in which all
things are immerfed, and the other from the internal combinations of
the particles, which form the various fluids, which are produced in
the extenfive fyftems of the glands.   When either the external heat,
which furrounds us, or the internal production of it, becomes leffened
to a certain degree, the pain of cold is perceived.

This pain of cold is experienced moft fenfibly by our teeth, when
ice is held in the mouth ; or by our whole fyftem after having been
previoufly accuftomed to much warmth.   It is probable, that this
pain does not arife from the mechanical or chemical effects of a deficiency of heat ; but that, like the organs of fenfe by which we perceive hunger and thirft, this fenfe of heat fuffers pain, when the ftimulus of its object is wanting to excite the irritative motions of the organ ; that is, when the fenforial power becomes too much accumulated in the quiefcent fibres.   See Sect. XII. 5. 3.   For as the periftaltic motions of the ftomach are leffened, when the pain of hunger
is great, fo the action of the cutaneous capillaries are leffened during
the pain of cold ; as appears by the palenefs of the fkin, as explained
in Sect. XIV. 6. on the production of ideas.

The pain in the fmall of the back and forehead in the cold fits of
the ague, in nervous hemicrania, and in hyfteric paroxyfms, when all
the irritative motions are much impaired, feems to arife from this
caufe ; the veffels of thefe membranes or mufcles become torpid by
their irritative affociations with other parts of the body, and thence

7   produce

produce lefs of their accuftomed fecretions, and in confequence lefs heat is evolved, and they experience the pain of cold; which coldnefs may often be felt by the hand applied upon the affected part.

2. The importance of a greater or lefs deduction of heat from the fyftem will be more eafy to comprehend, if we firft confider the great expenfe of fenforial power ufed in carrying on the vital motions; that is, which circulates, abforbs, fecretes, aerates, and elaborates the whole mafs of fluids with unceafing affiduity. The fenforial power, or fpirit of animation, ufed in giving perpetual and ftrong motion to the heart, which overcomes the elafticity and vis inertiæ of the whole arterial fyftem; next the expenfe of fenforial power in moving with great force and velocity the innumerable trunks and ramifications of the arterial fyftem; the expenfe of fenforial power in circulating the whole mafs of blood through the long and intricate intortions of the very fine veffels, which compofe the glands and capillaries; then the expenfe of fenforial power in the exertions of the abforbent extremities of all the lacteals, and of all the lymphatics, which open their mouths on the external furface of the fkin, and on the internal furfaces of every cell or interftice of the body; then the expenfe of fenforial power in the venous abforption, by which the blood is received from the capillary veffels, or glands, where the arterial power ceafes, and is drank up, and returned to the heart; next the expenfe of fenforial power ufed by the mufcles of refpiration in their office of perpetually expanding the bronchia, or air-veffels, of the lungs; and laftly in the unceafing periftaltic motions of the ftomach and whole fyftem of inteftines, and in all the fecretions of bile, gaftric juice, mucus, perfpirable matter, and the various excretions from the fyftem. If we confider the ceafelefs expenfe of fenforial power thus perpetually employed, it will appear to be much greater in a day than all the voluntary exertions of our mufcles and organs of fenfe confume in a week; and all this without any fenfible fatigue! Now, if but a part of thefe vital motions are impeded, or totally ftopped for but a fhort time, we

gain

gain an idea, that there muſt be a great accumulation of ſenſorial power ; as its production in theſe organs, which are ſubject to perpetual activity, is continued during their quieſcence, and is in conſequence accumulated.

While, on the contrary, where thoſe vital organs act too forcibly by increaſe of ſtimulus without a proportionally-increaſed production of ſenſorial power in the brain, it is evident, that a great deficiency of action, that is torpor, muſt ſoon follow, as in fevers ; whereas the locomotive muſcles, which act only by intervals, are neither liable to ſo great accumulation of ſenſorial power during their times of inactivity, nor to ſo great an exhauſtion of it during their times of action.

Thus, on going into a very cold bath, ſuppoſe at 33 degrees of heat on Fahrenheit's ſcale, the action of the ſubcutaneous capillaries, or glands, and of the mouths of the cutaneous abſorbents is diminiſhed, or ceaſes for a time. Hence leſs or no blood paſſes theſe capillaries, and paleneſs ſucceeds. But ſoon after emerging from the bath, a more florid colour and a greater degree of heat is generated on the ſkin than was poſſeſſed before immerſion ; for the capillary glands, after this quieſcent ſtate, occaſioned by the want of ſtimulus, become more irritable than uſual to their natural ſtimuli, owing to the accumulation of ſenſorial power, and hence a greater quantity of blood is tranſmitted through them, and a greater ſecretion of perſpirable matter ; and, in conſequence, a greater degree of heat ſucceeds. During the continuance in cold water the breath is cold, and the act of reſpiration quick and laborious ; which have generally been aſcribed to the obſtruction of the circulating fluid by a ſpaſm of the cutaneous veſſels, and by a conſequent accumulation of blood in the lungs, occaſioned by the preſſure as well as by the coldneſs of the water. This is not a ſatisfactory account of this curious phænomenon, ſince at this time the whole circulation is leſs, as appears from the ſmallneſs of the pulſe and coldneſs of the breath ; which ſhew that leſs blood paſſes through the lungs in a given time ; the ſame laborious breathing
imme-

immediately occurs when the palenefs of the fkin is produced by fear, where no external cold or preffure are applied.

The minute veffels of the bronchia, through which the blood paffes from the arterial to the venal fyftem, and which correfpond with the cutaneous capillaries, have frequently been expofed to cold air, and become quiefcent along with thofe of the fkin; and hence their motions are fo affociated together, that when one is affected either with quiefcence or exertion, the other fympathizes with it, according to the laws of irritative affociation.   See Sect. XXVII. 1. on hæmorrhages.

Befides the quiefcence of the minute veffels of the lungs, there are many other fyftems of veffels which become torpid from their irritative affociations with thofe of the fkin, as the abforbents of the bladder and inteftines; whence an evacuation of pale urine occurs, when the naked fkin is expofed only to the coldnefs of the atmofphere; and fprinkling the naked body with cold water is known to remove even pertinacious conftipation of the bowels.   From the quiefcence of fuch extenfive fyftems of veffels as the glands and capillaries of the fkin, and the minute veffels of the lungs, with their various abforbent feries of veffels, a great accumulation of fenforial powers is occafioned; part of which is again expended in the increafed exertion of all thefe veffels, with an univerfal glow of heat in confequence of this exertion, and the remainder of it adds vigour to both the vital and voluntary exertions of the whole day.

If the activity of the fubcutaneous veffels, and of thofe with which their actions are affociated, was too great before cold immerfion, as in the hot days of fummer, and by that means the fenforial power was previoufly diminifhed, we fee the caufe why the cold bath gives fuch prefent ftrength; namely, by ftopping the unneceffary activity of the fubcutaneous veffels, and thus preventing the too great exhauftion of fenforial power; which, in metaphorical language, has been called *bracing* the fyftem: which is, however, a mechanical term, only applicable

to

to drums, or mufical ftrings: as on the contrary the word *relaxation*, when applied to living animal bodies, can only mean too fmall a quantity of ftimulus, or too fmall a quantity of fenforial power; as explained in Sect. XII. 1.

3. This experiment of cold bathing prefents us with a fimple fever-fit; for the pulfe is weak, fmall, and quick during the cold immerfion; and becomes ftrong, full, and quick during the fubfequent glow of heat; till in a few minutes thefe fymptoms fubfide, and the temporary fever ceafes.

In thofe conftitutions where the degree of inirritability, or of debility, is greater than natural, the coldnefs and palenefs of the fkin with the quick and weak pulfe continue a long time after the patient leaves the bath; and the fubfequent heat approaches by unequal flufhings, and he feels himfelf difordered for many hours. Hence the bathing in a cold fpring of water, where the heat is but forty-eight degrees on Fahrenheit's thermometer, much difagrees with thofe of weak or inirritable habits of body; who poffefs fo little fenforial power, that they cannot without injury bear to have it diminifhed even for a fhort time; but who can neverthelefs bear the more temperate coldnefs of Buxton bath, which is about eighty degrees of heat, and which ftrengthens them, and makes them by habit lefs liable to great quiefcence from fmall variations of cold, and thence lefs liable to be difordered by the unavoidable accidents of life. Hence it appears, why people of thefe inirritable conftitutions, which is another expreffion for fenforial deficiency, are often much injured by bathing in a cold fpring of water; and why they fhould continue but a very fhort time in baths, which are colder than their bodies; and fhould gradually increafe both the degree of coldnefs of the water, and the time of their continuance in it, if they would obtain falutary effects from cold immerfions. See Sect. XII. 2. 1.

On the other hand in all cafes where the heat of the external furface of the body, or of the internal furface of the lungs, is greater than

7

natural,

natural, the ufe of expofure to cool air may be deduced.   In fever-
fits attended with ftrength, that is with great quantity of fenforial
power, it removes the additional ftimulus of heat from the furfaces
above mentioned, and thus prevents their excefs of ufelefs motion;
and in fever-fits attended with debility, that is with a deficiency of
the quantity of fenforial power, it prevents the great and dangerous
wafte of fenforial power expended in the unneceffary increafe of the
actions of the glands and capillaries of the fkin and lungs.

4. In the fame manner, when any one is long expofed to very cold
air, a quiefcence is produced of the cutaneous and pulmonary capillaries and abforbents, owing to the deficiency of their ufual ftimulus of heat; and this quiefcence of fo great a quantity of veffels affects, by irritative affociation, the whole abforbent and glandular fyftem, which becomes in a greater or lefs degree quiefcent, and a cold
fit of fever is produced.

If the deficiency of the ftimulus of heat is very great, the quiefcence becomes fo general as to extinguifh life, as in thofe who are frozen to death.

If the deficiency of heat be in lefs degree, but yet fo great as in
fome meafure to diforder the fyftem, and fhould occur the fucceeding
day, it will induce a greater degree of quiefcence than before, from
its acting in concurrence with the period of the diurnal circle of actions, explained in Sect. XXXVI.   Hence from a fmall beginning a
greater and greater degree of quiefcence may be induced, till a complete fever-fit is formed; and which will continue to recur at the periods by which it was produced.   See Sect. XVII. 3. 6.

If the degree of quiefcence occafioned by defect of the ftimulus of
heat be very great, it will recur a fecond time by a flighter caufe, than
that which firft induced it.   If the caufe, which induces the fecond
fit of quiefcence, recurs the fucceeding day, the quotidian fever is produced; if not till the alternate day, the tertian fever; and if not till
after feventy-two hours from the firft fit of quiefcence, the quartan

fever

fever is formed.   This laft kind of fever recurs lefs frequently than the other, as it is a difeafe only of thofe of the temperament of affociability, as mentioned in Sect. XXXI.; for in other conftitutions the capability of forming a habit ceafes, before the new caufe of quiefcence is again applied, if that does not occur fooner than in feventy-two hours.

And hence thofe fevers, whofe caufe is from cold air of the night or morning, are more liable to obferve the folar day in their periods; while thofe from other caufes frequently obferve the lunar day in their periods, their paroxyfms returning near an hour later every day, as explained in Sect. XXXVI.

IV. Another frequent caufe of the cold fits of fever is the defect of the ftimulus of diftention.   The whole arterial fyftem would appear, by the experiments of Haller, to be irritable by no other ftimulus, and the motions of the heart and alimentary canal are certainly in fome meafure dependant on the fame caufe. See Sect. XIV. 7.   Hence there can be no wonder, that the diminution of diftention fhould frequently induce the quiefcence, which conftitutes the beginning of fever-fits.

Monfieur Leiutaud has judicioufly mentioned the deficiency of the quantity of blood amongft the caufes of difeafes, which he fays is frequently evident in diffections : fevers are hence brought on by great hæmorrhages, diarrhœas, or other evacuations ; or from the continued ufe of diet, which contains but little nourifhment ; or from the exhauftion occafioned by violent fatigue, or by thofe chronic difeafes in which the digeftion is much impaired ; as where the ftomach has been long affected with the gout or fchirrus ; or in the paralyfis of the liver, as defcribed in Sect. XXX.   Hence a paroxyfm of gout is liable to recur on bleeding or purging ; as the torpor of fome vifcus, which precedes the inflammation of the foot, is thus induced by the want of the ftimulus of diftention.   And hence the extremities of the body, as the nofe and fingers, are more liable to become cold,

when

when we have long abftained from food ; and hence the pulfe is in-creafed both in ftrength and velocity above the natural ftandard after a full meal by the ftimulus of diftention.

However, this ftimulus of diftention, like the ftimulus of heat above defcribed, though it contributes much to the due action not only of the heart, arteries, and alimentary canal, but feems neceffary to the proper fecretion of all the various glands ; yet perhaps it is not the fole caufe of any of thefe numerous motions : for as the lacteals, cu-taneous abforbents, and the various glands appear to be ftimulated into action by the peculiar pungency of the fluids they abforb, fo in the inteftinal canal the pungency of the digefting aliment, or the acri-mony of the fæces, feem to contribute, as well as their bulk, to pro-mote the periftaltic motions ; and in the arterial fyftem, the mo-mentum of the particles of the circulating blood, and their acrimony, ftimulate the arteries, as well as the diftention occafioned by it. Where the pulfe is fmall this defect of diftention is prefent, and con-tributes much to produce the febris irritativa pulfu debili, or irritative fever with weak pulfe, called by modern writers nervous fever, as a predifponent caufe.   See Sect. XII. 1. 4.   Might not the transfufion of blood, fuppofe of four ounces daily from a ftrong man, or other healthful animal, as a fheep or an afs, be ufed in the early ftate of ner-vous or putrid fevers with great profpect of fuccefs ?

V. The defect of the momentum of the particles of the circu-lating blood is another caufe of the quiefcence, with which the cold fits of fever commence.   This ftimulus of the momentum of the progreffive particles of the blood does not act over the whole body like thofe of heat and diftention above defcribed, but is confined to the arterial fyftem ; and differs from the ftimulus of the diftention of the blood, as much as the vibration of the air does from the currents of it.   Thus are the different organs of our bodies ftimulated by four different mechanic properties of the external world : the fenfe of touch by the preffure of folid bodies fo as to diftinguifh their figure ;

the

the mufcular fyftem by the diftention, which they occafion ; the internal furface of the arteries, by the momentum of their moving particles ; and the auditory nerves, by the vibration of them : and thefe four mechanic properties are as different from each other as the various chemical ones, which are adapted to the numerous glands, and to the other organs of fenfe.

2. The momentum of the progreffive particles of blood is compounded of their velocity and their quantity of matter : hence whatever circumftances diminifh either of thefe without proportionally increafing the other, and without fuperadding either of the general ftimuli of heat or diftention, will tend to produce a quiefcence of the arterial fyftem, and from thence of all the other irritative motions, which are connected with it.

Hence in all thofe conftitutions or difeafes where the blood contains a greater proportion of ferum, which is the lighteft part of its compofition, the pulfations of the arteries are weaker, as in nervous fevers, chlorofis, and hyfteric complaints ; for in thefe cafes the momentum of the progreffive particles of blood is lefs : and hence, where the denfer parts of its compofition abound, as the red part of it, or the coagulable lymph, the arterial pulfations are ftronger ; as in thofe of robuft health, and in inflammatory difeafes.

That this ftimulus of the momentum of the particles of the circulating fluid is of the greateft confequence to the arterial action, appears from the experiment of injecting air into the blood veffels, which feems to deftroy animal life from the want of this ftimulus of momentum ; for the diftention of the arteries is not diminifhed by it, it poffeffes no corrofive acrimony, and is lefs liable to repafs the valves than the blood itfelf ; fince air-valves in all machinery require much lefs accuracy of conftruction than thofe which are oppofed to water.

3. One method of increafing the velocity of the blood, and in confequence the momentum of its particles, is by the exercife of the body, or by the friction of its furface : fo, on the contrary, too great

indolence

indolence contributes to decreafe this ftimulus of the momentum of the particles of the circulating blood, and thus tends to induce quiefcence ; as is feen in hyfteric cafes, and chlorofis, and the other difeafes of fedentary people.

4. The velocity of the particles of the blood in certain circumftances is increafed by venefection, which, by removing a part of it, diminifhes the refiftance to the motion of the other part, and hence the momentum of the particles of it is increafed. This may be eafily underftood by confidering it in the extreme, fince, if the refiftance was greatly increafed, fo as to overcome the propelling power, there could be no velocity, and in confequence no momentum at all. From this circumftance arifes that curious phænomenon, the truth of which I have been more than once witnefs to, that venefection will often inftantaneoufly relieve thofe nervous pains, which attend the cold periods of hyfteric, afthmatic, or epileptic difeafes ; and that even where large dofes of opium have been in vain exhibited. In thefe cafes the pulfe becomes ftronger after the bleeding, and the extremities regain their natural warmth ; and an opiate then given acts with much more certain effect.

VI. There is another caufe, which feems occafionally to induce quiefcence into fome part of our fyftem, I mean the influence of the fun and moon ; the attraction of thefe luminaries, by decreafing the gravity of the particles of the blood, cannot affect their momentum, as their vis inertiæ remains the fame ; but it may neverthelefs produce fome chemical change in them, becaufe whatever affects the general attractions of the particles of matter may be fuppofed from analogy to affect their fpecific attractions or affinities : and thus the ftimulus of the particles of blood may be diminifhed, though not their momentum. As the tides of the fea obey the fouthing and northing of the moon (allowing for the time neceffary for their motion, and the obftructions of the fhores), it is probable, that there are alfo atmofpheric tides on both fides of the earth, which to the inhabitants of another

ther

ther planet might fo deflect the light as to refemble the ring of Saturn. Now as thefe tides of water, or of air, are raifed by the diminution of their gravity, it follows, that their preffure on the furface of the earth is no greater than the preffure of the other parts of the ocean, or of the atmofphere, where no fuch tides exift; and therefore that they cannot affect the mercury in the barometer. In the fame manner, the gravity of all other terreftrial bodies is diminifhed at the times of the fouthing and northing of the moon, and that in a greater degree when this coincides with the fouthing and northing of the fun, and this in a ftill greater degree about the times of the equinoxes. This decreafe of the gravity of all bodies during the time the moon paffes our zenith or nadir might poffibly be fhewn by the flower vibrations of a pendulum, compared with a fpring clock, or with aftronomical obfervation. Since a pendulum of a certain length moves flower at the line than near the poles, becaufe the gravity being diminifhed and the vis inertiæ continuing the fame, the motive power is lefs, but the refiftance to be overcome continues the fame. The combined powers of the lunar and folar attraction is eftimated by Sir Ifaac Newton not to exceed one 7,868,850th part of the power of gravitation, which feems indeed but a fmall circumftance to produce any confiderable effect on the weight of fublunary bodies, and yet this is fufficient to raife the tides at the equator above ten feet high; and if it be confidered, what fmall impulfes of other bodies produce their effects on the organs of fenfe adapted to the perception of them, as of vibration on the auditory nerves, we fhall ceafe to be furprifed, that fo minute a diminution in the gravity of the particles of blood fhould fo far affect their chemical changes, or their ftimulating quality, as, joined with other caufes, fometimes to produce the beginnings of difeafes.

Add to this, that if the lunar influence produces a very fmall degree of quiefcence at firft, and if that recurs at certain periods even with lefs power to produce quiefcence than at firft, yet the quiefcence will

daily

daily increafe by the acquired habit acting at the fame time, till at length fo great a degree of quiefcence is induced as to produce phrenfy, canine madnefs, epilepfy, hyfteric pains, or cold fits of fever, inftances of many of which are to be found in Dr. Mead's work on this fubject.   The folar influence alfo appears daily in feveral difeafes; but as darknefs, filence, fleep, and our periodical meals mark the parts of the folar circle of actions, it is fometimes dubious to which of thefe the periodical returns of thefe difeafes are to be afcribed.

As far as I have been able to obferve, the periods of inflammatory difeafes obferve the folar day; as the gout and rheumatifm have their greateft quiefcence about noon and midnight, and their exacerbations fome hours after; as they have more frequently their immediate caufe from cold air, inanition, or fatigue, than from the effects of lunations: whilft the cold fits of hyfteric patients, and thofe in nervous fevers, more frequently occur twice a day, later by near half an hour each time, according to the lunar day; whilft fome fits of intermittents, which are undifturbed by medicines, return at regular folar periods, and others at lunar ones; which may, probably, be owing to the difference of the periods of thofe external circumftances of cold, inanition, or lunation, which immediately caufed them.

We muft, however, obferve, that the periods of quiefcence and exacerbation in difeafes do not always commence at the times of the fyzygies or quadratures of the moon and fun, or at the times of their paffing the zenith or nadir; but as it is probable, that the ftimulus of the particles of the circumfluent blood is gradually diminifhed from the time of the quadratures to that of the fyzygies, the quiefcence may commence at any hour, when, co-operating with other caufes of quiefcence, it becomes great enough to produce a difeafe: afterwards it will continue to recur at the fame period of the lunar or folar influence; the fame caufe operating conjointly with the acquired habit, that is with the catenation of this new motion with the diffevered links of the lunar or folar circles of animal action.

3 C   In

In this manner the periods of menſtruation obey the lunar month with great exactneſs in healthy patients (and perhaps the venereal orgaſm in brute animals does the ſame), yet theſe periods do not commence either at the ſyzygies or quadratures of the lunations, but at whatever time of the lunar periods they begin, they obſerve the ſame in their returns till ſome greater cauſe diſturbs them.

Hence, though the beſt way to calculate the time of the expected returns of the paroxyſms of periodical diſeaſes is to count the number of hours between the commencement of the two preceding fits, yet the following obſervations may be worth attending to, when we endeavour to prevent the returns of maniacal or epileptic diſeaſes; whoſe periods (at the beginning of them eſpecially) frequently obſerve the ſyzygies of the moon and ſun, and particularly about the equinox.

The greateſt of the two tides happening in every revolution of the moon, is that when the moon approaches neareſt to the zenith or nadir; for this reaſon, while the ſun is in the northern ſigns, that is during the vernal and ſummer months, the greater of the two diurnal tides in our latitude is that, when the moon is above the horizon; and when the ſun is in the ſouthern ſigns, or during the autumnal and winter months, the greater tide is that, which ariſes when the moon is below the horizon: and as the ſun approaches ſomewhat nearer the earth in winter than in ſummer, the greateſt equinoctial tides are obſerved to be a little before the vernal equinox, and a little after the autumnal one.

Do not the cold periods of lunar diſeaſes commence a few hours before the ſouthing of the moon during the vernal and ſummer months, and before the northing of the moon during the autumnal and winter months? Do not palſies and apoplexies, which occur about the equinoxes, happen a few days before the vernal equinoctial lunation, and after the autumnal one? Are not the periods of thoſe diurnal diſeaſes more obſtinate, that commence many hours before the ſouthing or northing of the moon, than of thoſe which commence at thoſe times? Are not thoſe palſies and apoplexies more dangerous which

3　　　　　　　　　　　　　　　　　　　　　　commence

commence many days before the fyzygies of the moon, than thofe which happen at thofe times ?   See Sect. XXXVI. on the periods of difeafes.

VII. Another very frequent caufe of the cold fit of fever is the quiefcence of fome of thofe large congeries of glands, which compofe the liver, fpleen, or pancreas; one or more of which are frequently fo enlarged in the autumnal intermittents as to be perceptible to the touch externally, and are called by the vulgar ague-cakes.   As thefe glands are ftimulated into action by the fpecific pungency of the fluids, which they abforb, the general caufe of their quiefcence feems to be the too great infipidity of the fluids of the body, co-operating perhaps at the fame time with other general caufes of quiefcence.

Hence, in marfhy countries at cold feafons, which have fucceeded hot ones, and amongft thofe, who have lived on innutritious and un-ftimulating diet, thefe agues are moft frequent.   The enlargement of thefe quiefcent vifcera, and the fwelling of the præcordia in many other fevers, is, moft probably, owing to the fame caufe ; which may confift in a general deficiency of the production of fenforial power, as well as in the diminifhed ftimulation of the fluids; and when the quiefcence of fo great a number of glands, as conftitute one of thofe large vifcera, commences, all the other irritative motions are affected by their connection with it, and the cold fit of fever is produced.

VIII. There are many other caufes, which produce quiefcence of fome part of the animal fyftem, as fatigue, hunger, thirft, bad diet, difappointed love, unwholefome air, exhauftion from evacuations, and many others ; but the laft caufe, that we fhall mention, as frequently productive of cold fits of fever, is fear or anxiety of mind. The pains, which we are firft and moft generally acquainted with, have been produced by defect of fome ftimulus ; thus, foon after our nativity we become acquainted with the pain from the coldnefs of the air, from the want of refpiration, and from the want of food.   Now all thefe pains occafioned by defect of ftimulus are attended with quiefcence

3 C 2   of

of the organ, and at the fame time with a greater or lefs degree of quiefcence of other parts of the fyftem : thus, if we even endure the pain of hunger fo as to mifs one meal inftead of our daily habit of re-pletion, not only the periftaltic motions of the ftomach and bowels are diminifhed, but we are more liable to coldnefs of our extremities, as of our nofes, and ears, and feet, than at other times.

Now, as fear is originally excited by our having experienced pain, and is itfelf a painful affection, the fame quiefcence of other fibrous motions accompany it, as have been moft frequently connected with this kind of pain, as explained in Sect. XVI. 8. 1. as the coldnefs and palenefs of the fkin, trembling, difficult refpiration, indigeftion, and other fymptoms, which contribute to form the cold fit of fevers. Anxiety is fear continued through a longer time, and, by producing chronical torpor of the fyftem, extinguifhes life flowly, by what is commonly termed a broken heart.

IX. 1. We now ftep forwards to confider the other fymptoms in confequence of the quiefcence which begins the fits of fever. If by any of the circumftances before defcribed, or by two or more of them acting at the fame time, a great degree of quiefcence is in-duced on any confiderable part of the circle of irritative motions, the whole clafs of them is more or lefs difturbed by their irritative affocia-tions. If this torpor be occafioned by a deficient fupply of fenforial power, and happens to any of thofe parts of the fyftem, which are ac-cuftomed to perpetual activity, as the vital motions, the torpor in-creafes rapidly, becaufe of the great expenditure of fenforial power by the inceffant activity of thofe parts of the fyftem, as fhewn in No. 3. 2. of this Section. Hence a deficiency of all the fecretions fucceeds, and as animal heat is produced in proportion to the quantity of thofe fecre-tions, the coldnefs of the fkin is the firft circumftance, which is at-tended to. Dr. Martin afferts, that fome parts of his body were warmer than natural in the cold fit of fever; but it is certain, that thofe, which are uncovered, as the fingers, and nofe, and ears, are

much.

much colder to the touch, and paler in appearance. It is possible, that his experiments were made at the beginning of the subsequent hot fits; which commence with partial distributions of heat, owing to some parts of the body regaining their natural irritability sooner than others.

From the quiescence of the anastomosing capillaries a paleness of the skin succeeds, and a less secretion of the perspirable matter; from the quiescence of the pulmonary capillaries a difficulty of respiration arises; and from the quiescence of the other glands less bile, less gastric and pancreatic juice, are secreted into the stomach and intestines, and less mucus and saliva are poured into the mouth; whence arises the dry tongue, costiveness, dry ulcers, and paucity of urine. From the quiescence of the absorbent system arises the great thirst, as less moisture is absorbed from the atmosphere. The absorption from the atmosphere was observed by Dr. Lyster to amount to eighteen ounces in one night, above what he had at the same time insensibly perspired. See Langrish. On the same account the urine is pale, though in small quantity, for the thinner part is not absorbed from it; and when repeated ague-fits continue long, the legs swell from the diminished absorption of the cellular absorbents.

From the quiescence of the intestinal canal a loss of appetite and flatulencies proceed. From the partial quiescence of the glandular viscera a swelling and tension about the præcordia becomes sensible to the touch; which is occasioned by the delay of the fluids from the defect of venous or lymphatic absorption. The pain of the forehead, and of the limbs, and of the small of the back, arises from the quiescence of the membranous fascia, or muscles of those parts, in the same manner as the skin becomes painful, when the vessels, of which it is composed, become quiescent from cold. The trembling in consequence of the pain of coldness, the restlessness, and the yawning, and stretching of the limbs, together with the shuddering, or rigours, are

convulsive

convulfive motions; and will be explained amongft the difeafes of volition, Sect. XXXIV.

Sicknefs and vomiting is a frequent fymptom in the beginnings of
fever-fits, the mufcular fibres of the ftomach fhare the general torpor
and debility of the fyftem; their motions become firft leffened, and
then ftop, and then become retrograde; for the act of vomiting, like
the globus hyftericus and the borborigmi of hypocondriafis, is always
a fymptom of debility, either from want of ftimulus, as in hunger;
or from want of fenforial power, as after intoxication; or from fympathy with fome other torpid irritative motions, as in the cold fits of
ague.   See Sect. XII. 5. 5. XXIX. 11. and XXXV. 1. 3. where this
act of vomiting is further explained.

The fmall pulfe, which is faid by fome writers to be flow at the
commencement of ague-fits, and which is frequently trembling and
intermittent, is owing to the quiefcence of the heart and arterial fyftem, and to the refiftance oppofed to the circulating fluid from the
inactivity of all the glands and capillaries.   The great weaknefs and
inability to voluntary motions, with the infenfibility of the extremities, are owing to the general quiefcence of the whole moving fyftem;
or, perhaps, fimply to the deficient production of fenforial power.

If all thefe fymptoms are further increafed, the quiefcence of all the
mufcles, including the heart and arteries, becomes complete, and death
enfues.   This is, moft probably, the cafe of thofe who are ftarved to
death with cold, and of thofe who are faid to die in Holland from long
fkaiting on their frozen canals.

2. As foon as this general quiefcence of the fyftem ceafes, either by
the diminution of the caufe, or by the accumulation of fenforial power,
(as in fyncope, Sect. XII. 7. 1.) which is the natural confequence of
previous quiefcence, the hot fit commences.   Every gland of the
body is now ftimulated into ftronger action than is natural, as its irritability is increafed by accumulation of fenforial power during its late
quiefcence,

quiefcence, a fuperabundance of all the fecretions is produced, and an increafe of heat in confequence of the increafe of thefe fecretions. The fkin becomes red, and the perfpiration great, owing to the increafed action of the capillaries during the hot part of the paroxyfm. The fecretion of perfpirable matter is perhaps greater during the hot fit than in the fweating fit which follows ; but as the abforption of it alfo is greater, it does not ftand on the fkin in vifible drops : add to this, that the evaporation of it alfo is greater, from the increafed heat of the fkin.   But at the decline of the hot fit, as the mouths of the abforbents of the fkin are expofed to the cooler air, or bed-clothes, thefe veffels fooner lofe their increafed activity, and ceafe to abforb more than their natural quantity : but the fecerning veffels for fome time longer, being kept warm by the circulating blood, continue to pour out an increafed quantity of perfpirable matter, which now ftands on the fkin in large vifible drops ; the exhalation of it alfo being leffened by the greater coolnefs of the fkin, as well as its abforption by the diminifhed action of the lymphatics.   See Clafs I. 1. 2. 3.

The increafed fecretion of bile and of other fluids poured into the inteftines frequently induce a purging at the decline of the hot fit ; for as the external abforbent veffels have their mouths expofed to the cold air, as above mentioned, they ceafe to be excited into unnatural activity fooner than the fecretory veffels, whofe mouths are expofed to the warmth of the blood : now, as the internal abforbents fympathize with the external ones, thefe alfo, which during the hot fit drank up the thinner part of the bile, or of other fecreted fluids, lofe their increafed activity before the gland lofes its increafed activity, at the decline of the hot fit ; and the loofe dejections are produced from the fame caufe, that the increafed perfpiration ftands on the furface of the fkin, from the increafed abforption ceafing fooner than the increafed fecretion.

The urine during the cold fit is in fmall quantity and pale, both from a deficiency of the fecretion and a deficiency of the abforption.

During

During the hot fit it is in its ufual quantity, but very high coloured
and turbid, becaufe a greater quantity had been fecreted by the increafed action of the kidnies, and alfo a greater quantity of its more
aqueous part had been abforbed from it in the bladder by the increafed
action of the abforbents; and laftly, at the decline of the hot fit it is
in large quantity and lefs coloured, or turbid, becaufe the abforbent
veffels of the bladder, as obferved above, lofe their increafed action by
fympathy with the cutaneous ones fooner than the fecretory veffels
of the kidnies lofe their increafed activity. Hence the quantity of
the fediment, and the colour of the urine, in fevers, depend much
on the quantity fecreted by the kidnies, and the quantity abforbed
from it again in the bladder: the kinds of fediment, as the lateri-
tious, purulent, mucous, or bloody fediments, depend on other
caufes. It fhould be obferved, that if the fweating be increafed by
the heat of the room, or of the bed-clothes, that a paucity of turbid
urine will continue to be produced, as the abforbents of the bladder
will have their activity increafed by their fympathy with the veffels
of the fkin, for the purpofe of fupplying the fluid expended in perfpiration.

The pulfe becomes ftrong and full owing to the increafed irritability of the heart and arteries, from the accumulation of fenforial
power during their quiefcence, and to the quicknefs of the return of
the blood from the various glands and capillaries. This increafed action of all the fecretory veffels does not occur very fuddenly, nor univerfally at the fame time. The heat feems to begin about the center,
and to be diffufed from thence irregularly to the other parts of the
fyftem. This may be owing to the fituation of the parts which firft
became quiefcent and caufed the fever-fit, efpecially when a hardnefs
or tumour about the præcordia can be felt by the hand; and hence
this part, in whatever vifcus it is feated, might be the firft to regain its
natural or increafed irritability.

It

3. It muft be here noted, that, by the increafed quantity of heat, and of the impulfe of the blood at the commencement of the hot fit, a great increafe of ftimulus is induced, and is now added to the in-creafed irritability of the fyftem, which was occafioned by its previ-ous quiefcence.   This additional ftimulus of heat and momentum of the blood augments the violence of the movements of the arterial and glandular fyftem in an increafing ratio.   Thefe violent exertions ftill producing more heat and greater momentum of the moving fluids, till at length the fenforial power becomes wafted by this great ftimu-lus beneath its natural quantity, and predifpofes the fyftem to a fe-cond cold fit.

At length all thefe unnatural exertions fpontaneoufly fubfide with the increafed irritability that produced them ; and which was itfelf produced by the preceding quiefcence, in the fame manner as the eye, on coming from darknefs into day-light, in a littie time ceafes to be dazzled and pained, and gradually recovers its natural degree of irri-tability.

4. But if the increafe of irritability, and the confequent increafe of the ftimulus of heat and momentum, produce more violent exer-tions than thofe above defcribed ; great pain arifes in fome part of the moving fyftem, as in the membranes of the brain, pleura, or joints ; and new motions of the veffels are produced in confequence of this pain, which are called inflammation ; or delirium or ftupor arifes ; as explained in Sect. XXI. and XXXIII.: for the immediate effect is the fame, whether the great energy of the moving organs arifes from an increafe of ftimulus or an increafe of irritability ; though in the former cafe the wafte of fenforial power leads to debility, and in the latter to health.

Recapitulation.

X. Thofe mufcles, which are lefs frequently exerted, and whofe actions are interrupted by fleep, acquire lefs accumulation of fenforial power during their quiefcent ftate, as the mufcles of locomotion. In thefe mufcles after great exertion, that is, after great exhauftion of fenforial power, the pain of fatigue enfues; and during reft there is a renovation of the natural quantity of fenforial power; but where the reft, or quiefcence of the mufcle, is long continued, a quantity of fenforial power becomes accumulated beyond what is neceffary; as appears by the uneafinefs occafioned by want of exercife; and which in young animals is one caufe exciting them into action, as is feen in the play of puppies and kittens.

But when thofe mufcles, which are habituated to perpetual action, as thofe of the ftomach by the ftimulus of food, thofe of the veffels of the fkin by the ftimulus of heat, and thofe which conftitute the arteries and glands by the ftimulus of the blood, become for a time quiefcent, from the want of their appropriated ftimuli, or by their affociations with other quiefcent parts of the fyftem; a greater accumulation of fenforial power is acquired during their quiefcence, and a greater or quicker exhauftion of it is produced during their increafed action.

This accumulation of fenforial power from deficient action, if it happens to the ftomach from want of food, occafions the pain of hunger; if it happens to the veffels of the fkin from want of heat, it occafions the pain of cold; and if to the arterial fyftem from the want of its adapted ftimuli, many difagreeable fenfations are occafioned, fuch as are experienced in the cold fits of intermittent fevers, and are as various, as there are glands or membranes in the fyftem, and are generally termed univerfal uneafinefs.

When the quiefcence of the arterial fyftem is not owing to defect of ftimulus as above, but to the defective quantity of fenforial power, as in the commencement of nervous fever, or irritative fever with weak pulfe, a great torpor of this fyftem is quickly induced; becaufe both the irritation from the ftimulus of the blood, and the affociation of the vafcular motions with each other, continue to excite the arteries into action, and thence quickly exhauft the ill-fupplied vafcular mufcles; for to reft is death; and therefore thofe vafcular mufcles continue to proceed, though with feebler action, to the extreme of wearinefs or faintnefs: while nothing fimilar to this affects the locomotive mufcles, whofe actions are generally caufed by volition, and not much fubject either to irritation or to other kinds of affociations befides the voluntary ones, except indeed when they are excited by the lafh of flavery.

In thefe vafcular mufcles, which are fubject to perpetual action, and thence liable to great accumulation of fenforial power during their quiefcence from want of ftimulus, a great increafe of activity occurs, either from the renewal of their accuftomed ftimulus, or even from much lefs quantities of ftimulus than ufual. This increafe of action conftitutes the hot fit of fever, which is attended with various increafed fecretions, with great concomitant heat, and general uneafinefs. The uneafinefs attending this hot paroxyfm of fever, or fit of exertion, is very different from that, which attends the previous cold fit, or fit of quiefcence, and is frequently the caufe of inflammation, as in pleurify, which is treated of in the next fection.

A fimilar effect occurs after the quiefcence of our organs of fenfe; thofe which are not fubject to perpetual action, as the tafte and fmell, are lefs liable to an exuberant accumulation of fenforial power after their having for a time been inactive; but the eye, which is in per-

petual

petual action during the day, becomes dazzled, and liable to inflammation after a temporary quiescence.

Where the previous quiescence has been owing to a defect of senforial power, and not to a defect of stimulus, as in the irritative fever with weak pulse, a similar increase of activity of the arterial system succeeds, either from the usual stimulus of the blood, or from a stimulus less than usual; but as there is in general in these cases of fever with weak pulse a deficiency of the quantity of the blood, the pulse in the hot fit is weaker than in health, though it is stronger than in the cold fit, as explained in No. 2. of this section. But at the same time in those fevers, where the defect of irritation is owing to the defect of the quantity of senforial power, as well as to the defect of stimulus, another circumstance occurs; which consists in the partial distribution of it, as appears in partial flushings, as of the face or bosom, while the extremities are cold; and in the increase of particular secretions, as of bile, saliva, insensible perspiration, with great heat of the skin, or with partial sweats, or diarrhœa.

There are also many uneasy sensations attending these increased actions, which, like those belonging to the hot fit of fever with strong pulse, are frequently followed by inflammation, as in scarlet fever; which inflammation is nevertheless accompanied with a pulse weaker, though quicker, than the pulse during the remission or intermission of the paroxysms, though stronger than that of the previous cold fit.

From hence I conclude, that both the cold and hot fits of fever are necessary consequences of the perpetual and incessant action of the arterial and glandular system; since those muscular fibres and those organs of sense, which are most frequently exerted, become necessarily most affected both with defect and accumulation of senforial power: and that hence *fever-fits are not an effort of nature to relieve herself*, and that therefore they should always be prevented or diminished as much as possible, by any means which decrease the general or partial

3   vascular

vafcular actions, when they are greater, or by increafing them when they are lefs than in health, as defcribed in Sect. XII. 6. 1.

Thus have I endeavoured to explain, and I hope to the fatisfaction of the candid and patient reader, the principal fymptoms or circum-ftances of fever without the introduction of the fupernatural power of fpafm. To the arguments in favour of the doctrine of fpafm it may be fufficient to reply, that in the evolution of medical as well as of dramatic cataftrophe,

Nec Deus interfit, nifi dignus vindice nodus inciderit.

HOR.

SECT.

SECT. XXXIII.

DISEASES OF SENSATION.

I. *Motions excited by sensation. Digestion. Generation. Pleasure of existence. Hypochondriacism.* 2. *Pain introduced. Sensitive fevers of two kinds.* 3. *Two sensorial powers exerted in sensitive fevers. Size of the blood. Nervous fevers distinguished from putrid ones. The septic and antiseptic theory.* 4. *Two kinds of delirium.* 5. *Other animals are less liable to delirium, cannot receive our contagious diseases, and are less liable to madness.* II. 1. *Sensitive motions generated.* 2. *Inflammation explained.* 3. *Its remote causes from excess of irritation, or of irritability, not from those pains which are owing to defect of irritation. New vessels produced, and much heat.* 4. *Purulent matter secreted.* 5. *Contagion explained.* 6. *Received but once.* 7. *If common matter be contagious?* 8. *Why some contagions are received but once.* 9. *Why others may be received frequently. Contagions of small-pox and measles do not act at the same times. Two cases of such patients.* 10. *The blood from patients in the small-pox will not infect others. Cases of children thus inoculated. The variolous contagion is not received into the blood. It acts by sensitive association between the stomach and skin.* III. 1. *Absorption of solids and fluids.* 2. *Art of healing ulcers.* 3. *Mortification attended with less pain in weak people.*

I. 1. AS many motions of the body are excited and continued by irritations, so others require, either conjunctly with these, or separately, the pleasurable or painful sensations, for the purpose of producing them with due energy. Amongst these the business of digestion supplies us with an instance: if the food, which we swallow, is not attended with agreeable sensation, it digests less perfectly; and if very disagreeable sensation accompanies it, such as a nauseous idea,

or

or very difguftful tafte, the digeftion becomes impeded ; or retrograde motions of the ftomach and oefophagus fucceed, and the food is ejected.

The bufinefs of generation depends fo much on agreeable fenfation, that, where the object is difguftful, neither voluntary exertion nor irritation can effect the purpofe ; which is alfo liable to be interrupted by the pain of fear or bafhfulnefs.

. Befides the pleafure, which attends the irritations produced by the objects of luft and hunger, there feems to be a fum of pleafurable affection accompanying the various fecretions of the numerous glands, which conftitutes the pleafure of life, in contradiftinction to the tedium vitæ.   This quantity or fum of pleafurable affection feems to contribute to the due or energetic performance of the whole moveable fyftem, as well that of the heart and arteries, as of digeftion and of abforption ; fince without the due quantity of pleafurable fenfation, flatulency and hypochondriacifm affect the inteftines, and a languor feizes the arterial pulfations and fecretions ; as occurs in great and continued anxiety of the mind.

2. Befides the febrile motions occafioned by irritation, defcribed in Sect. XXXII. and termed irritative fever, it frequently happens that pain is excited by the violence of the fibrous contractions ; and other new motions are then fuperadded, in confequence of fenfation, which we fhall term febris fenfitiva, or fenfitive fever. It muft be obferved, that moft irritative fevers begin with a decreafed exertion of irritation, owing to defect of ftimulus ; but that on the contrary the fenfitive fevers, or inflammations, generally begin with the increafed exertion of fenfation, as mentioned in Sect. XXXI. on temperaments : for though the cold fit, which introduces inflammation, commences with decreafed irritation, yet the inflammation itfelf commences in the hot fit during the increafe of fenfation.   Thus a common puftule, or phlegmon, in a part of little fenfibility does not excite an inflammatory fever ; but if the ftomach, inteftines, or the tender fubftance

5   beneath

beneath the nails, be injured, great fenfation is produced, and the whole fyftem is thrown into that kind of exertion, which conftitutes inflammation.

Thefe fenfitive fevers, like the irritative ones, refolve themfelves into thofe with arterial ftrength, and thofe with arterial debility, that is with excefs or defect of fenforial power; thefe may be termed the febris fenfitiva pulfu forti, fenfitive fever with ftrong pulfe, which is the fynocha, or inflammatory fever; and the febris fenfitiva pulfu debili, fenfitive fever with weak pulfe, which is the typhus gravior, or putrid fever of fome writers.

3. The inflammatory fevers, which are here termed fenfitive fevers with ftrong pulfe, are generally attended with fome topical inflammation, as pleurify, peripneumony, or rheumatifm, which diftinguifhes them from irritative fevers with ftrong pulfe. The pulfe is ftrong, quick, and full; for in this fever there is great irritation, as well as great fenfation, employed in moving the arterial fyftem. The fize, or coagulable lymph, which appears on the blood, is probably an increafed fecretion from the inflamed internal lining of the whole arterial fyftem, the thinner part being taken away by the increafed abforption of the inflamed lymphatics.

The fenfitive fevers with weak pulfe, which are termed putrid or malignant fevers, are diftinguifhed from irritative fevers with weak pulfe, called nervous fevers, defcribed in the laft fection, as the former confift of inflammation joined with debility, and the latter of debility alone. Hence there is greater heat and more florid colour of the fkin in the former, with petechiæ, or purple fpots, and aphthæ, or floughs in the throat, and generally with previous contagion.

When animal matter dies, as a flough in the throat, or the mortified part of a carbuncle, if it be kept moift and warm, as during its adhefion to a living body, it will foon putrify. This, and the origin of contagion from putrid animal fubftances, feem to have given rife to the feptic and antifeptic theory of thefe fevers.

The

SECT. XXXIII. I. DISEASES OF SENSATION. 393

The matter in puſtules and ulcers is thus liable to become putrid, and to produce microſcopic animalcula; the urine, if too long retained, may alſo gain a putreſcent ſmell, as well as the alvine feces; but ſome writers have gone ſo far as to believe, that the blood itſelf in theſe fevers has ſmelt putrid, when drawn from the arm of the patient: but this ſeems not well founded; ſince a ſingle particle of putrid matter taken into the blood can produce fever, how can we conceive that the whole maſs could continue a minute in a putrid ſtate without deſtroying life? Add to this, that putrid animal ſubſtances give up air, as in gangrenes; and that hence if the blood was putrid, air ſhould be given out, which in the blood-veſſels is known to occaſion immediate death.

In theſe ſenſitive fevers with ſtrong pulſe (or inflammations) there are two ſenſorial faculties concerned in producing the diſeaſe, viz. irritation and ſenſation; and hence, as their combined action is more violent, the general quantity of ſenſorial power becomes further exhauſted during the exacerbation, and the ſyſtem more rapidly weakened than in irritative fever with ſtrong pulſe; where the ſpirit of animation is weakened by but one mode of its exertion: ſo that this febris ſenſitiva pulſu forti (or inflammatory fever,) may be conſidered as the febris irritativa pulſu forti, with the addition of inflammation; and the febris ſenſitiva pulſu debili (or malignant fever) may be conſidered as the febris irritativa pulſu debili (or nervous fever), with the addition of inflammation.

4. In theſe putrid or malignant fevers a deficiency of irritability accompanies the increaſe of ſenſibility; and by this waſte of ſenſorial power by the exceſs of ſenſation, which was already too ſmall, ariſes the delirium and ſtupor which ſo perpetually attend theſe inflammatory fevers with arterial debility. In theſe caſes the voluntary power firſt ceaſes to act from deficiency of ſenſorial ſpirit; and the ſtimuli from external bodies have no effect on the exhauſted ſenſorial power, and a delirium like a dream is the conſequence. At length the in-

3 E   ternal

ternal ſtimuli ceaſe to excite ſufficient irritation, and the ſecretions are either not produced at all, or too parſimonious in quantity.   Amongſt theſe the ſecretion of the brain, or production of the ſenſorial power, becomes deficient, till at laſt all ſenſorial power ceaſes, except what is juſt neceſſary to perform the vital motions, and a ſtupor ſucceeds; which is thus owing to the ſame cauſe as the preceding delirium exerted in a greater degree.

This kind of delirium is owing to a ſuſpenſion of volition, and to the diſobedience of the ſenſes to external ſtimuli, and is always occaſioned by great debility, or paucity of ſenſorial power; it is therefore a bad ſign at the end of inflammatory fevers, which had previous arterial ſtrength, as rheumatiſm, or pleuriſy, as it ſhews the preſence of great exhauſtion of ſenſorial power in a ſyſtem, which having lately been expoſed to great excitement, is not ſo liable to be ſtimulated into its healthy action, either by additional ſtimulus of food and medicines, or by the accumulation of ſenſorial power during its preſent torpor. In inflammatory fevers with debility, as thoſe termed putrid fevers, delirium is ſometimes, as well as ſtupor, rather a favourable ſign; as leſs ſenſorial power is waſted during its continuance (ſee Claſs II. 1. 6. 8.), and the conſtitution not having been previouſly expoſed to exceſs of ſtimulation, is more liable to be excited after previous quieſcence.

When the ſum of general pleaſurable ſenſation becomes too great, another kind of delirium ſupervenes, and the ideas thus excited are miſtaken for the irritations of external objects: ſuch a delirium is produced for a time by intoxicating drugs, as fermented liquors, or opium: a permanent delirium of this kind is ſometimes induced by the pleaſures of inordinate vanity, or by the enthuſiaſtic hopes of heaven.   In theſe caſes the power of volition is incapable of exertion, and in a great degree the external ſenſes become incapable of perceiving their adapted ſtimuli, becauſe the whole ſenſorial power is employed or expended on the ideas excited by pleaſurable ſenſation.

This

This kind of delirium is diftinguifhed from that which attends the fevers above mentioned from its not being accompanied with general debility, but fimply with excefs of pleafureable fenfation ; and is therefore in fome meafure allied to madnefs or to reverie ; it differs from the delirium of dreams, as in this the power of volition is not totally fufpended, nor are the fenfes precluded from external ftimulation ; there is therefore a degree of confiftency, in this kind of delirium, and a degree of attention to external objects, neither of which exift in the delirium of fevers or in dreams.

5. It would appear, that the vafcular fyftem of other animals are lefs liable to be put into action by their general fum of pleafureable or painful fenfation ; and that the trains of their ideas, and the mufcular motions ufually affociated with them, are lefs powerfully connected than in the human fyftem. For other animals neither weep, nor fmile, nor laugh ; and are hence feldom fubject to delirium, as treated of in Sect. XVI. on Inftinct. Now as our epidemic and contagious difeafes are probably produced by difagreeable fenfation, and not fimply by irritation ; there appears a reafon, why brute animals are lefs liable to epidemic or contagious difeafes ; and fecondly, why none of our contagions, as the fmall-pox or meafles, can be communicated to them, though one of theirs, viz. the hydrophobia, as well as many of their poifons, as thofe of fnakes and of infects, communicate their deleterious or painful effects to mankind.

Where the quantity of general painful fenfation is too great in the fyftem, inordinate voluntary exertions are produced either of our ideas, as in melancholy and madnefs, or of our mufcles, as in convulfion. From thefe maladies alfo brute animals are much more exempt than mankind, owing to their greater inaptitude to voluntary exertion, as mentioned in Sect. XVI. on Inftinct.

II. 1. When any moving organ is excited into fuch violent motions, that a quantity of pleafureable or painful fenfation is produced, it frequently happens (but not always) that new motions of the af-

fected

fected organ are generated in confequence of the pain or pleafure, which are termed inflammation.

Thefe new motions are of a peculiar kind, tending to diftend the old, and to produce new fibres, and thence to elongate the ftraight mufcles, which ferve locomotion, and to form new veffels at the extremities or fides of the vafcular mufcles.

2. Thus the pleafureable fenfations produce an enlargement of the nipples of nurfes, of the papillæ of the tongue, of the penis, and probably produce the growth of the body from its embryon ftate to its maturity ; whilft the new motions in confequence of painful fenfation, with the growth of the fibres or veffels, which they occafion, are termed inflammation.

Hence when the ftraight mufcles are inflamed, part of their tendons at each extremity gain new life and fenfibility, and thus the mufcle is for a time elongated ; and inflamed bones become foft, vafcular, and fenfible. Thus new veffels fhoot over the cornea of inflamed eyes, and into fchirrous tumours, when they become inflamed ; and hence all inflamed parts grow together by intermixture, and inofculation of the new and old veffels.

The heat is occafioned from the increafed fecretions either of mucus, or of the fibres, which produce or elongate the veffels. The red colour is owing to the pellucidity of the newly formed veffels, and as the arterial parts of them are probably formed before their correfpondent venous parts.

3. Thefe new motions are excited either from the increafed quantity of fenfation in confequence of greater fibrous contractions, or from increafed fenfibility, that is, from the increafed quantity of fenforial power in the moving organ. Hence they are induced by great external ftimuli, as by wounds, broken bones ; and by acrid or infectious materials ; or by common ftimuli on thofe organs, which have been fome time quiefcent ; as the ufual light of the day inflames the eyes of thofe, who have been confined in dungeons ; and the

warmth

warmth of a common fire inflames thofe, who have been previoufly expofed to much cold.

But thefe new motions are never generated by that pain, which arifes from defect of ftimulus, as from hunger, thirft, cold, or in-anition, with all thofe pains, which are termed nervous. Where thefe pains exift, the motions of the affected part are leffened; and if inflammation fucceeds, it is in fome diftant parts; as coughs are caufed by coldnefs and moifture being long applied to the feet; or it is in confequence of the renewal of the ftimulus, as of heat or food, which excites our organs into ftronger action after their temporary quiefcence; as kibed heels after walking in fnow.

4. But when thefe new motions of the vafcular mufcles are ex-erted with greater violence, and thefe veffels are either elongated too much or too haftily, a new material is fecreted from their extremi-ties, which is of various kinds according to the peculiar animal mo-tions of this new kind of gland, which fecretes it; fuch is the pus laudabile or common matter, the variolous matter, venereal matter, catarrhous matter, and many others.

5. Thefe matters are the product of an animal procefs; they are fecreted or produced from the blood by certain difeafed motions of the extremities of the blood-veffels, and are on that account all of them contagious; for if a portion of any of thefe matters is tranfmitted into the circulation, or perhaps only inferted into the fkin, or beneath the cuticle of an healthy perfon, its ftimulus in a certain time produces the fame kind of morbid motions, by which itfelf was produced; and hence a fimilar matter is generated. See Sect. XXXIX. 6. 1.

6. It is remarkable, that many of thefe contagious matters are capable of producing a fimilar difeafe but once; as the fmall-pox and meafles; and I fuppofe this is true of all thofe contagious difeafes, which are fpontaneoufly cured by nature in a certain time; for if the body was capable of receiving the difeafe a fecond time, the patient muft perpetually infect himfelf by the very matter, which he has

himfelf

himfelf produced, and is lodged about him; and hence he could never become free from the difeafe. Something fimilar to this is feen in the fecondary fever of the confluent fmall-pox; there is a great abforption of variolous matter, a very minute part of which would give the genuine fmall-pox to another perfon; but here it only ftimulates the fyftem into common fever; like that which common pus, or any other acrid material might occafion.

7. In the pulmonary confumption, where common matter is daily abforbed, an irritative fever only, not an inflammatory one, is produced; which is terminated like other irritative fevers by fweats, or loofe ftools. Hence it does not appear, that this abforbed matter always acts as a contagious material producing frefh inflammation or new abfceffes. Though there is reafon to believe, that the firft time any common matter is abforbed, it has this effect, but not the fecond time, like the variolous matter above mentioned.

This accounts for the opinion, that the pulmonary confumption is fometimes infectious, which opinion was held by the ancients, and continues in Italy at prefent; and I have myfelf feen three or four inftances, where a hufband and wife, who have flept together, and have thus much received each other's breath, who have infected each other, and both died in confequence of the original taint of only one of them. This alfo accounts for the abfceffes in various parts of the body, that are fometimes produced after the inoculated fmall-pox is terminated; for this fecond abforption of variolous matter acts like common matter, and produces only irritative fever in thofe children, whofe conftitutions have already experienced the abforption of common matter; and inflammation with a tendency to produce new ab-fceffes in thofe, whofe conftitutions have not experienced the abforptions of common matter.

It is probable, that more certain proofs might have been found to fhew, that common matter is infectious the firft time it is abforbed,

tending

tending to produce fimilar abfceffes, but not the fecond time of its abforption, if this fubject had been attended to.

8. Thefe contagious difeafes are very numerous, as the plague, fmall-pox, chicken-pox, meafles, fcarlet-fever, pemphigus, catarrh, chincough, venereal difeafe, itch, trichoma, tinea. The infectious material does not feem to be diffolved by the air, but only mixed with it perhaps in fine powder, which foon fubfides; fince many of thefe contagions can only be received by actual contact; and others of them only at fmall diftances from the infected perfon; as is evident from many perfons having been near patients of the fmall-pox without acquiring the difeafe.

The reafon, why many of thefe difeafes are received but once, and others repeatedly, is not well underftood; it appears to me, that the conftitution becomes fo accuftomed to the ftimuli of thefe infectious materials, by having once experienced them, that though irritative motions, as hectic fevers, may again be produced by them, yet no fenfation, and in confequence no general inflammation fucceeds; as difagreeable fmells or taftes by habit ceafe to be perceived; they continue indeed to excite irritative ideas on the organs of fenfe, but thefe are not fucceeded by fenfation.

There are many irritative motions, which were at firft fucceeded by fenfation, but which by frequent repetition ceafe to excite fenfation, as explained in Sect. XX. on Vertigo. And, that this circumftance exifts in refpect to infectious matter appears from a known fact; that nurfes, who have had the fmall-pox, are liable to experience fmall ulcers on their arms by the contact of variolous matter in lifting their patients; and that when patients, who have formerly had the fmall-pox have been inoculated in the arm, a phlegmon, or inflamed fore, has fucceeded, but no fubfequent fever. Which fhews, that the contagious matter of the fmall-pox has not loft its power of ftimulating the part it is applied to, but that the general

fyftem

fyftem is not affected in confequence.   See Section XII. 7. 6;
XIX. 10.

9. From the accounts of the plague, virulent catarrh, and putrid
dyfentery, it feems uncertain, whether thefe difeafes are experienced
more than once ; but the venereal difeafe and itch are doubtlefs repeatedly infectious ; and as thefe difeafes are never cured fpontaneoufly,
but require medicines, which act without apparent operation, fome
have fufpected, that the contagious material produces fimilar matter
rather by a chemical change of the fluids, than by an animal procefs ;
and that the fpecific medicines deftroy their virus by chemically com-
bining with it.   This opinion is fuccefsfully combated by Mr. Hunter, in his Treatife on Venereal Difeafe, Part I. c. i.

But this opinion wants the fupport of analogy, as there is no
known procefs in animal bodies, which is purely chemical, not even
digeftion ; nor can any of thefe matters be produced by chemical proceffes.   Add to this, that it is probable, that the infects, obferved in
the puftules of the itch, and in the ftools of dyfenteric patients, are
the confequences, and not the caufes of thefe difeafes.   And that the
fpecific medicines, which cure the itch and lues venerea, as brimftone
and mercury, act only by increafing the abforption of the matter in
the ulcufcles of thofe difeafes, and thence difpofing them to heal ;
which would otherwife continue to fpread.

Why the venereal difeafe, and itch, and tenia, or fcald head, are
repeatedly contagious, while thofe contagions attended with fever can
be received but once, feems to depend on their being rather local difeafes than univerfal ones, and are hence not attended with fever, except the purulent fever in their laft ftages, when the patient is deftroyed by them.   On this account the whole of the fyftem does not become habituated to thefe morbid actions, fo as to ceafe to be affected
with fenfation by a repetition of the contagion.   Thus the contagious
matter of the venereal difeafe, and of the tenia, affects the lymphatic
glands,

5

glands, as the inquinal glands, and thofe about the roots of the hair and neck, where it is arrefted, but does not feem to affect the blood-veffels, fince no fever enfues.

Hence it would appear, that thefe kinds of contagion are propagated not by means of the circulation, but by fympathy of diftant parts with each other; fince if a diftant part, as the palate, fhould be excited by fenfitive affociation into the fame kind of motions, as the parts originally affected by the contact of infectious matter; that diftant part will produce the fame kind of infectious matter; for every fecretion from the blood is formed from it by the peculiar motions of the fine extremities of the gland, which fecretes it; the various fecreted fluids, as the bile, faliva, gaftric juice, not previoufly exifting, as fuch, in the blood-veffels.

And this peculiar fympathy between the genitals and the throat, owing to fenfitive affociation, appears not only in the production of venereal ulcers in the throat, but in variety of other inftances, as in the mumps, in the hydrophobia, fome coughs, ftrangulation, the production of the beard, change of voice at puberty. Which are further defcribed in Clafs IV. 2. 1. 7.

To evince that the production of fuch large quantities of contagious matter, as are feen in fome variolous patients, fo as to cover the whole fkin almoft with puftules, does not arife from any chemical fermentation in the blood, but that it is owing to morbid motions of the fine extremities of the capillaries, or glands, whether thefe be ruptured or not, appears from the quantity of this matter always correfponding with the quantity of the fever; that is, with the violent exertions of thofe glands and capillaries, which are the terminations of the arterial fyftem.

The truth of this theory is evinced further by a circumftance obferved by Mr. J. Hunter, in his Treatife on Venereal Difeafe; that in a patient, who was inoculated for the fmall-pox, and who appeared afterwards to have been previoufly infected with the meafles,

3 F

the

the progrefs of the fmall-pox was delayed till the meafles had run their courfe, and that then the fmall-pox went through its ufual periods.

Two fimilar cafes fell under my care, which I fhall here relate, as it confirms that of Mr. Hunter, and contributes to illuftrate this part of the theory of contagious difeafes. I have tranfcribed the particulars from a letter of Mr. Lightwood of Yoxal, the furgeon who daily attended them, and at my requeft, after I had feen them, kept a kind of journal of their cafes.

Mifs H. and Mifs L. two fifters, the one about four and the other about three years old, were inoculated Feb. 7, 1791. On the 10th there was a rednefs on both arms difcernible by a glafs. On the 11th their arms were fo much inflamed as to leave no doubt of the infection having taken place. On the 12th lefs appearance of inflammation on their arms. In the evening Mifs L. had an eruption, which refembled the meafles. On the 13th the eruption on Mifs L. was very full on the face and breaft, like the meafles, with confiderable fever. It was now known, that the meafles were in a farm houfe in the neighbourhood. Mifs H.'s arm lefs inflamed than yefterday. On the 14th Mifs L.'s fever great, and the eruption univerfal. The arm appears to be healed. Mifs H.'s arm fomewhat redder. They were now put into feparate rooms. On the 15th Mifs L.'s arms as yefterday. Eruption continues. Mifs H.'s arms have varied but little. 16th, the eruptions on Mifs L. are dying away, her fever gone. Begins to have a little rednefs in one arm at the place of inoculation. Mifs H.'s arms get redder, but fhe has no appearance of complaint. 20th, Mifs L.'s arms have advanced flowly till this day, and now a few puftules appear. Mifs H.'s arm has made little progrefs from the 16th to this day, and now fhe has fome fever. 21ft, Mifs L. as yefterday. Mifs H. has much inflammation, and an increafe of the red circle on one arm to the fize of half a crown, and had much fever at night, with fetid breath. 22d, Mifs L.'s puftules continue advancing.

ing.   Mifs H.'s inflammation of her arm and red circle increafes.   A few red fpots appear in different parts with fome degree of fever this morning.   23d. Mifs L. has a larger crop of puftules.   Mifs H. has fmall puftules and great inflammation of her arms, with but one puftule likely to fuppurate.   After this day they gradually got well, and the puftules difappeared.

In one of thefe cafes the meafles went through their common courfe with milder fymptoms than ufual, and in the other the meafly contagion feemed juft fufficient to ftop the progrefs of variolous contagion, but without itfelf throwing the conftitution into any diforder.   At the fame time both the meafles and fmall-pox feem to have been rendered milder.   Does not this give an idea, that if they were both inoculated at the fame time, that neither of them might affect the patient?

From thefe cafes I contend, that the contagious matter of thefe difeafes does not affect the conftitution by a fermentation, or chemical change of the blood, becaufe then they muft have proceeded together, and have produced a third fomething, not exactly fimilar to either of them : but that they produce new motions of the cutaneous terminations of the blood-veffels, which for a time proceed daily with increafing activity, like fome paroxyfms of fever, till they at length fecrete or form a fimilar poifon by thefe unnatural actions.

Now as in the meafles one kind of unnatural motion takes place, and in the fmall-pox another kind, it is eafy to conceive, that thefe different kinds of morbid motions cannot exift together ; and therefore, that that which has firft begun will continue till the fyftem becomes habituated to the ftimulus which occafions it, and has ceafed to be thrown into action by it ; and then the other kind of ftimulus will in its turn produce fever, and new kinds of motions peculiar to itfelf.

10. On further confidering the action of contagious matter, fince the former part of this work was fent to the prefs ; where I have afferted, in Sect. XII. 3. 6. that it is probable, that the variolous matter is diffufed through the blood ; I prevailed on my friend Mr. Power,

furgeon

furgeon at Bofworth in Leicefterfhire to try, whether the fmall-pox could be inoculated by ufing the blood of a variolous patient inftead of the matter from the puftules; as I thought fuch an experiment migh throw fome light at leaft on this interefting fubject. The following is an extract from his letter:—

" March 11, 1793. I inoculated two children, who had not had the fmall-pox, with blood; which was taken from a patient on the fecond day after the eruption commenced, and before it was completed. And at the fame time I inoculated myfelf with blood from the fame perfon, in order to compare the appearances, which might arife in a perfon liable to receive the infection, and in one not liable to receive it. On the fame day I inoculated four other children liable to receive the infection with blood taken from another perfon on the fourth day after the commencement of the eruption. The patients from whom the blood was taken had the difeafe mildly, but had the moft puftules of any I could felect from twenty inoculated patients; and as much of the blood was infinuated under the cuticle as I could introduce by elevating the fkin without drawing blood; and three or four fuch punctures were made in each of their arms, and the blood was ufed in its fluid ftate.

" As the appearances in all thefe patients, as well as in myfelf, were fimilar, I fhall only mention them in general terms. March 13. A flight fubcuticular difcoloration, with rather a livid appearance, without forenefs or pain, was vifible in them all, as well as in my own hand. 15. The difcoloration fomewhat lefs, without pain or forenefs. Some patients inoculated on the fame day with variolous matter have confiderable inflammation. 17. The difcoloration is quite gone in them all, and from my own hand, a dry mark only remaining. And they were all inoculated on the 18th, with variolous matter, which produced the difeafe in them all."

Mr. Power afterwards obferves, that, as the patients from whom the blood was taken had the difeafe mildly, it may be fuppofed, that

though

though the contagious matter might be mixed with the blood, it might ftill be in too dilute a ftate to convey the infection ; but adds at the fame time, that he has diluted recent matter with at leaft five times its quantity of water, and which has ftill given the infection ; though he has fometimes diluted it fo far as to fail.

The following experiments were inftituted at my requeft by my friend Mr. Hadley, furgeon in Derby, to afcertain whether the blood of a perfon in the fmall-pox be capable of communicating the difeafe. "Experiment 1ft. October 18th, 1793. I took fome blood from a vein in the arm of a perfon, who had the fmall-pox on the fecond day of the eruption, and introduced a fmall quantity of it immediately with the point of a lancet between the fcarf and true fkin of the right arm of a boy nine years old in two or three different places ; the other arm was inoculated with variolous matter at the fame time.

" 19th. The punctured parts of the right arm were furrounded with fome degree of fubcuticular inflammation. 20th. The inflammation more confiderable, with a flight degree of itching, but no pain upon preffure. 21ft. Upon examining the arm this day with a lens I found the inflammation lefs extenfive, and the rednefs changing to a deep yellow or orange-colour. 22d. Inflammation nearly gone. 23d. Nothing remained, except a flight difcoloration and a little fcurfy appearance on the punctures. At the fame time the inflammation of the arm inoculated with variolous matter was increafing faft, and he had the difeafe mildly at the ufual time.

" Experiment 2d. I inoculated another child at the fame time and in the fame manner, with blood taken on the firft day of the eruption ; but as the appearance and effects were fimilar to thofe in the preceding experiment, I fhall not relate them minutely.

Experiment 3d. October 20th. Blood was taken from a perfon who had the fmall-pox, on the third day of the eruption, and on the fixth from the commencement of the eruptive fever. I introduced fome of it in its fluid ftate into both arms of a boy feven years old.

21ft. There appeared to be fome inflammation under the cuticle, where the punctures were made.   22d. Inflammation more confiderable.   23d. On this day the inflammation was fomewhat greater, and the cuticle rather elevated.

24th. Inflammation much lefs, and only a brown or orange-colour remained.   25th. Scarcely any difcoloration left.   On this day he was inoculated with variolous matter, the progrefs of the infection went on in the ufual way, and he had the fmall-pox very favourably.

" At this time I was requefted to inoculate a young perfon, who was thought to have had the fmall-pox, but his parents were not quite certain; in one arm I introduced variolous matter, and in the other blood, taken as in experiment 3d.   On the fecond day after the operation, the punctured parts were inflamed, though I think the arm in which I had inferted variolous matter was rather more fo than the other.   On the third the inflammation was increafed, and looked much the fame as in the preceding experiment.   4th. The inflammation was much diminifhed, and on the 5th almoft gone.   He was expofed at the fame time to the natural infection, but has continued perfectly well.

" I have frequently obferved (and believe moft practitioners have done the fame), that if variolous matter be inferted in the arm of a perfon who has previoufly had the fmall-pox, that the inflammation on the fecond or third days is much greater, than if they had not had the difeafe, but on the fourth or fifth it difappears.

" On the 23d I introduced blood into the arms of three more children, taken on the third and fourth days of the eruption.   The appearances were much the fame as mentioned in experiments firft and third.   They were afterwards inoculated with variolous matter, and had the difeafe in the regular way.

" The above experiments were made with blood taken from a fmall vein in the hand or foot of three or four different patients,

whom

whom I had at that time under inoculation.   They were felected from 160, as having the greateft number of puftules.   The part was wafhed with warm water before the blood was taken, to prevent the poffibility of any matter being mixed with it from the furface."

Shall we conclude from hence, that the variolous matter never enters the blood-veffels? but that the morbid motions of the veffels of the fkin around the infertion of it continue to increafe in a larger and larger circle for fix or feven days; that then their quantity of morbid action becomes great enough to produce a fever-fit, and to affect the ftomach by affociation of motions? and finally, that a fecond affociation of motions is produced between the ftomach and the other parts of the fkin, inducing them into morbid actions fimilar to thofe of the circle round the infertion of the variolous matter?   Many more experiments and obfervations are required before this important queftion can be fatisfactorily anfwered.

It may be adduced, that as the matter inferted into the fkin of the arm frequently fwells the lymphatic in the axilla, that in that circumftance it feems to be there arrefted in its progrefs, and cannot be imagined to enter the blood by that lymphatic gland till the fwelling of it fubfides.   Some other phænomena of the difeafe are more eafily reconcileable to this theory of fympathetic motions than to that of abforption; as the time taken up between the infertion of the matter, and the operation of it on the fyftem, as mentioned above.   For the circle around the infertion is feen to increafe, and to inflame; and I believe, undergoes a kind of diurnal paroxyfm of torpor and palenefs with a fucceeding increafe of action and colour, like a topical fever-fit.   Whereas if the matter is conceived to circulate for fix or feven days with the blood, without producing diforder, it ought to be rendered milder, or the blood-veffels more familiarized to its acrimony.

It is much eafier to conceive from this doctrine of affociated or fympathetic motions of diftant parts of the fyftem, how it happens, that the variolous infection can be received but once, as before explained;

plained ; than by fuppofing, that a change is effected in the mafs of blood by any kind of fermentative procefs.

The curious circumftance of the two contagions of fmall-pox and meafles not acting at the fame time, but one of them refting or fuf-pending its action till that of the other ceafes, may be much eafier explained from fympathetic or affociated actions of the infected part with other parts of the fyftem, than it can from fuppofing the two contagions to enter the circulation.

The fkin of the face is fubject to more frequent viciffitudes of heat and cold, from its expofure to the open air, and is in confequence more liable to fenfitive affociation with the ftomach than any other part of the furface of the body, becaufe their actions have been more fre-quently thus affociated. Thus in a furfeit from drinking cold water, when a perfon is very hot and fatigued, an eruption is liable to appear on the face in confequence of this fympathy. In the fame manner the rofy eruption on the faces of drunkards more probably arifes from the fympathy of the face with the ftomach, rather than between the face and the liver, as is generally fuppofed.

This fympathy between the ftomach and the fkin of the face is ap-parent in the eruption of the fmall-pox ; fince, where the difeafe is in confiderable quantity, the eruption on the face firft fucceeds the fick-nefs of the ftomach. In the natural difeafe the ftomach feems to be frequently primarily affected, either alone or along with the tonfils, as the matter feems to be only diffufed in the air, and by being mixed with the faliva, or mucus of the tonfils, to be fwallowed into the ftomach.

After fome days the irritative circles of motions become difordered by this new ftimulus, which acts upon the mucous lining of the fto-mach ; and ficknefs, vertigo, and a diurnal fever fucceed. Thefe difordered irritative motions become daily increafed for two or three days, and then by their increafed action certain fenfitive motions, or inflammation, is produced ; and at the next cold fit of fever, when the

the ftomach recovers from its torpor, an inflammation of the external fkin is formed in points (which afterwards fuppurate), by fenfitive
affociation, in the fame manner as a cough is produced in confequence
of expofing the feet to cold, as defcribed in Sect. XXV. 1. 1. and
Clafs IV. 2. 2. 4.   If the inoculated fkin of the arm, as far as it appears inflamed, was to be cut out, or deftroyed by cauftic, before the
fever commenced, as fuppofe on the fourth day after inoculation,
would this prevent the difeafe? as it is fuppofed to prevent the hydrophobia.

III.   1. Where the new veffels, and enlarged old ones, which
conftitute inflammation, are not fo haftily diftended as to burft, and
form a new kind of gland for the fecretion of matter, as above mentioned; if fuch circumftances happen as diminifh the painful fenfation, the tendency to growth ceafes, and by and by an abforption
commences, not only of the fuperabundant quantity of fluids depofited in the inflamed part, but of the folids likewife, and this even
of the hardeft kind.

Thus during the growth of the fecond fet of teeth in children,
the roots of the firft fet are totally abforbed, till at length nothing of
them remains but the crown; though a few weeks before, if they are
drawn immaturely, their roots are found complete.   Similar to this
Mr. Hunter has obferved, that where a dead piece of bone is to exfo-
liate, or to feparate from a living one, that the dead part does not putrify, but remains perfectly found, while the furface of the living
part of the bone, which is in contact with the dead part, becomes
abforbed, and thus effects its feparation.   Med. Comment. Edinb. V.
1. 425.   In the fame manner the calcareous matter of gouty concretions, the coagulable lymph depofited on inflamed membranes in rheumatifm and extravafated blood become abforbed; which are all as folid and as indiffoluble materials as the new veffels produced in inflammation.

This abforption of the new veffels and depofited fluids of inflamed parts is called refolution : it is produced by firft ufing fuch internal means as decreafe the pain of the part, and in confequence its new motions, as repeated bleeding, cathartics, diluent potations, and warm bath.

After the veffels are thus emptied, and the abforption of the new veffels and depofited fluids is evidently begun, it is much promoted by ftimulating the part externally by folutions of lead, or other metals, and internally by the bark, and fmall dofes of opium. ⁻Hence when an ophthalmy begins to become paler, any acrid eye-water, as a folution of fix grains of white vitriol in an ounce of water, haftens the abforption, and clears the eye in a very fhort time. But the fame application ufed a few days fooner would have increafed the inflammation. Hence after evacuation opium in fmall dofes may contribute to promote the abforption of fluids depofited on the brain, as obferved by Mr. Bromfield in his treatife of furgery.

2. Where an abfcefs is formed by the rupture of thefe new veffels, the violence of inflammation ceafes, and a new gland feparates a material called pus : at the fame time a lefs degree of inflammation produces new veffels called vulgarly proud flefh ; which, if no bandage confines its growth, nor any other circumftance promotes abforption in the wound, would rife to a great height above the ufual fize of the part.

Hence the art of healing ulcers confifts in producing a tendency to abforption in the wound greater than the depofition. Thus when an ill-conditioned ulcer feparates a copious and thin difcharge, by the ufe of any ftimulus, as of falts of lead, or mercury, or copper externally applied, the difcharge becomes diminifhed in quantity, and becomes thicker, as the thinner parts are firft abforbed.

But nothing fo much contributes to increafe the abforption in a wound as covering the whole limb above the fore with a bandage, which fhould be fpread with fome plafter, as with emplaftrum de minio, to prevent it from flipping. By this artificial tight-

.nefs

nefs of the fkin, the arterial pulfations act with double their ufual power in promoting the afcending current of the fluid in the valvular lymphatics.

Internally the abforption from ulcers fhould be promoted firft by evacuation, then by opium, bark, mercury, fteel.

3. Where the inflammation proceeds with greater violence or rapidity, that is, when by the painful fenfation a more inordinate activity of the organ is produced, and by this great activity an additional quantity of painful fenfation follows in an increafing ratio, till the whole of the fenforial power, or fpirit of animation, in the part becomes exhaufted, a mortification enfues, as in a carbuncle, in inflammations of the bowels, in the extremities of old people, or in the limbs of thofe who are brought near a fire after having been much benumbed with cold.   And from hence it appears, why weak people are more fubject to mortification than ftrong ones, and why in weak perfons lefs pain will produce mortification, namely, becaufe the fenforial power is fooner exhaufted by any excefs of activity.   I remember feeing a gentleman who had the preceding day travelled two ftages in a chaife with what he termed a bearable pain in his bowels ; which when I faw him had ceafed rather fuddenly, and without a paffage through him ; his pulfe was then weak, though not very quick ; but as nothing which he fwallowed would continue in his ftomach many minutes, I concluded that the bowel was mortified ; he died on the next day.   It is ufual for patients finking under the fmall-pox with mortified puftules, and with purple fpots intermixed, to complain of no pain, but to fay they are pretty well to the laft moment.

Recapitulation.

IV. When the motions of any part of the fyftem, in confequence of previous torpor, are performed with more energy than in the irritative fevers, a difagreeable fenfation is produced, and new actions of fome part of the fyftem commence in confequence of this fenfation conjointly with the irritation; which motions conftitute inflammation. If the fever be attended with a ftrong pulfe, as in pleurify, or rheumatifm, it is termed fynocha fenfitiva, or fenfitive fever with ftrong pulfe; which is ufually termed inflammatory fever. If it be attended with weak pulfe, it is termed typhus fenfitivus, or fenfitive fever with weak pulfe, or typhus gravior, or putrid malignant fever.

The fynocha fenfitiva, or fenfitive fever with ftrong pulfe, is generally attended with fome topical inflammation, as in peripneumony, hepatitis, and is accompanied with much coagulable lymph, or fize; which rifes to the furface of the blood, when taken into a bafon, as it cools; and which is believed to be the increafed mucous fecretion from the coats of the arteries, infpiffated by a greater abforption of its aqueous and faline part, and perhaps changed by its delay in the circulation.

The typhus fenfitivus, or fenfitive fever with weak pulfe, is frequently attended with delirium, which is caufed by the deficiency of the quantity of fenforial power, and with variety of cutaneous eruptions.

Inflammation is caufed by the pains occafioned by excefs of action, and not by thofe pains which are occafioned by defect of action. Thefe morbid actions, which are thus produced by two fenforial powers, viz. by irritation and fenfation, fecrete new living fibres,
which

which elongate the old veffels, or form new ones, and at the fame time much heat is evolved from thefe combinations. By the rupture of thefe veffels, or by a new conftruction of their apertures, purulent matters are fecreted of various kinds; which are infectious the firft time they are applied to the fkin beneath the cuticle, or fwallowed with the faliva into the ftomach. This contagion acts not by its being abforbed into the circulation, but by the fympathies, or affociated actions, between the part firft ftimulated by the contagious matter and the other parts of the fyftem. Thus in the natural fmall-pox the contagion is fwallowed with the faliva, and by its ftimulus inflames the ftomach; this variolous inflammation of the ftomach increafes every day, like the circle round the puncture of an inoculated arm, till it becomes great enough to diforder the circles of irritative and fenfitive motions, and thus produces fever-fits, with ficknefs and vomiting. Laftly, after the cold paroxyfm, or fit of torpor, of the ftomach has increafed for two or three fucceffive days, an inflammation of the fkin commences in points; which generally firft appear upon the face, as the affociated actions between the fkin of the face and that of the ftomach have been more frequently exerted together than thofe of any other parts of the external furface.

Contagious matters, as thofe of the meafles and fmall-pox, do not act upon the fyftem at the fame time; but the progrefs of that which was laft received is delayed, till the action of the former infection ceafes. All kinds of matter, even that from common ulcers, are probably contagious the firft time they are inferted beneath the cuticle or fwallowed into the ftomach; that is, as they were formed by certain morbid actions of the extremities of the veffels, they have the power to excite fimilar morbid actions in the extremities of other veffels, to which they are applied; and thefe by fympathy, or affociations of motion, excite fimilar morbid actions in diftant parts of the fyftem, without entering the circulation; and hence the blood of a patient in the fmall-pox will not give that difeafe by inoculation to others.

When

When the new fibres or veffels become again abforbed into the circulation, the inflammation ceafes ; which is promoted, after fufficient
evacuations, by external ftimulants and bandages : but where the action of the veffels is very great, a mortification of the part is liable to
enfue, owing to the exhauftion of fenforial power ; which however
occurs in weak people without much pain, and without very violent
previous inflammation ; and, like partial paralyfis, may be efteemed one
mode of natural death of old people, a part dying before the whole.

SECT.

SECT. XXXIV.

DISEASES OF VOLITION.

I. 1. *Volition defined. Motions termed involuntary are caused by volition. Defires oppofed to each other. Deliberation. Afs between two hay-cocks. Saliva fwallowed againft one's defire. Voluntary motions diftinguifhed from thofe affociated with fenfitive motions. 2. Pains from excefs, and from defect of motion. No pain is felt during vehement voluntary exertion; as in cold fits of ague, labour-pains, ftrangury, tenefmus, vomiting, reftleffnefs in fevers, convulfion of a wounded mufcle. 3. Of holding the breath and fcreaming in pain; why fwine and dogs cry out in pain, and not fheep and horfes. Of grinning and biting in pain; why mad animals bite others. 4. Epileptic convulfions explained, why the fits begin with quivering of the under jaw, biting the tongue, and fetting the teeth; why the convulfive motions are alternately relaxed. The phænomenon of laughter explained. Why children cannot tickle themfelves. How fome have died from immoderate laughter. 5. Of cataleptic fpafms, of the locked jaw, of painful cramps. 6. Syncope explained. Why no external objects are perceived in fyncope. 7. Of palfy and apoplexy from violent exertions. Cafe of Mrs. Scot. From dancing, fkating, fwimming. Cafe of Mr. Nairn. Why palfies are not always immediately preceded by violent exertions. Palfy and epilepfy from difeafed livers. Why the right arm more frequently paralytic than the left. How paralytic limbs regain their motions. II. Difeafes of the fenfual motions from excefs or defect of voluntary exertion. 1. Madnefs. 2. Diftinguifhed from delirium. 3. Why mankind more liable to infanity than brutes. 4. Sufpicion. Want of fhame, and of cleanlinefs. 5. They bear cold, hunger, and fatigue. Charles XII. of Sweden. 6. Pleafureable delirium, and infanity. Child riding on a ftick. Pains of martyrdom not felt. 7. Dropfy. 8. Inflammation cured by infanity. III. 1. Pain relieved by reverie. Reverie*

I. 1. BEFORE we commence this Section on Difeafed Voluntary Motions, it may be neceffary to premife, that the word volition is not ufed in this work exactly in its common acceptation.   Volition is faid in Section V. to bear the fame analogy to defire and averfion, which fenfation does to pleafure and pain.   And hence that, when defire or averfion produces any action of the mufcular fibres, or of the organs of fenfe, they are termed volition; and the actions produced in confequence are termed voluntary actions.   Whence it appears, that motions of our mufcles or ideas may be produced in confequence of defire or averfion without our having the power to prevent them, and yet thefe motions may be termed voluntary, according to our definition of the word; though in common language they would be called involuntary.

The objects of defire and averfion are generally at a diftance, whereas thofe of pleafure and pain are immediately acting upon our organs. Hence, before defire or averfion are exerted, fo as to caufe any actions, there is generally time for deliberation; which confifts in difcovering the means to obtain the object of defire, or to avoid the object of averfion; or in examining the good or bad confequences, which may refult from them.   In this cafe it is evident, that we have a power to delay the propofed action, or to perform it; and this power of choofing, whether we fhall act or not, is in common language expreffed by the word volition, or will.   Whereas in this work the word volition means fimply the active ftate of the fenforial faculty in producing motion in confequence of defire or averfion; whether we have the power of reftraining that action, or not; that is, whether we exert any actions in confequence of oppofite defires or averfions, or not.

For

For if the objects of defire or averfion are prefent, there is no neceffity to inveftigate or compare the *means* of obtaining them, nor do
we always deliberate about their confequences ; that is, no deliberation neceffarily intervenes, and in confequence the power of choofing
to act or not is not exerted. It is probable, that this twofold ufe of
the word volition in all languages has confounded the metaphyficians,
who have difputed about free will and neceffity. Whereas from the
above analyfis it would appear, that during our fleep, we ufe no voluntary exertions at all ; and in our waking hours, that they are the
confequence of defire or averfion.

  To will is to act in confequence of defire ; but to defire means to
defire fomething, even if that fomething be only to become free from
the pain, which caufes the defire ; for to defire nothing is not to defire ; the word defire, therefore, includes both the action and the object or motive ; for the object and motive of defire are the fame thing.
Hence to defire without an object, that is, without a motive, is a
folecifm in language. As if one fhould afk, if you could eat without
food, or breathe without air.

  From this account of volition it appears, that convulfions of the
mufcles, as in epileptic fits, may in the common fenfe of that word
be termed involuntary ; becaufe no deliberation is interpofed between
the defire or averfion and the confequent action ; but in the fenfe of
the word, as above defined, they belong to the clafs of voluntary motions, as delivered in Vol. II. Clafs III. If this ufe of the word be
difcordant to the ear of the reader, the term morbid voluntary motions, or motions in confequence of averfion, may be fubftituted in its
ftead.

  If a perfon has a defire to be cured of the ague, and has at the fame
time an averfion (or contrary defire) to fwallowing an ounce of Peruvian bark ; he balances defire againft defire, or averfion againft
averfion ; and thus he acquires the power of choofing, which is the
common acceptation of the word *willing*. But in the cold fit of ague,

3 H   after

after having difcovered that the act of fhuddering, or exerting the fubcutaneous mufcles, relieves the pain of cold; he immediately exerts
this act of volition, and fhudders, as foon as the pain and confequent
averfion return, without any deliberation intervening; yet is this act,
as well as that of fwallowing an ounce of the bark, caufed by volition; and that even though he endeavours in vain to prevent it by a
weaker contrary volition. This recalls to our minds the ftory of the
hungry afs between two hay-ftacks, where the two defires are fuppofed fo exactly to counteract each other, that he goes to neither of
the ftacks, but perifhes by want. Now as two equal and oppofite defires are thus fuppofed to balance each other, and prevent all action,
it follows, that if one of thefe hayftacks was fuddenly removed, that
the afs would irrefiftibly be hurried to the other, which in the common ufe of the word might be called an involuntary act; but which,
in our acceptation of it, would be claffed amongft voluntary actions,
as above explained.

Hence to deliberate is to compare oppofing defires or averfions, and
that which is the moft interefting at length prevails, and produces action. Similar to this, where two pains oppofe each other, the ftronger
or more interefting one produces action; as in pleurify the pain from
fuffocation would produce expanfion of the lungs, but the pain occafioned by extending the inflamed membrane, which lines the cheft,
oppofes this expanfion, and one or the other alternately prevails.

When any one moves his hand quickly near another perfon's eyes,
the eye-lids inftantly clofe; this act in common language is termed
involuntary, as we have not time to deliberate or to exert any contrary defire or averfion, but in this work it would be termed a voluntary act, becaufe it is caufed by the faculty of volition, and after a
few trials the nictitation can be prevented by a contrary or oppofing
volition.

The power of oppofing volitions is beft-exemplified in the ftory of
Mutius Scævola, who is faid to have thruft his hand into the fire be-

fore

fore Porcenna, and to have fuffered it to be confumed for having failed him in his attempt on the life of that general.   Here the averfion for the lofs of fame, or the unfatisfied defire to ferve his country, the two prevalent enthufiafms at that time, were more powerful than the defire of withdrawing his hand, which muft be occafioned by tho pain of combuftion; of thefe oppofing volitions

Vincit amor patriæ, laudumque immenfa cupido.

If any one is told not to fwallow his faliva for a minute, he foon fwallows it contrary to his will, in the common fenfe of that word; but this alfo is a voluntary action, as it is performed by the faculty of volition, and is thus to be underftood.   When the power of volition is exerted on any of our fenfes, they become more acute, as in our attempts to hear fmall noifes in the night.   As explained in Section XIX. 6.   Hence by our attention to the fauces from our defire not to fwallow our faliva; the fauces become more fenfible; and the ftimulus of the faliva is followed by greater fenfation, and confequent defire of fwallowing it.   So that the defire or volition in confequence of the increafed fenfation of the faliva is more powerful, than the previous defire not to fwallow it.   See Vol. II. Deglutitio invita.   In the fame manner if a modeft man wifhes not to want to make water, when he is confined with ladies in a coach or an affembly-room; that very act of volition induces the circumftance, which he wifhes to avoid, as above explained; infomuch that I once faw a partial infanity, which might be called a voluntary diabetes, which was occafioned by the fear (and confequent averfion) of not being able to make water at all.

It is further neceffary to obferve here, to prevent any confufion of voluntary, with fenfitive, or affociate motions, that in all the inftances of violent efforts to relieve pain, thofe efforts are at firft voluntary exertions; but after they have been frequently repeated for the pur-

pofe of relieving certain pains, they become affociated with thofe
pains, and ceafe at thofe times to be fubfervient to the will; as in
coughing, fneezing, and ftrangury.   Of thefe motions thofe which
contribute to remove or diflodge the offending caufe, as the actions of
the abdominal mufcles in parturition or in vomiting, though they
were originally excited by volition, are in this work termed fenfitive
motions; but thofe actions of the mufcles or organs of fenfe, which
do not contribute to remove the offending caufe, as in general convulfions or in madnefs, are in this work termed voluntary motions, or
motions in confequence of averfion, though in common language they
are called involuntary ones.   Thofe fenfitive unreftrainable actions,
which contribute to remove the caufe of pain are uniformly and invariably exerted, as in coughing or fneezing; but thofe motions which
are exerted in confequence of averfion without contributing to remove
the painful caufe, but only to prevent the fenfation of it, as in epileptic, or cataleptic fits, are not uniformly and invariably exerted,
but change from one fet of mufcles to another, as will be further explained; and may by this criterion alfo be diftinguifhed from the
former.

At the fame time thofe motions, which are excited by perpetual
ftimulus, or by affociation with each other, or immediately by pleafureable or painful fenfation, may properly be termed involuntary motions, as thofe of the heart and arteries; as the faculty of volition feldom affects thofe, except when it exifts in unnatural quantity, as in
maniacal people.

2. It was obferved in Section XIV. on the Production of Ideas,
that thofe parts of the fyftem, which are ufually termed the organs
of fenfe, are liable to be excited into pain by the excefs of the ftimulus of thofe objects, which are by nature adapted to affect them; as of
too great light, found, or preffure.   But that thefe organs receive no
pain from the defect or abfence of thefe ftimuli, as in darknefs or
filence.   But that our other organs of perception, which have generally

rally been called appetites, as of hunger, thirft, want of heat, want of frefh air, are liable to be affected with pain by the defect, as well as by the excefs of their appropriated ftimuli.

This excefs or defect of ftimulus is however to be confidered only as the remote caufe of the pain, the immediate caufe being the excefs or defect of the natural action of the affected part, according to Sect. IV. 5. Hence all the pains of the body may be divided into thofe from excefs of motion, and thofe from defect of motion ; which diftinction is of great importance in the knowledge and the cure of many difeafes. For as the pains from excefs of motion either gradually fubfide, or are in general fucceeded by inflammation ; fo thofe from defect of motion either gradually fubfide, or are in general fucceeded by convulfion, or madnefs. Thefe pains are eafily diftinguifhable from each other by this circumftance, that the former are attended with heat of the pained part, or of the whole body ; whereas the latter exifts without increafe of heat in the pained part, and is generally attended with coldnefs of the extremities of the body ; which is the true criterion of what have been called nervous pains.

Thus when any acrid material, as fnuff or lime, falls into the eye, pain and inflammation and heat are produced from the excefs of ftimulus ; but violent hunger, hemicrania, or the clavus hyftericus, are attended with coldnefs of the extremities, and defect of circulation. When we are expofed to great cold, the pain we experience from the deficiency of heat is attended with a quiefcence of the motions of the vafcular fyftem ; fo that no inflammation is produced, but a great defire of heat, and a tremulous motion of the fubcutaneous mufcles, which is properly a convulfion in confequence of this pain from defect of the ftimulus of heat.

It was before mentioned, that as fenfation confifts in certain movements of the fenforium, beginning at fome of the extremities of it, and propagated to the central parts of it ; fo volition confifts of certain other movements of the fenforium, commencing in the central

parts

parts of it, and propagated to fome of its extremities. This idea of thefe two great powers of motion in the animal machine is confirmed from obferving, that they never exift in a great degree or univerfally at the fame time; for while we ftrongly exert our voluntary motions, we ceafe to feel the pains or uneafineffes, which occafioned us to exert them.

Hence during the time of fighting with fifts or fwords no pain is felt by the combatants, till they ceafe to exert themfelves. Thus in the beginning of ague-fits the painful fenfation of cold is diminifhed, while the patient exerts himfelf in the fhivering and gnafhing of his teeth. He then ceafes to exert himfelf, and the pain of cold returns; and he is thus perpetually induced to reiterate thefe exertions, from which he experiences a temporary relief. The fame occurs in labour-pains, the exertion of the parturient woman relieves the violence of the pains for a time, which recur again foon after fhe has ceafed to ufe thofe exertions. The fame is true in many other painful difeafes, as in the ftrangury, tenefmus, and the efforts of vomiting; all thefe difagreeable fenfations are diminifhed or removed for a time by the various exertions they occafion, and recur alternately with thofe exertions.

The reftleffnefs in fome fevers is an almoft perpetual exertion of this kind, excited to relieve fome difagreeable fenfations; the reciprocal oppofite exertions of a wounded worm, the alternate emproftho-tonos and opifthotonos of fome fpafmodic difeafes, and the intervals of all convulfions, from whatever caufe, feem to be owing to this cir-cumftance of the laws of animation; that great or univerfal exertion cannot exift at the fame time with great or univerfal fenfation, though they can exift reciprocally; which is probably refolvable into the more general law, that the whole fenforial power being expanded in one mode of exertion, there is none to fpare for any other. Whence fyncope, or temporary apoplexy, fucceeds to epileptic convul-fions.

3. Hence

3. Hence when any violent pain afflicts us, of which we can neither avoid nor remove the caufe, we foon learn to endeavour to alle-
viate it, by exerting fome violent voluntary effort, as mentioned
above; and are naturally induced to ufe thofe mufcles for this purpofe, which have been in the early periods of our lives moft frequently or moft powerfully exerted.

Now the firft mufcles, which infants ufe moft frequently, are
thofe of refpiration; and on this account we gain a habit of holding
our breath, at the fame time that we ufe great efforts to exclude it,
for this purpofe of alleviating unavoidable pain; or we prefs out our
breath through a fmall aperture of the larinx, and fcream violently,
when the pain is greater than is relievable by the former mode of exertion.   Thus children fcream to relieve any pain either of body or
mind, as from anger, or fear of being beaten.

Hence it is curious to obferve, that thofe animals, who have more
frequently exerted their mufcles of refpiration violently, as in talking,
barking, or grunting, as children, dogs, hogs, fcream much more,
when they are in pain, than thofe other animals, who ufe little or
no language in their common modes of life; as horfes, fheep, and
cows.

The next moft frequent or moft powerful efforts, which infants are
firft tempted to produce, are thofe with the mufcles in biting hard
fubftances; indeed the exertion of thefe mufcles is very powerful in
common maftication, as appears from the pain we receive, if a bit of
bone is unexpectedly found amongft our fofter food; and further appears from their acting to fo great mechanical difadvantage, particularly when we bite with the incifores, or canine teeth; which are
firft formed, and thence are firft ufed to violent exertion.

Hence when a perfon is in great pain, the caufe of which he cannot remove, he fets his teeth firmly together, or bites fome fubftance
between them with great vehemence, as another mode of violent exertion to produce a temporary relief.   Thus we have a proverb where

no help can be had in pain, " to grin and abide ;" and the tortures of hell are faid to be attended with " gnafhing of teeth."

Hence in violent fpafmodic pains I have feen people bite not only their tongues, but their arms or fingers, or thofe of the attendants, or any object which was near them ; and alfo ftrike, pinch, or tear, others or themfelves, particularly the part of their own body, which is painful at the time. Soldiers, who die of painful wounds in battle, are faid by Homer to bite the ground. Thus alfo in the bellon, or colica faturnina, the patients are faid to bite their own flefh, and dogs in this difeafe to bite up the ground they lie upon. It is probable that the great endeavours to bite in mad dogs, and the violence of other mad animals, is owing to the fame caufe.

4. If the efforts of our voluntary motions are exerted with ftill greater energy for the relief of fome difagreeable fenfation, convulfions are produced ; as the various kinds of epilepfy, and in fome hyfteric paroxifms. In all thefe difeafes a pain or difagreeable fenfation is produced, frequently by worms, or acidity in the bowels, or by a difeafed nerve in the fide, or head, or by the pain of a difeafed liver.

In fome conftitutions a more intolerable degree of pain is produced in fome part at a diftance from the caufe by fenfitive affociation, as before explained ; thefe pains in fuch conftitutions arife to fo great a degree, that I verily believe no artificial tortures could equal fome, which I have witneffed ; and am confident life would not have long been preferved, unlefs they had been foon diminifhed or removed by the univerfal convulfion of the voluntary motions, or by temporary madnefs.

In fome of the unfortunate patients I have obferved, the pain has rifen to an inexpreffible degree, as above defcribed, before the convulfions have fupervened ; and which were preceded by fcreaming, and grinning ; in others, as in the common epilepfy, the convulfion has immediately fucceeded the commencement of the difagreeable

5   fenfations ;

fenfations; and as a ftupor frequently fucceeds the convulfions, they only feemed to remember that a pain at the ftomach preceded the fit, or fome other uneafy feel; or more frequently retained no memory at all of the immediate caufe of the paroxyfm.   But even in this kind of epilepfy, where the patient does not recollect any preceding pain, the paroxifms generally are preceded by a quivering motion of the under jaw, with a biting of the tongue; the teeth afterwards become preffed together with vehemence, and the eyes are then convulfed, before the commencement of the univerfal convulfion; which are all efforts to relieve pain.

The reafon why thefe convulfive motions are alternately exerted and remitted was mentioned above, and in Sect. XII. 1. 3. when the exertions are fuch as give a temporary relief to the pain, which excites them, they ceafe for a time, till the pain is again perceived; and then new exertions are produced for its relief.   We fee daily examples of this in the loud reiterated laughter of fome people; the pleafure-able fenfation, which excites this laughter, arifes for a time fo high as to change its name and become painful: the convulfive motions of the refpiratory mufcles relieve the pain for a time; we are, however, unwilling to lofe the pleafure, and prefently put a ftop to this ex-ertion, and immediately the pleafure recurs, and again as inftantly rifes into pain.   All of us have felt the pain of immoderate laughter; children have been tickled into convulfions of the whole body; and others have died in the act of laughing; probably from a paralyfis fucceeding the long continued actions of the mufcles of refpira-tion.

Hence we learn the reafon, why children, who are fo eafily ex-cited to laugh by the tickling of other people's fingers, cannot tickle themfelves into laughter.   The exertion of their hands in the en-deavour to tickle themfelves prevents the neceffity of any exertion of the refpiratory mufcles to relieve the excefs of pleafureable affection. See Sect. XVII. 3. 5.

3 I   Chryfippus

Chryſippus is recorded to have died laughing, when an aſs was in-vited to ſup with him.   The ſame is related of one of the popes, who, when he was ill, ſaw a tame monkey at his bedſide put on the holy thiara.   Hall. Phyſ. T. III. p. 306.

There are inſtances of epilepſy being produced by laughing recorded by Van Swieton, T. III. 402 and 308.   And it is well known, that many people have died inſtantaneouſly from the painful exceſs of joy, which probably might have been prevented by the exertions of laughter.

Every combination of ideas, which we attend to, occaſions pain or pleaſure ; thoſe which occaſion pleaſure, furniſh either ſocial or ſelfiſh pleaſure, either malicious or friendly, or laſcivious, or ſublime plea-ſure; that is, they give us pleaſure mixed with other emotions, or they give us unmixed pleaſure, without occaſioning any other emo-tions or exertions at the ſame time.   This unmixed pleaſure, if it be great, becomes painful, like all other animal motions from ſtimuli of every kind ; and if no other exertions are occaſioned at the ſame time, we uſe the exertion of laughter to relieve this pain.   Hence laughter is occaſioned by ſuch wit as excites ſimple pleaſure without any other emotion, ſuch as pity, love, reverence.   For ſublime ideas are mixed with admiration, beautiful ones with love, new ones with ſurpriſe ; and theſe exertions of our ideas prevent the action of laugh-ter from being neceſſary to relieve the painful pleaſure above deſcribed. Whence laughable wit conſiſts of frivolous ideas, without connections of any conſequence, ſuch as puns on words, or on phraſes, incon-gruous junctions of ideas; on which account laughter is ſo frequent in children.

Unmixed pleaſure leſs than that, which cauſes laughter, cauſes ſleep, as in ſinging children to ſleep, or in ſlight intoxication from wine or food.   See Sect. XVIII. 12.

5. If the pains, or diſagreeable ſenſations, above deſcribed do not obtain a temporary relief from theſe convulſive exertions of the muſ-

cles,

cles, thofe convulfive exertions continue without remiffion, and one kind of catalepfy is produced. Thus when a nerve or tendon pro- duces great pain by its being inflamed or wounded, the patient fets his teeth firmly together, and grins violently, to diminifh the pain; and if the pain is not relieved by this exertion, no relaxation of the maxillary mufcles takes place, as in the convulfions above defcribed, but the jaws remain firmly fixed together. This locked jaw is the moft frequent inftance of cataleptic fpafm, becaufe we are more in- clined to exert the mufcles fubfervient to maftication from their early obedience to violent efforts of volition.

But in the cafe related in Sect. XIX. on Reverie, the cataleptic lady had pain in her upper teeth; and preffing one of her hands ve- hemently againft her cheek-bone to diminifh this pain, it remained in that attitude for about half an hour twice a day, till the painful pa- roxyfm was over.

I have this very day feen a young lady in this difeafe, (with which fhe has frequently been afflicted,) fhe began to-day with violent pain fhooting from one fide of the forehead to the occiput, and after various ftruggles lay on the bed with her fingers and wrifts bent and ftiff for about two hours; in other refpects fhe feemed in a fyncope with a natural pulfe. She then had intervals of pain and of fpafm, and took three grains of opium every hour till fhe had taken nine grains, before the pains and fpafm ceafed.

There is, however, another fpecies of fixed fpafm, which differs from the former, as the pain exifts in the contracted mufcle, and would feem rather to be the confequence than the caufe of the con- traction, as in the cramp in the calf of the leg, and in many other parts of the body.

In thefe fpafms it fhould feem, that the mufcle itfelf is firft thrown into contraction by fome difagreeable fenfation, as of cold; and that then the violent pain is produced by the great contraction of the muf-

3 I 2

cular

cular fibres extending its own tendons, which are faid to be fenfible to extenfion only; and is further explained in Sect. XVIII. 15.

6. Many inftances have been given in this work, where after violent motions excited by irritation, the organ has become quiefcent to lefs, and even to the great irritation, which induced it into violent motion; as after looking long at the fun or any bright colour, they ceafe to be feen; and after removing from bright day-light into a gloomy room, the eye cannot at firft perceive the objects, which ftimulate it lefs. Similar to this is the fyncope, which fucceeds after the violent exertions of our voluntary motions, as after epileptic fits, for the power of volition acts in this cafe as the ftimulus in the other. This fyncope is a temporary palfy, or apoplexy, which ceafes after a time, the mufcles recovering their power of being excited into action by the efforts of volition; as the eye in the circumftance above mentioned recovers in a little time its power of feeing objects in a gloomy room; which were invifible immediately after coming out of a ftronger light. This is owing to an accumulation of fenforial power during the inaction of thofe fibres, which were before accuftomed to perpetual exertions, as explained in Sect. XII. 7. 1. A flighter degree of this difeafe is experienced by every one after great fatigue, when the mufcles gain fuch inability to further action, that we are obliged to reft them for a while, or to fummon a greater power of volition to continue their motions.

In all the fyncopes, which I have feen induced after convulfive fits, the pulfe has continued natural, though the organs of fenfe as well as the locomotive mufcles, have ceafed to perform their functions; for it is neceffary for the perception of objects, that the external organs of fenfe fhould be properly excited by the voluntary power, as the eye-lids muft be open, and perhaps the mufcles of the eye put into action to diftend, and thence give greater pellucidity to the cornea, which in fyncope, as in death, appears flat and lefs tranfparent.

The

The tympanum of the ear alfo feems to require a voluntary exertion of its mufcles, to gain its due tenfion, and it is probable the other external organs of fenfe require a fimilar voluntary exertion to adapt them to the diftinct perception of objects. Hence in fyncope as in fleep, as the power of volition is fufpended, no external objects are perceived. See Sect. XVIII. 5. During the time which the patient lies in a fainting fit, the fpirit of animation becomes accumulated; and hence the mufcles in a while become irritable by their ufual ftimulation, and the fainting fit ceafes. See Sect. XII. 7. 1.

7. If the exertion of the voluntary motions has been ftill more energetic, the quiefcence, which fucceeds, is fo complete, that they cannot again be excited into action by the efforts of the will. In this manner the palfy, and apoplexy (which is an univerfal palfy) are frequently produced after convulfions, or other violent exertions; of this I fhall add a few inftances.

Platernus mentions fome, who have died apoplectic from violent exertions in dancing; and Dr. Mead, in his Effay on Poifons, records a patient in the hydrophobia, who at one effort broke the cords which bound him, and at the fame inftant expired. And it is probable, that thofe, who have expired from immoderate laughter, have died from this paralyfis confequent to violent exertion. Mrs. Scott of Stafford was walking in her garden in perfect health with her neighbour Mrs. ———; the latter accidentally fell into a muddy rivulet, and tried in vain to difengage herfelf by the affiftance of Mrs. Scott's hand. Mrs. Scott exerted her utmoft power for many minutes, firft to affift her friend, and next to prevent herfelf from being pulled into the morafs, as her diftreffed companion would not difengage her hand. After other affiftance was procured by their united fcreams, Mrs. Scott walked to a chair about twenty yards from the brook, and was feized with an apoplectic ftroke; which continued many days, and terminated in a total lofs of her right arm, and her fpeech; neither of which fhe ever after perfectly recovered.

It

It is faid, that many people in Holland have died after fkating too long or too violently on their frozen canals ; it is probable the death of thefe, and of others, who have died fuddenly in fwimming, has been owing to this great quiefcence or paralyfis ; which has fucceeded very violent exertions, added to the concomitant cold, which has had greater effect after the fufferers had been heated and exhaufted by previous exercife.

I remember a young man of the name of Nairne at Cambridge, who walking on the edge of a barge fell into the river. His coufin and fellow-ftudent of the fame name, knowing the other could not fwim, plunged into the water after him, caught him by his clothes, and approaching the bank by a vehement exertion propelled him fafe to the land, but that inftant, feized, as was fuppofed, by the cramp, or paralyfis, funk to rife no more. The reafon why the cramp of the mufcles, which compofe the calf of the leg, is fo liable to affect fwimmers, is, becaufe thefe mufcles have very weak antagonifts, and are in walking generally elongated again after their contraction by the weight of the body on the ball of the toe, which is very much greater than the refiftance of the water in fwimming. See Section XVIII. 15.

It does not follow that every apoplectic or paralytic attack is immediately preceded by vehement exertion ; the quiefcence, which fucceeds exertion, and which is not fo great as to be termed paralyfis, frequently recurs afterwards at certain periods ; and by other caufes of quiefcence, occurring with thofe periods, as was explained in treating of the paroxyfms of intermitting fevers ; the quiefcence at length becomes fo great as to be incapable of again being removed by the efforts of volition, and complete paralyfis is formed. See Section XXXII. 3. 2.

Many of the paralytic patients, whom I have feen, have evidently had difeafed livers from the too frequent potation of fpirituous liquors ; fome of them have had the gutta rofea on their faces and breafts ;
which

which has in some degree receded either spontaneously, or by the use of external remedies, and the paralytic stroke has succeeded; and as in several persons, who have drank much vinous spirits, I have observed epileptic fits to commence at about forty or fifty years of age, without any hereditary taint, from the stimulus, as I believed, of a diseased liver; I was induced to ascribe many paralytic cases to the same source; which were not evidently the effect of age, or of un-acquired debility.   And the account given before of dropsies, which very frequently are owing to a paralysis of the absorbent system, and are generally attendant on free drinkers of spirituous liquors, confirmed me in this opinion.

The disagreeable irritation of a diseased liver produces exertions and consequent quiescence; these by the accidental concurrence of other causes of quiescence, as cold, solar or lunar periods, inanition, the want of their usual portion of spirit of wine, at length produces paralysis.

This is further confirmed by observing, that the muscles, we most frequently, or most powerfully exert, are most liable to palsy; as those of the voice and of articulation, and of those paralytics which I have seen, a much greater proportion have lost the use of their right arm; which is so much more generally exerted than the left.

I cannot dismiss this subject without observing, that after a paralytic stroke, if the vital powers are not much injured, that the patient has all the movements of the affected limb to learn over again, just as in early infancy; the limb is first moved by the irritation of its muscles, as in stretching, (of which a case was related in Section VII. 1. 3.) or by the electric concussion; afterwards it becomes obedient to sensation, as in violent danger or fear; and lastly, the muscles become again associated with volition, and gradually acquire their usual habits of acting together.

Another phænomenon in palsies is, that when the limbs of one side

are

are difabled, thofe of the other are in perpetual motion.   This can only be explained from conceiving that the power of motion, whatever it is, or wherever it refides, and which is capable of being exhaufted by fatigue, and accumulated in reft, is now lefs expended, whilft one half of the body is incapable of receiving its ufual proportion of it, and is hence derived with greater eafe or in greater abundance into the limbs, which remain unaffected.

II. 1. The excefs or defect of voluntary exertion produces fimilar effects upon the fenfual motions, or ideas of the mind, as thofe already mentioned upon the mufcular fibres.   Thus when any violent pain, arifing from the defect of fome peculiar ftimulus, exifts either in the mufcular or fenfual fyftems of fibres, and which cannot be removed by acquiring the defective ftimulus; as in fome conftitutions convulfions of the mufcles are produced to procure a temporary relief, fo in other conftitutions vehement voluntary exertions of the ideas of the mind are produced for the fame purpofe; for during this exertion, like that of the mufcles, the pain either vanifhes or is diminifhed: this violent exertion conftitutes madnefs; and in many cafes I have feen the madnefs take place, and the convulfions ceafe, and reciprocally the madnefs ceafe, and the convulfions fupervene.   See Section III. 5. 8.

2. Madnefs is diftinguifhable from delirium, as in the latter the patient knows not the place where he refides, nor the perfons of his friends or attendants, nor is confcious of any external objects, except when fpoken to with a louder voice, or ftimulated with unufual force, and even then he foon relapfes into a ftate of inattention to every thing about him.   Whilft in the former he is perfectly fenfible to every thing external, but has the voluntary powers of his mind intenfely exerted on fome particular object of his defire or averfion, he harbours in his thoughts a fufpicion of all mankind, left they fhould counteract his defigns; and while he keeps his intentions, and the

motives

motives of his actions profoundly secret; he is perpetually studying
the means of acquiring the object of his wish, or of preventing or re-
venging the injuries he suspects.

3. A late French philosopher, Mr. Helvetius, has deduced almost
all our actions from this principle of their relieving us from the ennui
or tædium vitæ; and true it is, that our desires or aversions are the
motives of all our voluntary actions; and human nature seems to excel other animals in the more facil use of this voluntary power, and
on that account is more liable to insanity than other animals. But
in mania this violent exertion of volition is expended on mistaken objects, and would not be relieved, though we were to gain or escape
the objects, that excite it. Thus I have seen two instances of mad-
men, who conceived that they had the itch, and several have believed
they had the venereal infection, without in reality having a symptom
of either of them. They have been perpetually thinking upon this
subject, and some of them were in vain salivated with design of con-
vincing them to the contrary.

4. In the minds of mad people those volitions alone exist, which
are unmixed with sensation; immoderate suspicion is generally the
first symptom, and want of shame, and want of delicacy about cleanliness. Suspicion is a voluntary exertion of the mind arising from the
pain of fear, which it is exerted to relieve: shame is the name of a
peculiar disagreeable sensation, see Fable of the Bees, and delicacy
about cleanliness arises from another disagreeable sensation. And
therefore are not found in the minds of maniacs, which are employed
solely in voluntary exertions. Hence the most modest women in
this disease walk naked amongst men without any kind of concern,
use obscene discourse, and have no delicacy about their natural evacuations.

5. Nor are maniacal people more attentive to their natural appetites, or to the irritations which surround them, except as far as may
respect their suspicions or designs; for the violent and perpetual ex-

3 K   ertions

ertions of their voluntary powers of mind prevents their perception of almoft every other object, either of irritation or of fenfation. Hence it is that they bear cold, hunger, and fatigue, with much greater pertinacity than in their fober hours, and are lefs injured by them in refpect to their general health. Thus it is afferted by hiftorians, that Charles the Twelfth of Sweden flept on the fnow, wrapped only in his cloak, at the fiege of Frederickftad, and bore extremes of cold, and hunger, and fatigue, under which numbers of his foldiers perifh-ed; becaufe the king was infane with ambition, but the foldier had no fuch powerful ftimulus to preferve his fyftem from debility and death.

6. Befides the infanities arifing from exertions in confequence of pain, there is alfo a pleafureable infanity, as well as a pleafureable delirium; as the infanity of perfonal vanity, and that of religious fa-naticifm. When agreeable ideas excite into motion the fenforial power of fenfation, and this again caufes other trains of agreeable ideas, a conftant ftream of pleafureable ideas fucceeds, and produces pleafureable delirium. So when the fenforial power of volition ex-cites agreeable ideas, and the pleafure thus produced excites more vo-lition in its turn, a conftant flow of agreeable voluntary ideas fuc-ceeds; which when thus exerted in the extreme conftitutes in-fanity.

Thus when our mufcular actions are excited by our fenfations of pleafure, it is termed play; when they are excited by our volition, it is termed work; and the former of thefe is attended with lefs fa-tigue, becaufe the mufcular actions in play produce in their turn more pleafureable fenfation; which again has the property of pro-ducing more mufcular action. An agreeable inftance of this I faw this morning. A little boy, who was tired with walking, begged of his papa to carry him. " Here," fays the reverend doctor, " ride upon my gold-headed cane;" and the pleafed child, putting it be-tween his legs, gallopped away with delight, and complained no

more

more of his fatigue.   Here the aid of another fenforial power, that of pleafureable fenfation, fuperadded vigour to the exertion of exhaufted volition.   Which could otherwife only have been excited by additional pain, as by the lafh of flavery.   On this account where the whole fenforial power has been exerted on the contemplation of the promifed joys of heaven, the faints of all perfecuted religions have borne the tortures of martyrdom with otherwife unaccountable firmnefs.

7.   There are fome difeafes, which obtain at leaft a temporary relief from the exertions of infanity; many inftances of dropfies being thus for a time cured are recorded.   An elderly woman labouring with afcites I twice faw relieved for fome weeks by infanity, the dropfy ceafed for feveral weeks, and recurred again alternating with the infanity.   A man afflicted with difficult refpiration on lying down, with very irregular pulfe, and œdematous legs, whom I faw this day, has for above a week been much relieved in refpect to all thofe fymptoms by the acceffion of infanity, which is fhewn by inordinate fufpicion, and great anger.

In cafes of common temporary anger the increafed action of the arterial fyftem is feen by the red fkin, and increafed pulfe, with the immediate increafe of mufcular activity.   A friend of mine, when he was painfully fatigued by riding on horfeback, was accuftomed to call up ideas into his mind, which ufed to excite his anger or indignation, and thus for a time at leaft relieved the pain of fatigue.   By this temporary infanity, the effect of the voluntary power upon the whole of his fyftem was increafed; as in the cafes of dropfy above mentioned, it would appear, that the increafed action of the voluntary faculty of the fenforium affected the abforbent fyftem, as well as the fecerning one.

8.   In refpect to relieving inflammatory pains, and removing fever, I have feen many inftances, as mentioned in Sect. XII. 2. 4.   One lady, whom I attended, had twice at fome years interval a locked

jaw,

jaw, which relieved a pain on her ſternum with peripneumony. Two
other ladies I ſaw, who towards the end of violent peripneumony, in
which they frequently loſt blood, were at length cured by inſanity
ſupervening. In the former the increaſed voluntary exertion of the
muſcles of the jaw, in the latter that of the organs of ſenſe, removed
the diſeaſe; that is, the diſagreeable ſenſation, which had produced
the inflammation, now excited the voluntary power, and theſe new
voluntary exertions employed or expended the ſuperabundant ſenſorial
power, which had previouſly been exerted on the arterial ſyſtem, and
cauſed inflammation.

Another caſe, which I think worth relating, was of a young man
about twenty; he had laboured under an irritative fever with debility
for three or four weeks, with very quick and very feeble pulſe, and
other uſual ſymptoms of that ſpecies of typhus, but at this time complained much and frequently of pain of his legs and feet. When thoſe
who attended him were nearly in deſpair of his recovery, I obſerved
with pleaſure an inſanity of mind ſupervene: which was totally different from delirium, as he knew his friends, calling them by their
names, and the room in which he lay, but became violently ſuſpici-
ous of his attendants, and calumniated with vehement oaths his tender mother, who ſat weeping by his bed. On this his pulſe became
ſlower and firmer, but the quickneſs did not for ſome time intirely
ceaſe, and he gradually recovered. In this caſe the introduction of an
increaſed quantity of the power of volition gave vigour to thoſe movements of the ſyſtem, which are generally only actuated by the power
of irritation, and of aſſociation.

Another caſe I recollect of a young man, about twenty-five, who
had the ſcarlet-fever, with very quick pulſe, and an univerſal eruption on his ſkin, and was not without reaſon eſteemed to be in great
danger of his life. After a few days an inſanity ſupervened, which
his friends miſtook for delirium, and he gradually recovered, and the

cuticle

cuticle peeled off.   From thefe and a few other cafes I have always efteemed infanity to be a favourable fign in fevers, and have cautioufly diftinguifhed it from delirium.

III. Another mode of mental exertion to relieve pain, is by producing a train of ideas not only by the efforts of volition, as in infanity; but by thofe of fenfation likewife, as in delirium and fleep. This mental effort is termed reverie, or fomnambulation, and is defcribed more at large in Sect. XIX. on that fubject.   But I fhall here relate another cafe of that wonderful difeafe, which fell yefterday under my eye, and to which I have feen many analogous alienations of mind, though not exactly fimilar in all circumftances.   But as all of them either began or terminated with pain or convulfion, there can be no doubt but that they are of epileptic origin, and conftitute another mode of mental exertion to relieve fome painful fenfation.

1. Mafter A. about nine years old, had been feized at feven every morning for ten days with uncommon fits, and had had flight returns in the afternoon.   They were fuppofed to originate from worms, and had been in vain attempted to be removed by vermifuge purges.   As his fit was expected at feven yefterday morning, I faw him before that hour; he was afleep, feemed free from pain, and his pulfe natural.   About feven he began to complain of pain about his navel, or more to the left fide, and in a few minutes had exertions of his arms and legs like fwimming.   He then for half an hour hunted a pack of hounds; as appeared by his hallooing, and calling the dogs by their names, and difcourfing with the attendants of the chafe, defcribing exactly a day of hunting, which (I was informed) he had witnefled a year before, going through all the moft minute circumftances of it; calling to people, who were then prefent, and lamenting the abfence of others, who were then alfo abfent.   After this fcene he imitated, as he lay in bed,

some

some of the plays of boys, as swimming and jumping. He then sung an English and then an Italian song; part of which with his eyes open, and part with them closed, but could not be awakened or excited by any violence, which it was proper to use.

After about an hour he came suddenly to himself with apparent surprise, and seemed quite ignorant of any part of what had passed, and after being apparently well for half an hour, he suddenly fell into a great stupor, with slower pulse than natural, and a slow moaning respiration, in which he continued about another half hour, and then recovered.

The sequel of this disease was favourable; he was directed one grain of opium at six every morning, and then to rise out of bed; at half past six he was directed fifteen drops of laudanum in a glass of wine and water. The first day the paroxysm became shorter, and less violent. The dose of opium was increased to one-half more, and in three or four days the fits left him. The bark and filings of iron were also exhibited twice a day; and I believe the complaint returned no more.

2. In this paroxysm it must be observed, that he began with pain, and ended with stupor, in both circumstances resembling a fit of epilepsy. And that therefore the exertions both of mind and body, both the voluntary ones, and those immediately excited by pleasureable sensation, were exertions to relieve pain.

The hunting scene appeared to be rather an act of memory than of imagination, and was therefore rather a voluntary exertion, though attended with the pleasureable eagerness, which was the consequence of those ideas recalled by recollection, and not the cause of them.

These ideas thus voluntarily recollected were succeeded by sensations of pleasure, though his senses were unaffected by the sti-

muli

muli of vifible or audible objects; or fo weakly excited by them as not to produce fenfation or attention. And the pleafure thus excited by volition produced other ideas and other motions in confequence of the fenforial power of fenfation. Whence the mixed catenations of voluntary and fenfitive ideas and mufcular motions in reverie; which, like every other kind of vehement exertion, contribute to relieve pain, by expending a large quantity of fenforial power.

Thofe fits generally commence during fleep, from whence I fuppofe they have been thought to have fome connection with fleep, and have thence been termed Somnambalifm; but their commencement during fleep is owing to our increafed excitability by internal fenfations at that time, as explained in Sect. XVIII. 14 and 15, and not to any fimilitude between reverie and fleep.

3. I was once concerned for a very elegant and ingenious young lady, who had a reverie on alternate days, which continued nearly the whole day; and as in her days of difeafe fhe took up the fame kind of ideas, which fhe had converfed about on the alternate day before, and could recollect nothing of them on her well-day; fhe appeared to her friends to poffefs two minds. This cafe alfo was of epileptic kind, and was cured, with fome relapfes, by opium adminiftered before the commencement of the paroxyfm.

4 Whence it appears, that the methods of relieving inflammatory pains, is by removing all ftimulus, as by venefection, cool air, mucilaginous diet, aqueous potation, filence, darknefs.

The methods of relieving pains from defect of ftimulus is by fupplying the peculiar ftimulus required, as of food, or warmth.

And the general method of relieving pain is by exciting into action fome great part of the fyftem for the purpofe of expending a part of the fenforial power. This is done either by exertion of the voluntary ideas and mufcles, as in infanity and convulfion; or by

exerting

exerting both voluntary and fenfitive motions, as in reverie ; or by exciting the irritative motions by wine or opium internally, and by the warm bath or blifters externally ; or laftly, by exciting the fenfitive ideas by good news, affecting ftories, or agreeable paffions.

SECT.

SECT. XXXV.

DISEASES OF ASSOCIATION.

I.   1. *Sympathy or confent of parts.   Primary and fecondary parts of an affociated train of motions reciprocally affect each other.   Parts of irritative trains of motion affect each other in four ways.   Sympathies of the fkin and ftomach.   Flufh-ing of the face after a meal.   Eruption of the fmall-pox on the face.   Chilnefs after a meal.   2. Vertigo from intoxication.   3. Abforption from the lungs and pericardium by emetics.   In vomiting the actions of the ftomach are decreafed, not increafed.   Digeftion ftrengthened after an emetic.   Vomitting from deficiency of fenforial power.   4. Dyfpnœa from cold bathing.   Slow pulfe from digitalis.   Death from gout in the ftomach.   II.   1. Primary and fecondary parts of fen-fitive affociations affect each other.   Pain from gall-ftone, from urinary ftone.   Hemicrania.   Painful epilepfy.   2. Gout and red face from inflamed liver.   Shingles from inflamed kidney.   3. Coryza from cold applied to the feet.   Pleu-rify.   Hepatitis.   4. Pain of fhoulders from inflamed liver.   III. Difeafes from the affociations of ideas.*

I.   1. MANY fynchronous and fucceffive motions of our mufcular fibres, and of our organs of fenfe, or ideas, become affociated fo as to form indiffoluble tribes or trains of action, as fhewn in Section X. on Affociate Motions.   Some conftitutions more eafily eftablifh thefe af-fociations, whether by voluntary, fenfitive, or irritative repetitions, and fome more eafily lofe them again, as fhewn in Section XXXI. on Temperaments.

When the beginning of fuch a train of actions becomes by any means difordered, the fucceeding part is liable to become difturbed in confequence, and this is commonly termed fympathy or confent of parts by the writers of medicine.   For the more clear underftanding of thefe fympathies we muft confider a tribe or train of actions as divided into two parts, and call one of them the primary or original motions, and the other the fecondary or fympathetic ones.

The primary and fecondary parts of a train of irritative actions may reciprocally affect each other in four different manners.   1. They may both be exerted with greater energy than natural.   2. The former may act with greater, and the latter with lefs energy.   3. The former may act with lefs, and the latter with greater energy.   4. They may both act with lefs energy than natural.   I fhall now give an example of each kind of thefe modes of action, and endeavour to fhew, that though the primary and fecondary parts of thefe trains or tribes of motion are connected by irritative affociation, or their previous habits of acting together, as defcribed in Sect. XX. on Vertigo.   Yet that their acting with fimilar or diffimilar degrees of energy, depends on the greater or lefs quantity of fenforial power, which the primary part of the train expends in its exertions.

The actions of the ftomach conftitute fo important a part of the affociations of both irritative and fenfitive motions, that it is faid to fympathize with almoft every part of the body; the firft example, which I fhall adduce to fhew that both the primary and fecondary parts of a train of irritative affociations of motion act with increafed energy, is taken from the confent of the fkin with this organ.   When the action of the fibres of the ftomach is increafed, as by the ftimulus of a full meal, the exertions of the cutaneous arteries of the face become increafed by their irritative affociations with thofe of the ftomach, and a glow or flufhing of the face fucceeds.   For the fmall veffels of the fkin of the face having been more accuftomed to the varieties of action, from their frequent expofure to various degrees of cold and heat become more eafily excited into increafed action, than thofe of the covered parts of our bodies, and thus act with more

3   energy

energy from their irritative or fenfitive affociations with the ftomach. On this account in fmall-pox the eruption in confequence of the previous affection of the ftomach breaks out a day fooner on the face than on the hands, and two days fooner than on the trunk, and recedes in fimilar times after maturation.

But fecondly, in weaker conftitutions, that is, in thofe who poffefs lefs fenforial power, fo much of it is expended in the increafed actions of the fibres of the ftomach excited by the ftimulus of a meal, that a fenfe of chilnefs fucceeds inftead of the univerfal glow above mentioned; and thus the fecondary part of the affociated train of motions is diminifhed in energy, in confequence of the increafed activity of the primary part of it.

2. Another inftance of a fimilar kind, where the fecondary part of the train acts with lefs energy in confequence of the greater exertions of the primary part, is the vertigo attending intoxication; in this circumftance fo much fenforial power is expended on the ftomach, and on its neareft or more ftrongly affociated motions, as thofe of the fubcutaneous veffels, and probably of the membranes of fome internal vifcera, that the irritative motions of the retina become imperfectly exerted from deficiency of fenforial power, as explained in Sect. XX. and XXI. on Vertigo and on Drunkennefs, and hence the ftaggering inebriate cannot completely balance himfelf by fuch indiftinct vifion.

3. An inftance of the third circumftance, where the primary part of a train of irritative motions acts with lefs, and the fecondary part with greater energy, may be obferved by making the following experiment. If a perfon lies with his arms and fhoulders out of bed, till they become cold, a temporary coryza or catarrh is produced; fo that the paffage of the noftrils becomes totally obftructed; at leaft this happens to many people; and then on covering the arms and fhoulders, till they become warm, the paffage of the noftrils ceafes

3 L 2

again

again to be obſtructed, and a quantity of mucus is diſcharged from them. In this caſe the quieſcence of the veſſels of the ſkin of the arms and ſhoulders, occaſioned by expoſure to cold air, produces by irritative aſſociation an increaſed action of the veſſels of the membrane of the noſtrils; and the accumulation of ſenſorial power during the torpor of the arms and ſhoulders is thus expended in producing a temporary coryza or catarrh.

Another inſtance may be adduced from the ſympathy or conſent of the motions of the ſtomach with other more diſtant links of the very extenſive tribes or trains of irritative motions aſſociated with them, deſcribed in Sect. XX. on Vertigo. When the actions of the fibres of the ſtomach are diminiſhed or inverted, the actions of the abſorbent veſſels, which take up the mucus from the lungs, pericardium, and other cells of the body, become increaſed, and abſorb the fluids accumulated in them with greater avidity, as appears from the exhibition of foxglove, antimony, or other emetics in caſes of anaſarca, attended with unequal pulſe and difficult reſpiration.

That the act of nauſea and vomiting is a decreaſed exertion of the fibres of the ſtomach may be thus deduced; when an emetic medicine is adminiſtered, it produces the pain of ſickneſs, as a diſagreeable taſte in the mouth produces the pain of nauſea; theſe pains, like that of hunger, or of cold, or like thoſe, which are uſually termed nervous, as the head-ach or hemicrania, do not excite the organ into greater action; but in this caſe I imagine the pains of ſickneſs or of nauſea counteract or deſtroy the pleaſureable ſenſation, which ſeems neceſſary to digeſtion, as ſhewn in Sect. XXXIII. r. 1. The periſtaltic motions of the fibres of the ſtomach become enfeebled by the want of this ſtimulus of pleaſureable ſenſation, and in conſequence ſtop for a time, and then become inverted; for they cannot become inverted without being previouſly ſtopped. Now that this inverſion of the trains of motion of the fibres of the ſtomach is owing to the de-

6

ficiency

ficiency of pleafureable fenfation is evinced from this circumftance, that a naufeous idea excited by words will produce vomiting as effectually as a naufeous drug.

Hence it appears, that the act of naufea or vomiting expends lefs fenforial power than the ufual periftaltic motions of the ftomach in the digeftion of our aliment; and that hence there is a greater quantity of fenforial power becomes accumulated in the fibres of the ftomach, and more of it in confequence to fpare for the action of thofe parts of the fyftem, which are thus affociated with the ftomach, as of the whole abforbent feries of veffels, and which are at the fame time excited by their ufual ftimuli.

From this we can underftand, how after the operation of an emetic the ftomach becomes more irritable and fenfible to the ftimulus, and the pleafure of food; fince as the fenforial power becomes accumulated during the naufea and vomiting, the digeftive power is afterwards exerted more forceably for a time. It fhould, however, be here remarked, that though vomiting is in general produced by the defect of this ftimulus of pleafureable fenfation, as when a naufeous drug is adminiftered; yet in long continued vomiting, as in fea-ficknefs, or from habitual dram-drinking, it arifes from deficiency of fenforial power, which in the former cafe is exhaufted by the increafed exertion of the irritative ideas of vifion, and in the latter by the frequent application of an unnatural ftimulus.

4. An example of the fourth circumftance above mentioned, where both the primary and fecondary parts of a train of motions proceed with energy lefs than natural, may be obferved in the dyfpnœa, which occurs in going into a very cold bath, and which has been defcribed and explained in Sect. XXXII. 3. 2.

And by the increafed debility of the pulfations of the heart and arteries during the operation of an emetic. Secondly, from the flownefs and intermiffion of the pulfations of the heart from the inceffant efforts to vomit occafioned by an overdofe of digitalis. And thirdly,

from the total ftoppage of the motions of the heart, or death, in confequence of the torpor of the ftomach, when affected with the commencement or cold paroxyfm of the gout.   See Sect. XXV. 17.

II.   1. The primary and fecondary parts of the trains of fenfitive
affociation reciprocally affect each other in different manners.   1. The
increafed fenfation of the primary part may ceafe, when that of the
fecondary part commences.   2. The increafed action of the primary
part may ceafe, when that of the fecondary part commences.   3.
The primary part may have increafed fenfation, and the fecondary
part increafed action.   4. The primary part may have increafed action, and the fecondary part increafed fenfation.

Examples of the firft mode, where the increafed fenfation of the
primary part of a train of fenfitive affociation ceafes, when that of
the fecondary part commences, are not unfrequent; as this is the
general origin of thofe pains, which continue fome time without
being attended with inflammation, fuch as the pain at the pit of the
ftomach from a ftone at the neck of the gall-bladder, and the pain
of ftrangury in the glans penis from a ftone at the neck of the urinary
bladder.   In both thefe cafes the part, which is affected fecondarily,
is believed to be much more fenfible than the part primarily affected,
as defcribed in the catalogue of difeafes, Clafs II. 1. 1. 10. and IV. 2.
1. 1. and IV. 2. 1. 2.

The hemicrania, or nervous headach, as it is called, when it ori-
ginates from a decaying tooth, is another difeafe of this kind; as the
pain of the carious tooth always ceafes, when the pain over one eye
and temple commences.   And it is probable, that the violent pains,
which induce convulfions in painful epilepfies, are produced in the
fame manner, from a more fenfible part fympathizing with a difeafed
one of lefs fenfibility.   See Catalogue of Difeafes, Clafs IV. 2. 1. 5.
and III. 1. 1. 7.

The laft tooth, or dens fapientiæ, of the upper jaw moft frequently decays firft, and is liable to produce pain over the eye and temple

of

of that fide.   The laft tooth of the under-jaw is alfo liable to produce a fimilar hemicrania, when it begins to decay.   When a tooth in the upper-jaw is the caufe of the headach, a flighter pain is fometimes perceived on the cheek-bone.   And when a tooth in the lower-jaw is the caufe of headach, a pain fometimes affects the tendons of the mufcles of the neck, which are attached near the jaws.   But the clavus hyftericus, or pain about the middle of the parietal bone on one fide of the head, I have feen produced by the fecond of the molares, or grinders, of the under-jaw; of which I fhall relate the following cafe.   See Clafs II. 1. 1. 4. and IV. 2. 1. 5.

Mrs. ——, about 30 years of age, was feized with great pain about the middle of the right parietal bone, which had continued a whole day before I faw her, and was fo violent as to threaten to occafion convulfions.   Not being able to detect a decaying tooth, or a tender one, by examination with my eye, or by ftriking them with a tea-fpoon, and fearing bad confequences from her tendency to convulfion, I advifed her to extract the laft tooth of the under-jaw on the affected fide; which was done without any good effect.   She was then directed to lofe blood, and to take a brifk cathartic; and after that had operated, about 60 drops of laudanum were given her, with large dofes of bark; by which the pain was removed.   In about a fortnight fhe took a cathartic medicine by ill advice, and the pain returned with greater violence in the fame place; and, before I could arrive, as fhe lived 30 miles from me, fhe fuffered a paralytic ftroke; which affected her limbs and her face on one fide, and relieved the pain of her head.

About a year afterwards I was again called to her on account of a pain as violent as before exactly on the fame part of the other parietal bone.   On examining her mouth I found the fecond molaris of the under-jaw on the fide before affected was now decayed, and concluded, that this tooth had occafioned the ftroke of the palfy by the pain and confequent exertion it had caufed.   On this account I earneftly

neftly entreated her to allow the found molaris of the fame jaw oppofite to the decayed one to be extracted; which was forthwith done,
and the pain of her head immediately ceafed, to the aftonifhment of
her attendants.

In the cafes above related of the pain exifting in a part diftant from
the feat of the difeafe, the pain is owing to defect of the ufual motions of the painful part. This appears from the coldnefs, palenefs,
and emptinefs of the affected veffels, or of the extremities of the body
in general, and from there being no tendency to inflammation. The
increafed action of the primary part of thefe affociated motions, as of
the hepatic termination of the bile-duct from the ftimulus of a gall-
ftone, or of the interior termination of the urethra from the ftimulus
of a ftone in the bladder, or laftly, of a decaying tooth in hemicrania,
deprives the fecondary part of thefe affociated motions, namely, the
exterior terminations of the bile-duct or urethra, or the pained membranes of the head in hemicrania, of their natural fhare of fenforial
power: and hence the fecondary parts of thefe fenfitive trains of affociation become pained from the deficiency of their ufual motions,
which is accompanied with deficiency of fecretions and of heat. See
Sect. IV. 5. XII. 5. 3. XXXIV. 1.

Why does the pain of the primary part of the affociation ceafe,
when that of the fecondary part commences? This is a queftion of
intricacy, but perhaps not inexplicable. The pain of the primary
part of thefe affociated trains of motion was owing to too great ftimulus, as of the ftone at the neck of the bladder, and was confequently
caufed by too great action of the pained part. This greater action
than natural of the primary part of thefe affociated motions, by employing or expending the fenforial power of irritation belonging to the
whole affociated train of motions, occafioned torpor, and confequent
pain, in the fecondary part of the affociated trai ; which was poffeffed of greater fenfibility than the primary part of it. Now the
great pain of the fecondary part of the train, as foon as it commences,

employs

employs or expends the fenforial power of fenfation belonging to the whole affociated train of motions; and in confequence the motions of the primary part, though increafed by the ftimulus of an extraneous body, ceafe to be accompanied with pain or fenfation.

If this mode of reafoning be juft it explains a curious fact, why when two parts of the body are ftrongly ftimulated, the pain is felt only in one of them, though it is poffible by voluntary attention it may be alternately perceived in them both. In the fame manner, when two new ideas are prefented to us from the ftimulus of external bodies, we attend to but one of them at a time. In other words, when one fet of fibres, whether of the mufcles or organs of fenfe, contract fo ftrongly as to excite much fenfation; another fet of fibres contracting more weakly do not excite fenfation at all, becaufe the fenforial power of fenfation is pre-occupied by the firft fet of fibres. So we cannot will more then one effect at once, though by affociations previoufly formed we can move many fibres in combination.

Thus in the inftances above related, the termination of the bile duct in the duodenum, and the exterior extremity of the urethra, are more fenfible than their other terminations. When thefe parts are deprived of their ufual motions by deficiency of the fenforial power of irritation, they become painful according to law the fifth in Section IV. and the lefs pain originally excited by the ftimulus of concreted bile, or of a ftone at their other extremities ceafes to be perceived. Afterwards, however, when the concretions of bile, or the ftone on the urinary bladder, become more numerous or larger, the pain from their increafed ftimulus becomes greater than the affociated pain; and is then felt at the neck of the gall bladder or urinary bladder; and the pain of the glans penis, or at the pit of the ftomach, ceafes to be perceived.

2. Examples of the fecond mode, where the increafed action of the primary part of a train of fenfitive affociation ceafes, when that of the fecondary part commences, are alfo not unfrequent; as this is

the

the ufual manner of the tranflation of inflammations from internal to external parts of the fyftem, fuch as when an inflammation of the liver or ftomach is tranflated to the membranes of the foot, and forms the gout; or to the fkin of the face, and forms the rofy drop, or when an inflammation of the membranes of the kidneys is tranflated to the fkin of the loins, and forms one kind of herpes, called fhingles; in thefe cafes by whatever caufe the original inflammation may have been produced, as the fecondary part of the train of fenfitive affociation is more fenfible, it becomes exerted with greater violence than the firft part of it; and by both its increafed pain, and the increafed motion of its fibres, fo far diminifhes or exhaufts the fenforial power of fenfation; that the primary part of the train being lefs fenfible ceafes both to feel pain, and to act with unnatural energy.

3. Examples of the third mode, where the primary part of a train of fenfitive affociation of motions may experience increafed fenfation, and the fecondary part increafed action, are likewife not unfrequent; as it is in this manner that moft inflammations commence. Thus, after ftanding fome time in fnow, the feet become affected with the pain of cold, and a common coryza, or inflammation of the membrane of the noftrils, fucceeds. It is probable that the internal inflammations, as pleurifies, or hepatitis, which are produced after the cold paroxyfm of fever, originate in the fame manner from the fympathy of thofe parts with fome others, which were previoufly pained from quiefcence; as happens to various parts of the fyftem during the cold fits of fevers. In thefe cafes it would feem, that the fenforial power of fenfation becomes accumulated during the pain of cold, as the torpor of the veffels occafioned by the defect of heat contributes to the increafe or accumulation of the fenforial power of irritation, and that both thefe become exerted on fome internal part, which was not rendered torpid by the cold which affected the external

parts,

parts, nor by its affociation with them; or which fooner recovered its fenfibility.

4. An example of the fourth mode, or where the primary part of a fenfitive affociation of motions may have increafed action, and the fecondary part increafed fenfation, may be taken from the pain of the fhoulder, which attends inflammation of the membranes of the liver, fee Hepatitis, Clafs IV. 2. 1. 6. in this circumftance fo much fenforial power feems to be expended in the violent actions and fenfations of the inflamed membranes of the liver, that the membranes affociated with them become quiefcent to their ufual ftimuli, and painful in confequence.

There may be other modes in which the primary and fecondary parts of the trains of affociated fenfitive motions may reciprocally affect each other, as may be feen by looking over Clafs IV. in the catalogue of difeafes; all which may probably be refolved into the plus and minus of fenforial power, but we have not yet had fufficient obfervations made upon them with a view to this doctrine.

III. The affociated trains of our ideas may have fympathies, and their primary and fecondary parts affect each other in fome manner fimilar to thofe above defcribed; and may thus occafion various curious phenomena not yet adverted to, befides thofe explained in the Sections on Dreams, Reveries, Vertigo, and Drunkennefs; and may thus difturb the deductions of our reafonings, as well as the ftreams of our imaginations; prefent us with falfe degrees of fear, attach unfounded value to trivial circumftances; give occafion to our early prejudices and antipathies; and thus embarrafs the happinefs of our lives. A copious and curious harveft might be reaped from this province of fcience, in which, however, I fhall not at prefent wield my fickle.

SECT.

SECT. XXXVI.

OF THE PERIODS OF DISEASES.

I. *Muscles excited by volition soon cease to contract, or by sensation, or by irritation, owing to the exhaustion of sensorial power. Muscles subjected to less stimulus have their sensorial power accumulated. Hence the periods of some fevers. Want of irritability after intoxication.* II. 1. *Natural actions catenated with daily habits of life.* 2. *With solar periods. Periods of sleep. Of evacuating the bowels.* 3. *Natural actions catenated with lunar periods. Menstruation. Venereal orgasm of animals. Barrenness.* III. *Periods of diseased animal actions from stated returns of nocturnal cold, from solar and lunar influence. Periods of diurnal fever, hectic fever, quotidian, tertian, quartan fever. Periods of gout, pleurisy, of fevers with arterial debility, and with arterial strength. Periods of rhaphania, of nervous cough, hemicrania, arterial hæmorrhages, hæmorrhoids, hæmoptoe, epilepsy, palsy, apoplexy, madness.* IV. *Critical days depend on lunar periods. Lunar periods in the small pox.*

I. IF any of our muscles be made to contract violently by the power of volition, as those of the fingers, when any one hangs by his hands on a swing, fatigue soon ensues; and the muscles cease to act owing to the temporary exhaustion of the spirit of animation; as soon as this is again accumulated in the muscles, they are ready to contract again by the efforts of volition.

Those violent muscular actions induced by pain become in the same manner intermitted and recurrent; as in labour-pains, vomiting, tenesmus, strangury; owing likewise to the temporary exhaustion of the spirit of animation, as above mentioned.

When

When any ſtimulus continues long to act with unnatural violence, ſo as to produce too energetic action of any of our moving organs, thoſe motions ſoon ceaſe, though the ſtimulus continues to act; as in looking long on a bright object, as on an inch-ſquare of red ſilk laid on white paper in the ſunſhine. See Plate I. in Sect. III. 1.

On the contrary, where leſs of the ſtimulus of volition, ſenſation, or irritation, have been applied to a muſcle than uſual; there appears to be an accumulation of the ſpirit of animation in the moving organ; by which it is liable to act with greater energy from leſs quantity of ſtimulus, than was previouſly neceſſary to excite it into ſo great action; as after having been immerſed in ſnow the cutaneous veſſels of our hands are excited into ſtronger action by the ſtimulus of a leſs degree of heat, than would previouſly have produced that effect.

From hence the periods of ſome fever-fits may take their origin, either ſimply, or by their accidental coincidence with lunar and ſolar periods, or with the diurnal periods of heat and cold, to be treated of below; for during the cold fit at the commencement of a fever, from whatever cauſe that cold fit may have been induced, it follows, 1. That the ſpirit of animation muſt become accumulated in the parts, which exert during this cold fit leſs than their natural quantity of action. 2. If the cauſe producing the cold fit does not increaſe, or becomes diminiſhed; the parts before benumbed or inactive become now excitable by ſmaller ſtimulus, and are thence thrown into more violent action than is natural; that is a hot fit ſucceeds the cold one. 3. By the energetic action of the ſyſtem during the hot fit, if it continues long, an exhauſtion of the ſpirit of animation takes place; and another cold fit is liable to ſucceed, from the moving ſyſtem not being excitable into action from its uſual ſtimulus. This inirritability of the ſyſtem from a too great previous ſtimulus, and conſequent exhauſtion of ſenſorial power, is the cauſe of the general debility, and ſickneſs, and head-ach, ſome hours after intoxication.

toxication. And hence we fee one of the caufes of the periods of fever-fits; which however are frequently combined with the periods of our diurnal habits, or of heat and cold, or of folar or lunar periods.

When befides the tendency to quiefcence occafioned by the expenditure of fenforial power during the hot fit of fever, fome other caufe of torpor, as the folar or lunar periods, is neceffary to the introduction of a fecond cold fit; the fever becomes of the intermittent kind; that is, there is a fpace of time intervenes between the end of the hot fit, and the commencement of the next cold one. But where no exteriour caufe is neceffary to the introduction of the fecond cold fit; no fuch interval of health intervenes; but the fecond cold fit commences, as foon as the fenforial power is fufficiently exhaufted by the hot fit; and the fever becomes continual.

II. 1. The following are natural animal actions, which are frequently catenated with our daily habits of life, as well as excited by their natural irritations. The periods of hunger and thirft become catenated with certain portions of time, or degrees of exhauftion, cr other diurnal habits of life. And if the pain of hunger be not relieved by taking food at the ufual time, it is liable to ceafe till the next period of time or other habits recur; this is not only true in refpect to our general defire of food, but the kinds of it alfo are governed by this periodical habit; infomuch that beer taken to breakfaft will difturb the digeftion of thofe, who have been accuftomed to tea; and tea taken at dinner will difagree with thofe, who have been accuftomed to beer. Whence it happens, that thofe, who have weak ftomachs, will be able to digeft more food, if they take their meals at regular hours; becaufe they have both the ftimulus of the aliment they take, and the periodical habit, to affift their digeftion.

The periods of emptying the bladder are not only dependent on the acrimony or diftention of the water in it, but are frequently

3

catenated

catenated with external cold applied to the fkin, as in cold bathing, or wafhing the hands; or with other habits of life, as many are accuftomed to empty the bladder before going to bed, or into the houfe after a journey, and this whether it be full or not.

Our times of refpiration are not only governed by the ftimulus of the blood in the lungs, or our defire of frefh air, but alfo by our attention to the hourly objects before us.   Hence when a perfon is earneftly contemplating an idea of grief, he forgets to breathe, till the fenfation in his lungs becomes very urgent; and then a figh fucceeds for the purpofe of more forceably pufhing forwards the blood, which is accumulated in the lungs.

Our times of refpiration are alfo frequently governed in part by our want of a fteady fupport for the actions of our arms, and hands, as in threading a needle, or hewing wood, or in fwimming; when we are intent upon thefe objects, we breathe at the intervals of the exertion of the pectoral mufcles.

2. The following natural animal actions are influenced by folar periods.   The periods of fleep and of waking depends much on the folar period, for we are inclined to fleep at a certain hour, and to awake at a certain hour, whether we have had more or lefs fatigue during the day, if within certain limits; and are liable to wake at a certain hour, whether we went to bed earlier or later within certain limits.   Hence it appears, that thofe who complain of want of fleep, will be liable to fleep better or longer, if they accuftom themfelves to go to reft, and to rife, at certain hours.

The periods of evacuating the bowels are generally connected with fome part of the folar day, as well as with the acrimony or diftention occafioned by the feces.   Hence one method of correcting coftivenefs is by endeavouring to eftablifh a habit of evacuation at a certain hour of the day, as recommended by Mr. Locke, which may be accomplifhed by ufing daily voluntary efforts at thofe times, joined with the ufual ftimulus of the material to be evacuated.

3. The

3. The following natural animal actions are connected with lunar periods. 1. The periods of female menstruation are connected with lunar periods to great exactnefs, in fome inftances even to a few hours. Thefe do not commence or terminate at the full or change, or at any other particular part of the lunation, but after they have commenced at any part of it, they continue to recur at that part with great regularity, unlefs difturbed by fome violent circumftance, as explained in Sect. XXXII. No. 6. their return is immediately caufed by deficient venous abforption, which is owing to the want of the ftimulus, defigned by nature, of amatorial copulation, or of the growing fetus. When the catamenia returns fooner than the period of lunation, it fhews a tendency of the conftitution to inirritability; that is to debility, or deficiency of fenforial power, and is to be relieved by fmall dofes of fteel and opium.

The venereal orgafm of birds and quadrupeds feems to commence, or return about the moft powerful lunations at the vernal or autumnal equinoxes; but if it be difappointed of its object, it is faid to recur at monthly periods; in this refpect refembling the female catamenia. Whence it is believed, that women are more liable to become pregnant at or about the time of their catamenia, than at the intermediate times; and on this account they are feldom much miftaken in their reckoning of nine lunar periods from the laft menftruation; the inattention to this may fometimes have been the caufe of fuppofed barrennefs, and is therefore worth the obfervation of thofe, who wifh to have children.

III. We now come to the periods of difeafed animal actions. The periods of fever-fits, which depend on the ftated returns of nocturnal cold, are difcuffed in Sect. XXXII. 3. Thofe, which originate or recur at folar or lunar periods, are alfo explained in Section XXXII. 6. Thefe we fhall here enumerate; obferving, however, that it is not more furprifing, that the influence of the varying attractions of the fun and moon, fhould raife the ocean into mountains, than

that

that it fhould affect the nice fenfibilities of animal bodies; though the manner of its operation on them is difficult to be underftood. It is probable however, that as this influence gradually leffens during the courfe of the day, or of the lunation, or of the year, fome actions of our fyftem become lefs and lefs; till at length a total quiefcence of fome part is induced; which is the commencement of the paroxyfms of fever, of menftruation, of pain with decreafed action of the affected organ, and of confequent convulfion.

1. A diurnal fever in fome weak people is diftinctly obferved to come on towards evening, and to ceafe with a moift fkin early in the morning, obeying the folar periods. Perfons of weak conftitutions are liable to get into better fpirits at the accefs of the hot fit of this evening fever; and are thence inclined to fit up late; which by further enfeebling them increafes the difeafe; whence they lofe their ftrength and their colour.

2. The periods of hectic fever, fuppofed to arife from abforption of matter, obeys the diurnal periods like the above, having the exacerbefcence towards evening, and its remiffion early in the morning, with fweats, or diarrhœa, or urine with white fediment.

3. The periods of quotidian fever are either catenated with folar time, and return at the intervals of twenty-four hours; or with lunar time, recurring at the intervals of about twenty-five hours. There is great ufe in knowing with what circumftances the periodical return of new morbid motions are conjoined, as the moft effectual times of exhibiting the proper medicines are thus determined. So if the torpor, which ufhers in an ague fit, is catenated with the lunar day: it is know, when the bark or opium muft be given, fo as to exert its principal effect about the time of the expected return. Solid opium fhould be given about an hour before the expected cold fit; liquid opium and wine about half an hour; the bark repeatedly for fix or eight hours previous to the expected return.

4. The

4. The periods of tertian fevers, reckoned from the commencement of one cold fit to the commencement of the next cold fit, recur
with folar intervals of forty-eight hours, or with lunar ones of about
fifty hours. When thefe times of recurrence begin one or two hours
earlier than the folar period, it fhews, that the torpor or cold fit is
produced by lefs external influence; and therefore that it is more liable to degenerate into a fever with only remiffions; fo when menftruation recurs fooner than the period of lunation, it fhews a tendency of the habit to torpor or inirritability.

5. The periods of quartan fevers return at folar intervals of feventy-
two hours, or at lunar ones of about feventy-four hours and an half.
This kind of ague appears moft in moift cold autumns, and in cold
countries replete with marfhes. It is attended with greater debility,
and its cold accefs more difficult to prevent. For where there is previoufly a deficiency of fenforial power, the conftitution is liable to
run into greater torpor from any further diminution of it; two
ounces of bark and fome fteel fhould be given on the day before the
return of the cold paroxyfm, and a pint of wine by degrees a few
hours before its return, and thirty drops of laudanum one hour before
the expected cold fit.

6. The periods of the gout generally commence about an hour before fun-rife, which is ufually the coldeft part of the twenty-four
hours. The greater periods of the gout feem alfo to obferve the folar
influence, returning about the fame feafon of the year.

7. The periods of the pleurify recur with exacerbation of the pain
and fever about fun-fet, at which time venefection is of moft fervice.
The fame may be obferved of the inflammatory rheumatifm, and
other fevers with arterial ftrength, which feem to obey folar periods;
and thofe with debility feem to obey lunar ones.

8. The periods of fevers with arterial debility feem to obey the
lunar day, having their accefs daily nearly an hour later; and have
fometimes

fometimes two acceffes in a day, refembling the lunar effects upon the tides.

9. The periods of rhaphania, or convulfions of the limbs from rheumatic pains, feem to be connected with folar influence, returning at nearly the fame hour for weeks together, unlefs difturbed by the exhibition of powerful dofes of opium.

So the periods of Tuffis ferina, or violent cough with flow pulfe, called nervous cough, recurs by folar periods. Five grains of opium given at the time the cough commenced difturbed the period, from feven in the evening to eleven, at which time it regularly returned for fome days, during which time the opium was gradually omitted. Then 120 drops of laudanum were given an hour before the accefs of the cough, and it totally ceafed. The laudanum was continued a fortnight, and then gradually difcontinued.

10. The periods of hemicrania, and of painful epilepfy, are liable to obey lunar periods, both in their diurnal returns, and in their greater periods of weeks, but are alfo induced by other exciting caufes.

11. The periods of arterial hæmorrhages feem to return at folar periods about the fame hour of the evening or morning. Perhaps the venous hæmorrhages obey the lunar periods, as the catamenia, and hæmorrhoids.

12. The periods of the hæmorrhoids, or piles, in fome recur monthly, in others only at the greater lunar influence about the equinoxes.

13. The periods of hæmoptoe fometimes obey folar influence, recurring early in the morning for feveral days; and fometimes lunar periods, recurring monthly; and fometimes depend on our hours of fleep. See Clafs I. 2. 1. 9.

14. Many of the firft periods of epileptic fits obey the monthly lunation with fome degree of accuracy; others recur only at the moft powerful lunations before the vernal equinox, and after the autumnal

one; but when the conftitution has gained a habit of relieving difagreeable fenfations by this kind of exertion, the fit recurs from any flight caufe.

15. The attack of palfy and apoplexy are known to recur with great frequency about the equinoxes.

16. There are numerous inftances of the effect of the lunations upon the periods of infanity, whence the name of lunatic has been given to thofe afflicted with this difeafe.

IV. The critical days, in which fevers are fuppofed to terminate, have employed the attention of medical philofophers from the days of Hippocrates to the prefent time. In whatever part of a lunation a fever commences, which owes either its whole caufe to folar and lunar influence, or to this in conjunction with other caufes; it would feem, that the effect would be the greateft at the full and new moon, as the tides rife higheft at thofe times, and would be the leaft at the quadratures; thus if a fever-fit fhould commence at the new or full moon, occafioned by the folar and lunar attraction diminishing fome chemical affinity of the particles of blood, and thence decreafing their ftimulus on our fanguiferous fyftem, as mentioned in Sect. XXXII. 6. this effect will daily decreafe for the firft feven days, and will then increafe till about the fourteenth day, and will again decreafe till about the twenty-firft day, and increafe again till the end of the lunation. If a fever-fit from the above caufe fhould commence on the feventh day after either lunation, the reverfe of the above circumftances would happen. Now it is probable, that thofe fevers, whofe crifis or terminations are influenced by lunations, may begin at one or other of the above times, namely at the changes or quadratures; though fufficient obfervations have not been made to afcertain this circumftance. Hence I conclude, that the fmall-pox and meafles have their critical days, not governed by the times required for certain chemical changes in the blood, which affect or alter the ftimulus of the contagious matter, but from the daily increafing or decreafing effect of this lunar

3

link

link of catenation, as explained in Section XVII. 3. 3.   And as other fevers terminate moft frequently about the feventh, fourteenth, twenty-firft, or about the end of four weeks, when no medical af-fiftance has difturbed their periods, I conclude, that thefe crifes, or terminations, are governed by periods of the lunations, though we are ftill ignorant of their manner of operation.

In the diftinct fmall-pox the veftiges of lunation are very apparent; after inoculation a quarter of a lunation precedes the commencement of the fever, another quarter terminates with the complete eruption, another quarter with the complete maturation; and another quarter terminates the complete abforption of a material now rendered inoffen-five to the conftitution.

SECT.

SECT. XXXVII.

OF DIGESTION, SECRETION, NUTRITION.

I. Cryftals increafe by the greater attraction of their fides. Accretion by chemical precipitations, by welding, by preffure, by agglutination. II. Hunger, digeftion, why it cannot be imitated out of the body. Lacteals abforb by animal felection, or appetency. III. The glands and pores abforb nutritious particles by animal felection. Organic particles of Buffon. Nutrition applied at the time of elongation of fibres. Like inflammation. IV. It feems eafier to have preferved animals than to reproduce them. Old age and death from inirritability. Three caufes of this. Original fibres of the organs of fenfe and mufcles unchanged. V. Art of producing long life.

I. THE larger cryftals of faline bodies may be conceived to arife from the combination of fmaller cryftals of the fame form, owing to the greater attractions of their fides than of their angles. Thus if four cubes were floating in a fluid, whofe friction or refiftance is nothing, it is certain the fides of thefe cubes would attract each other ftronger than their angles; and hence that thefe four fmaller cubes would fo arrange themfelves as to produce one larger one.

There are other means of chemical accretion, fuch as the depofitions of diffolved calcareous or filiceous particles, as are feen in the formation of the ftalactites of limeftone in Derbyfhire, or of calcedone in Cornwall. Other means of adhefion are produced by heat and preffure, as in the welding of iron-bars; and other means by fimple preffure, as in forcing two pieces of caoutchou, or elaftic gum, to adhere;

here; and laftly, by the agglutination of a third fubftance penetrating the pores of the other two, as in the agglutination of wood by means of animal gluten. Though the ultimate particles of animal bodies are held together during life, as well as after death, by their fpecific attraction of cohefion, like all other matter; yet it does not appear, that their original organization was produced by chemical laws, and their production and increafe muft therefore only be looked for from the laws of animation.

II. When the pain of hunger requires relief, certain parts of the material world, which furround us, when applied to our palates, excite into action the mufcles of deglutition; and the material is fwallowed into the ftomach. Here the new aliment becomes mixed with certain animal fluids, and undergoes a chemical procefs, termed digeftion; which however chemiftry has not yet learnt to imitate out of the bodies of living animals or vegetables. This procefs feems very fimilar to the faccharine procefs in the lobes of farinaceous feeds, as of barley, when it begins to germinate; except that, along with the fugar, oil and mucilage are alfo produced; which form the chyle of animals, which is very fimilar to their milk.

The reafon, I imagine, why this chyle-making, or faccharine procefs, has not yet been imitated by chemical operations, is owing to the materials being in fuch a fituation in refpect to warmth, moifture, and motion; that they will immediately change into the vinous or acetous fermentation; except the new fugar be abforbed by the numerous lacteal or lymphatic veffels, as foon as it is produced; which is not eafy to imitate in the laboratory.

Thefe lacteal veffels have mouths, which are irritated into action by the ftimulus of the fluid, which furrounds them; and by animal felection, or appetency, they abforb fuch part of the fluid as is agreeable to their palate; thofe parts, for inftance, which are already converted into chyle, before they have time to undergo another change by a vinous or acetous fermentation. This animal abforption of fluid

S

is

is almoſt viſible to the naked eye in the action of the puncta lacrymalia; which imbibe the tears from the eye, and diſcharge them
again into the noſtrils.

III. The arteries conſtitute another reſervoir of a changeful fluid;
from which, after its recent oxygenation in the lungs, a further animal ſelection of various fluids is abſorbed by the numerous glands;
theſe ſelect their reſpective fluids from the blood, which is perpetually undergoing a chemical change; but the ſelection by theſe glands,
like that of the lacteals, which open their mouths into the digeſting
aliment in the ſtomach, is from animal appetency, not from chemical
affinity; ſecretion cannot therefore be imitated in the laboratory, as
it conſiſts in a ſelection of part of a fluid during the chemical change
of that fluid.

The mouths of the lacteals, and lymphatics, and the ultimate terminations of the glands, are finer than can eaſily be conceived; yet
it is probable, that the pores, or interſtices of the parts, or coats,
which conſtitute theſe ultimate veſſels, may ſtill have greater tenuity;
and that theſe pores from the above analogy muſt poſſeſs a ſimilar
power of irritability, and abſorb by their living energy the particles of
fluid adapted to their purpoſes, whether to replace the parts abraded
or diſſolved, or to elongate and enlarge themſelves. Not only every
kind of gland is thus endued with its peculiar appetency, and ſelects
the material agreeable to its taſte from the blood, but every individual
pore acquires by animal ſelection the material, which it wants; and
thus nutrition ſeems to be performed in a manner ſo ſimilar to ſecretion; that they only differ in the one retaining, and the other
parting again with the particles, which they have ſelected from the
blood.

This way of accounting for nutrition from ſtimulus, and the conſequent animal ſelection of particles, is much more analogous to other
phenomena of the animal microcoſm, than by having recourſe to the
microſcopic animalcula, or organic particles of Buffon and Needham;

which

which being already compounded muſt themſelves require nutritive particles to continue their own exiſtence. And muſt be liable to undergo a change by our digeſtive or ſecretory organs; otherwiſe mankind would ſoon reſemble by their theory the animals, which they feed upon. He, who is nouriſhed by beef or veniſon, would in time become horned; and he, who feeds on pork or bacon, would gain a noſe proper for rooting into the earth, as well as for the perception of odours.

The whole animal ſyſtem may be conſidered as conſiſting of the extremities of the nerves, or of having been produced from them; if we except perhaps the medullary part of the brain reſiding in the head and ſpine, and in the trunks of the nerves. Theſe extremities of the nerves are either of thoſe of locomotion, which are termed muſcular fibres; or of thoſe of ſenſation, which conſtitute the immediate organs of ſenſe, and which have alſo their peculiar motions. Now as the fibres, which conſtitute the bones and membranes, poſſeſſed originally ſenſation and motion; and are liable again to poſſeſs them, when they become inflamed; it follows, that thoſe were, when firſt formed, appendages to the nerves of ſenſation or locomotion, or were formed from them. And that hence all theſe ſolid parts of the body, as they have originally conſiſted of extremities of nerves, require an appoſition of nutritive particles of a ſimilar kind, contrary to the opinion of Buffon and Needham above recited.

Laſtly, as all theſe filaments have poſſeſſed, or do poſſeſs, the power of contraction, and of conſequent inertion or elongation; it ſeems probable, that the nutritive particles are applied during their times of elongation; when their original conſtituent particles are removed to a greater diſtance from each other. For each muſcular or ſenſual fibre may be conſidered as a row or ſtring of beads; which approach, when in contraction, and recede during its reſt or elongation; and our daily experience ſhews us, that great action emaciates the ſyſtem, and that it is repaired during reſt.

3 O

Something

Something like this is seen out of the body; for if a hair, or a single untwisted fibre of flax or silk, be soaked in water; it becomes longer and thicker by the water, which is abforbed into its pores. Now if a hair could be suppofed to be thus immerfed in a folution of particles fimilar to thofe, which compofe it; one may imagine, that it might be thus increafed in weight and magnitude; as the particles of oak-bark increafe the fubftance of the hides of beafts in the procefs of making leather.   I mention thefe not as philofophic analogies, but as fimiles to facilitate our ideas, how an accretion of parts may be ef-fected by animal appetences, or felections, in a manner fomewhat fimilar to mechanical or chemical attractions.

If thofe new particles of matter, previoufly prepared by digeftion and fanguification, only fupply the places of thofe, which have been abraded by the actions of the fyftem, it is properly termed nutrition. If they are applied to the extremities of the nervous fibrils, or in fuch quantity as to increafe the length or craffitude of them, the body be-comes at the fame time enlarged, and its growth is increafed, as well as its deficiences repaired.

In this laft cafe fomething more than a fimple appofition or felec-tion of particles feems to be neceffary; as many parts of the fyftem during its growth are caufed to recede from thofe, with which they were before in contact; as the ends of the bones, or cartilages, recede from each other, as their growth advances: this procefs refembles inflammation, as appears in ophthalmy, or in the production of new flefh in ulcers, where old veffels are enlarged, and new ones pro-duced; and like that is attended with fenfation.   In this fituation the veffels become diftended with blood, and acquire greater fenfibility, and may thus be compared to the erection of the penis, or of the nipples of the breafts of women; while new particles become added at the fame time; as in the procefs of nurition above defcribed.

When only the natural growth of the various parts of the body are produced, a pleafureable fenfation attends it, as in youth, and

perhaps

perhaps in thofe, who are in the progrefs of becoming fat.   When an unnatural growth is the confequence, as in inflammatory difeafes, a painful fenfation attends the enlargement of the fyftem.

IV. This appofition of new parts, as the old ones difappear, fe-lected from the aliment we take, firft enlarges and ftrengthens our bodies for twenty years, for another twenty years it keeps us in health and vigour, and adds ftrength and folidity to the fyftem ; and then gradually ceafes to nourifh us properly, and for another twenty years we gradually fink into decay, and finally ceafe to act, and to exift.

On confidering this fubject one fhould have imagined at firft view, that it might have been eafier for nature to have fupported her pro-geny for ever in health and life, than to have perpetually reproduced them by the wonderful and myfterious procefs of generation.   But it feems our bodies by long habit ceafe to obey the ftimulus of the aliment, which fhould fupport us.   After we have acquired our height and folidity we make no more new parts, and the fyftem obeys the irritations, fenfations, volitions, and affociations, with lefs and lefs energy, till the whole finks into inaction.

Three caufes may confpire to render our nerves lefs excitable, which have been already mentioned.   1. If a ftimulus be greater than natural, it produces too great an exertion of the ftimulated organ, and in confequence exhaufts the fpirit of animation ; and the moving organ ceafes to act, even though the ftimulus be con-tinued.   And though reft will recruit this exhauftion, yet fome de-gree of permanent injury remains, as is evident after expofing the eyes long to too ftrong a light.   2. If excitations weaker than na-tural be applied, fo as not to excite the organ into action, (as when fmall dofes of aloe or rhubarb are exhibited,) they may be gradually increafed, without exciting the organ into action ; which will thus acquire a habit of difobedience to the ftimulus ; thus by increafing the

3 O 2

dofe

dose by degrees, great quantities of opium or wine may be taken without intoxication.   See Sect. XII. 3. 1.

3. Another mode, by which life is gradually undermined, is when irritative motions continue to be produced in consequence of stimulus, but are not succeeded by sensation; hence the stimulus of contagious matter is not capable of producing fever a second time, because it is not succeeded by sensation.   See Sect. XII. 3. 6.   And hence, owing to the want of the general pleasureable sensation, which ought to attend digestion and glandular secretion, an irksomeness of life ensues; and, where this is in greater excess, the melancholy of old age occurs, with torpor or debility.

From hence I conclude, that it is probable that the fibrillæ, or moving filaments at the extremities of the nerves of sense, and the fibres which constitute the muscles (which are perhaps the only parts of the system that are endued with contractile life) are not changed, as we advance in years, like the other parts of the body; but only enlarged or elongated with our growth; and in consequence they become less and less excitable into action.   Whence, instead of gradually changing the old animal, the generation of a totally new one becomes necessary with undiminished excitability; which many years will continue to acquire new parts, or new solidity, and then losing its excitability in time, perish like its parent.

V. From this idea the art of preserving long health and life may be deduced; which must consist in using no greater stimulus, whether of the quantity or kind of our food and drink, or of external circumstances, such as heat, and exercise, and wakefulness, than is sufficient to preserve us in vigour; and gradually, as we grow old to increase the stimulus of our aliment, as the inirritability of our system increases.

The debilitating effects ascribed by the poet Martial to the excessive use of warm bathing in Italy, may with equal propriety be ap-

plied

plied to the warm rooms of England; which, with the general excessive stimulus of spirituous or fermented liquors, and in some instances of immoderate venery, contribute to shorten our lives.

Balnea, vina, venus, corrumpunt corpora nostra,
At faciunt vitam balnea, vina, venus!

Wine, women, warmth, against our lives combine;
But what is life without warmth, women, wine!

SECT. XXXVIII.

OF THE OXYGENATION OF THE BLOOD IN THE LUNGS, AND IN THE PLACENTA.

I. *Blood abforbs oxygene from the air, whence phofphoric acid, changes its colour, gives out heat, and fome phlogiftic material, and acquires an etherial fpirit, which is diffipated in fibrous motion.* II. *The placenta is a pulmonary organ like the gills of fifh. Oxygenation of the blood from air, from water, by lungs, by gills, by the placenta; neceffity of this oxygenation to quadrupeds, to fifh, to the fœtus in utero. Placental veffels inferted into the arteries of the mother. Ufe of cotyledons in cows. Why quadrupeds have not fanguiferous lochia. Oxygenation of the chick in the egg, of feeds.* III. *The liquor amnii is not excrementitious. It is nutrititious. It is found in the efophagus and ftomach, and forms the meconium. Monftrous births without heads. Queftion of Dr. Harvey.*

I. FROM the recent difcoveries of many ingenious philofophers it appears, that during refpiration the blood imbibes the vital part of the air, called oxygene, through the membranes of the lungs; and that hence refpiration may be aptly compared to a flow combuftion. As in combuftion the oxygene of the atmofphere unites with fome phlogiftic or inflammable body, and forms an acid (as in the production of vitriolic acid from fulphur, or carbonic acid from charcoal,) giving out at the fame time a quantity of the matter of heat; fo in refpiration the oxygene of the air unites with the phlogiftic part of the blood, and probably produces phofphoric or animal acid, changing the colour

of

of the blood from a dark to a bright red; and probably fome of the matter of heat is at the fame time given out according to the theory of Dr. Crawford. But as the evolution of heat attends almoft all chemical combinations, it is probable, that it alfo attends the fecretions of the various fluids from the blood; and that the conftant combinations or productions of new fluids by means of the glands conftitute the more general fource of animal heat; this feems evinced by the univerfal evolution of the matter of heat in the blufh of fhame or of anger; in which at the fame time an increafed fecretion of the perfpirable matter occurs; and the partial evolution of it from topical inflammations, as in gout or rheumatifm, in which there is a fecretion of new blood-veffels.

Some medical philofophers have afcribed the heat of animal bodies to the friction of the particles of the blood againft the fides of the veffels. But no perceptible heat has ever been produced by the agitation of water, or oil, or quickfilver, or other fluids; except thofe fluids have undergone at the fame time fome chemical change, as in agitating milk or wine, till they become four.

Befides the fuppofed production of phofphoric acid, and change of colour of the blood, and the production of carbonic acid, there would appear to be fomething of a more fubtile nature perpetually acquired from the atmofphere; which is too fine to be long contained in animal veffels, and therefore requires perpetual renovation; and without which life cannot continue longer than a minute or two; this ethereal fluid is probably fecreted from the blood by the brain, and perpetually diffipated in the actions of the mufcles and organs of fenfe.

That the blood acquires fomething from the air, which is immediately neceffary to life, appears from an experiment of Dr. Hare (Philof. Tranfact. abridged, Vol. III. p. 239.) who found, " that birds, mice, &c. would live as long again in a veffel, where he had crowded in double the quantity of air by a condenfing engine, than

3   they

they did when confined in air of the common denfity." Whereas if some kind of deleterious vapour only was exhaled from the blood in respiration; the air, when condensed into half its compass, could not be supposed to receive so much of it.

II. Sir Edward Hulse, a physician of reputation at the beginning of the present century, was of opinion, that the placenta was a respiratory organ, like the gills of fish; and not an organ to supply nutriment to the fœtus; as mentioned in Derham's Physico-theology. Many other physicians seem to have espoused the same opinion, as noticed by Haller. Elem. Physiologiæ, T. 1. Dr. Gipson published a defence of this theory in the Medical Essays of Edinburgh, Vol. I. and II. which doctrine is there controverted at large by the late Alexander Monro; and since that time the general opinion has been, that the placenta is an organ of nutrition only, owing perhaps rather to the authority of so great a name, than to the validity of the arguments adduced in its support. The subject has lately been resumed by Dr. James Jeffray, and by Dr. Forester French, in their inaugural dissertations at Edinburgh and at Cambridge; who have defended the contrary opinion in an able and ingenious manner; and from whose Theses I have extracted many of the following remarks.

First, by the late discoveries of Dr. Priestley, M. Lavoisier, and other philosophers, it appears, that the basis of atmospherical air, called oxygene, is received by the blood through the membranes of the lungs; and that by this addition the colour of the blood is changed from a dark to a light red. Secondly, that water possesses oxygene also as a part of its composition, and contains air likewise in its pores; whence the blood of fish receives oxygene from the water, or from the air it contains, by means of their gills, in the same manner as the blood is oxygenated in the lungs of air-breathing animals; it changes its colour at the same time from a dark to a light red in the vessels of their gills, which constitute a pulmonary organ adapted to the medium in which they live. Thirdly, that the placenta consists of arteries

teries carrying the blood to its extremities, and a vein bringing it back, refembling exactly in ftructure the lungs and gills above mentioned ; and that the blood changes its colour from a dark to a light red in paffing through thefe veffels.

This analogy between the lungs and gills of animals, and the placenta of the fetus, extends through a great variety of other circumftances ; thus air-breathing creatures and fifh can live but a few minutes without air or water ; or when they are confined in fuch air or water, as has been fpoiled by their own refpiration ; the fame happens to the fetus, which, as foon as the placenta is feparated from the uterus, muft either expand its lungs, and receive air, or die.   Hence from the ftructure, as well as the ufe of the placenta, it appears to be a refpiratory organ, like the gills of fifh, by which the blood in the fetus becomes oxygenated.

From the terminations of the placental veffels not being obferved to bleed after being torn from the uterus, while thofe of the uterus effufe a great quantity of florid arterial blood, the terminations of the placental veffels would feem to be inferted into the arterial ones of the mother ; and to receive oxygenation from the paffing currents of her blood through their coats or membranes ; which oxygenation is proved by the change of the colour of the blood from dark to light red in its paffage from the placental arteries to the placental vein.

The curious ftructure of the cavities or lacunæ of the placenta, demonftrated by Mr. J. Hunter, explain this circumftance.   That ingenious philofopher has fhewn, that there are numerous cavities or lacunæ formed on that fide of the placenta, which is in contact with the uterus ; thofe cavities or cells are filled with blood from the maternal arteries, which open into them ; which blood is again taken up by the maternal veins, and is thus perpetually changed.   While the terminations of the placental arteries and veins are fpread in fine reticulation on the fides of thefe cells.   And thus, as the growing fetus

requires

requires greater oxygenation, an apparatus is produced refembling exactly the air-cells of the lungs.

In cows, and other ruminating animals, the internal furface of the uterus is unequal like hollow cups, which have been called cotyledons ; and into thefe cavities the prominencies of the numerous placentas, with which the fetus of thofe animals is furnifhed, are inferted, and ftrictly adhere ; though they may be extracted without effufion of blood. Thefe inequalities of the uterus, and the numerous placentas in confequence, feem to be defigned for the purpofe of expanding a greater furface for the terminations of the placental veffels for the purpofe of receiving oxygenation from the uterine ones ; as the progeny of this clafs of animals are more completely formed before their nativity, than that of the carnivorous claffes, and muft thence in the latter weeks of pregnancy require greater oxygenation. Thus calves and lambs can walk about in a few minutes after their birth ; while puppies and kittens remain many days without opening their eyes. And though on the feparation of the cotyledons of ruminating animals no blood is effufed, yet this is owing clearly to the greater power of contraction of their uterine lacunæ or alveoli. See Medical Effays, Vol. V. page 144. And from the fame caufe they are not liable to a fanguiferous menftruation.

The neceffity of the oxygenation of the blood in the fetus is farther illuftrated by the analogy of the chick in the egg ; which appears to have its blood oxygenated at the extremities of the veffels furrounding the yolk ; which are fpread on the air-bag at the broad end of the egg, and may abforb oxygene through that moift membrane from the air confined behind it ; and which is fhewn by experiments in the exhaufted receiver to be changeable through the fhell.

This analogy may even be extended to the growing feeds of vegetables ; which were fhewn by Mr. Scheele to require a renovation of the air over the water, in which they were confined. Many vegetable

5

table

table feeds are furrounded with air in their pods or receptacles, as peas, the fruit of ftaphylea, and lichnis veficaria ; but it is probable, that thofe feeds, after they are fhed, as well as the fpawn of fifh, by the fituation of the former on or near the moift and aerated furface of the earth, and of the latter in the ever-changing and ventilated water, may not be in need of an apparatus for the oxygenation of their firft blood, before the leaves of one, and the gills of the other, are produced for this purpofe.

III. 1. There are many arguments, befides the ftrict analogy between the liquor amnii and the albumen ovi, which fhew the former to be a nutritive fluid ; and that the fetus in the latter months of pregnancy takes it into its ftomach ; and that in confequence the placenta is produced for fome other important purpofe.

Firft, that the liquor amnii is not an excrementitious fluid is evinced, becaufe it is found in greater quantity, when the fetus is young, decreafing after a certain period till birth. Haller afferts, " that in fome animals but a fmall quantity of this fluid remains at the birth. In the eggs of hens it is confumed on the eighteenth day, fo that at the exclufion of the chick fcarcely any remains. In rabbits before birth there is none." Elem. Phyfiol. Had this been an excrementitious fluid, the contrary would probably have occurred. Secondly, the fkin of the fetus is covered with a whitifh cruft or pellicle, which would feem to preclude any idea of the liquor amnii being produced by any exfudation of perfpirable matter. And it cannot confift of urine, becaufe in brute animals the urachus paffes from the bladder to the alantois for the exprefs purpofe of carrying off that fluid ; which however in the human fetus feems to be retained in the diftended bladder, as the feces are accumulated in the bowels of all animals.

2. The nutritious quality of the liquid, which furrounds the fetus, appears from the following confiderations. 1. It is coagulable by heat, by nitrous acid, and by fpirit of wine, like milk, ferum of blood, and

other

other fluids, which daily experience evinces to be nutritious.　2. It has a faltifh tafte, according to the accurate Baron Haller, not unlike the whey of milk, which it even refembles in fmell.　3. The white of the egg which conftitutes the food of the chick, is fhewn to be nutritious by our daily experience; befides the experiment of its nutritious effects mentioned by Dr. Fordyce in his late Treatife on Digeftion, p. 178; who adds, that it much refembles the effential parts of the ferum of blood.

3. A fluid fimilar to the fluid, with which the fetus is furrounded, except what little change may be produced by a beginning digeftion, is found in the ftomach of the fetus; and the white of the egg is found in the fame manner in the ftomach of the chick.

Numerous hairs, fimilar to thofe of its fkin, are perpetually found among the contents of the ftomach in new-born calves; which muft therefore have licked themfelves before their nativity.　Blafii Anatom. See Sect. XVI. 2. on Inftinct.

The chick in the egg is feen gently to move in its furrounding fluid, and to open and fhut its mouth alternately.　The fame has been obferved in puppies.　Haller's El. Phyf. I. 8. p. 201.

A column of ice has been feen to reach down the œfophagus from the mouth to the ftomach in a frozen fetus; and this ice was the liquor amnii frozen.

The meconium, or firft fæces, in the bowels of new-born infants evince, that fomething has been digefted; and what could this be but the liquor amnii together with the recrements of the gaftric juice and gall, which were neceffary for its digeftion?

There have been recorded fome monftrous births of animals without heads, and confequently without mouths, which feem to have been delivered on doubtful authority, or from inaccurate obfervation. There are two of fuch monftrous productions however better attefted; one of a human fetus, mentioned by Gipfon in the Scots Medical Effays; which having the gula impervious was furnifhed with an aperture into

the

the wind-pipe, which communicated below into the gullet; by means of which the liquor amnii might be taken into the ftomach before nativity without danger of fuffocation, while the fetus had no occafion to breathe. The other monftrous fetus is defcribed by Vander Wiel, who afferts, that he faw a monftrous lamb, which had no mouth; but inftead of it was furnifhed with an opening in the lower part of the neck into the ftomach. Both thefe inftances evidently favour the doctrine of the fetus being nourifhed by the mouth; as otherwife there had been no neceffity for new or unnatural apertures into the ftomach, when the natural ones were deficient?

From thefe facts and obfervations we may fafely infer, that the fetus in the womb is nourifhed by the fluid which furrounds it; which during the firft period of geftation is abforbed by the naked lacteals; and is afterwards fwallowed into the ftomach and bowels, when thefe organs are perfected; and laftly that the placenta is an organ for the purpofe of giving due oxygenation to the blood of the fetus; which is more neceffary, or at leaft more frequently neceffary, than even the fupply of food.

The queftion of the great Harvey becomes thus eafily anfwered. " Why is not the fetus in the womb fuffocated for want of air, when it remains there even to the tenth month without refpiration: yet if it be born in the feventh or eighth month, and has once refpired, it becomes immediately fuffocated for want of air, if its refpiration be obftructed?"

For further information on this fubject, the reader is referred to the Tentamen Medicum of Dr. Jeffray, printed at Edinburgh in 1786. And it is hoped that Dr. French will fome time give his thefes on this fubject to the public.

SECT.

SECT. XXXIX.

OF GENERATION.

Felix, qui caufas altâ caligine merfas
Pandit, et evolvit tenuiffima vincula rerum.   ANON.

I. *Habits of acting and feeling of individuals attend the foul into a future life, and attend the new embryon at the time of its production. The new speck of entity abforbs nutriment, and receives oxygene. Spreads the terminations of its veffels on cells, which communicate with the arteries of the uterus; fometimes with thofe of the peritoneum. Afterwards it fwallows the liquor amnii, which it produces by its irritation from the uterus, or peritoneum. Like infects in the heads of calves and fheep. Why the white of egg is of two confiftencies. Why nothing is found in quadrupeds fimilar to the yolk, nor in moft vegetable feeds. II. 1. Eggs of frogs and fifh impregnated out of their bodies. Eggs of fowls which are not fecundated, contain only the nutriment for the embryon. The embryon is produced by the male, and the nutriment by the female. Animalcula in femine. Profufion of nature's births. 2. Vegetables viviparous. Buds and bulbs have each a father but no mother. Veffels of the leaf and bud inofculate. The paternal offspring exactly refembles the parent. 3. Infects impregnated for fix generations. Polypus branches like buds. Creeping roots. Viviparous flowers. Tænia, volvox. Eve from Adam's rib. Semen not a ftimulus to the egg. III. 1. Embryons not originally created within other embryons. Organized matter is not fo minute. 2. All the parts of the embryon are not formed in the male parent. Crabs produce their legs, worms produce their heads and tails. In wens, cancers, and inflammations, new veffels are formed. Mules partake of the forms of both parents. Hair and nails grow by elongation, not by diftention. 3. Organic particles of Buffon. IV. 1. Rudiment of the embryon a fimple living filament, becomes a living ring, and then a living tube. 2. It acquires new irritabilities, and fenfibili-*

ties

ties with new organizations, as in wounded snails, polypi, moths, gnats, tadpoles. Hence new parts are acquired by addition not by distention.   3. All parts of the body grow if not confined.   4. Fetuses deficient at their extremities, or have a duplicature of parts.   Monstrous births.   Double parts of vegetables.   5. Mules cannot be formed by distention of the seminal ens.   6. Families of animals from a mixture of their orders.   Mules imperfect.   7. Animal appetency like chemical affinity.   Vis fabricatrix and medicatrix of nature.   8. The changes of animals before and after nativity.   Similarity of their structure.   Changes in them from lust, hunger, and danger.   All warm-blooded animals derived from one living filament.   Cold-blooded animals, insects, worms, vegetables, derived also from one living filament.   Male animals have teats.   Male pidgeon gives milk.   The world itself generated.   The cause of causes.   A state of probation and responsibility.   V. 1. Efficient cause of the colours of birds eggs, and of hair and feathers, which become white in snowy countries.   Imagination of the female colours the egg.   Ideas or motions of the retina imitated by the extremities of the nerves of touch, or rete mucosum.   2. Nutriment supplied by the female of three kinds. Her imagination can only affect the first kind.   Mules how produced, and mulattoes.   Organs of reproduction why deficient in mules.   Eggs with double yolks. VI. 1. Various secretions produced by the extremities of the vessels, as in the glands.   Contageous matter.   Many glands affected by pleasurable ideas, as those which secrete the semen.   2. Snails and worms are hermaphrodite, yet cannot impregnate themselves.   Final cause of this.   3. The imagination of the male forms the sex.   Ideas, or motions of the nerves of vision or of touch, are imitated by the ultimate extremities of the glands of the testes, which mark the sex.   This effect of the imagination belongs only to the male.   The sex of the embryon is not owing to accident.   4. Causes of the changes in animals from imagination as in monsters. From the male.   From the female.   5. Miscarriages from fear.   6. Power of the imagination of the male over the colour, form, and sex of the progeny.   An instance of.   7. Act of generation accompanied with ideas of the male or female form.   Art of begetting beautiful children of either sex.   VII. Recapitulation. VIII. Conclusion.   Of cause and effect.   The atomic philosophy leads to a first cause.

I. THE

I. THE ingenious Dr. Hartley in his work on man, and some other philosophers, have been of opinion, that our immortal part acquires during this life certain habits of action or of sentiment, which become for ever indissoluble, continuing after death in a future state of existence; and add, that if these habits are of the malevolent kind, they must render the possessor miserable even in heaven. I would apply this ingenious idea to the generation or production of the embryon, or new animal, which partakes so much of the form and propensities of the parent.

Owing to the imperfection of language the offspring is termed a *new* animal, but is in truth a branch or elongation of the parent; since a part of the embryon-animal is, or was, a part of the parent; and therefore in strict language it cannot be said to be entirely *new* at the time of its production; and therefore it may retain some of the habits of the parent-system.

At the earliest period of its existence the embryon, as secreted from the blood of the male, would seem to consist of a living filament with certain capabilities of irritation, sensation, volition, and association; and also with some acquired habits or propensities peculiar to the parent: the former of these are in common with other animals; the latter seem to distinguish or produce the kind of animal, whether man or quadruped, with the similarity of feature or form to the parent. It is difficult to be conceived, that a living entity can be separated or produced from the blood by the action of a gland; and which shall afterwards become an animal similar to that in whose vessels it is formed; even though we should suppose with some modern theorists, that the blood is alive; yet every other hypothesis concerning generation rests on principles still more difficult to our comprehension.

At the time of procreation this speck of entity is received into an appropriated nidus, in which it must acquire two circumstances necessary to its life and growth; one of these is food or sustenance, which is to be received by the absorbent mouths of its vessels; and

the

the other is that part of atmofpherical air, or of water, which by the new chemiftry is termed oxygene, and which affects the blood by paffing through the coats of the veffels which contain it.   The fluid furrounding the embryon in its new habitation, which is called liquor amnii, fupplies it with nourifhment; and as fome air cannot but be introduced into the uterus along with the new embryon, it would feem that this fame fluid would for a fhort time, fuppofe for a few hours, fupply likewife a fufficient quantity of the oxygene for its immediate exiftence.

On this account the vegetable impregnation of aquatic plants is performed in the air; and it is probable that the honey-cup or nectary of vegetables requires to be open to the air, that the anthers and ftigmas of the flower may have food of a more oxygenated kind than the common vegetable fap-juice.

On the introduction of this primordium of entity into the uterus the irritation of the liquor amnii, which furrounds it, excites the abforbent mouths of the new veffels into action; they drink up a part of it, and a pleafurable fenfation accompanies this new action; at the fame time the chemical affinity of the oxygene acts through the veffels of the rubefcent blood; and a previous want, or difagreeable fenfation, is relieved by this procefs.

As the want of this oxygenation of the blood is perpetual, (as appears from the inceffant neceffity of breathing by lungs or gills,) the veffels become extended by the efforts of pain or defire to feek this neceffary object of oxygenation, and to remove the difagreeable fenfation, which that want occafions.   At the fame time new particles of matter are abforbed, or applied to thefe extended veffels, and they become permanently elongated, as the fluid in contact with them foon loofes the oxygenous part, which it at firft poffeffed, which was owing to the introduction of air along with the embryon.   Thefe new blood-veffels approach the fides of the uterus, and penetrate with their fine terminations into the veffels of the mother; or adhere to them,

3 Q   acquiring

acquiring oxygene through their coats from the paſſing currents of the arterial blood of the mother.   See Sect. XXXVIII. 2.

This attachment of the placental veſſels to the internal ſide of the uterus by their own proper efforts appears further illuſtrated by the many inſtances of extra-uterine fetuſes, which have thus attached or inſerted their veſſels into the peritoneum ; or on the viſcera, exactly in the ſame manner as they naturally inſert or attach them to the uterus.

The abſorbent veſſels of the embryon continue to drink up nouriſhment from the fluid in which they ſwim, or liquor amnii ; and which at firſt needs no previous digeſtive preparation ; but which, when the whole apparatus of digeſtion becomes complete, is ſwallowed by the mouth into the ſtomach, and being mixed with ſaliva, gaſtric juice, bile, pancreatic juice, and mucus of the inteſtines, becomes digeſted, and leaves a recrement, which produces the firſt feces of the infant, called meconium.

The liquor amnii is ſecreted into the uterus, as the fetus requires it, and may probably be produced by the irritation of the fetus as an extraneous body ; ſince a ſimilar fluid is acquired from the peritoneum in caſes of extra-uterine geſtation. The young caterpillars of the gadfly placed in the ſkins of cows, and the young of the ichneumon-fly placed in the backs of the caterpillars on cabbages, ſeem to produce their nouriſhment by their irritating the ſides of their nidus. A vegetable ſecretion and concretion is thus produced on oak-leaves by the gall-inſect, and by the cynips in the bedeguar of the roſe ; and by the young graſshopper on many plants, by which the animal ſurrounds itſelf with froth. But in no circumſtance is extra-uterine geſtation ſo exactly reſembled as by the eggs of a fly, which are depoſited in the frontal ſinus of ſheep and calves. Theſe eggs float in ſome ounces of fluid collected in a thin pellicle or hydatide. This bag of fluid compreſſes the optic nerve on one ſide, by which the viſion being leſs diſtinct in that eye, the animal turns in perpetual

8   circles

circles towards the fide affected, in order to get a more accurate view
of objects ; for the fame reafon as in fquinting the affected eye is
turned away from the object contemplated. Sheep in the warm
months keep their nofes clofe to the ground to prevent this fly from
fo readily getting into their noftrils.

The liquor amnii is fecreted into the womb as it is required, not
only in refpect to quantity, but, as the digeftive powers of the fetus
become formed, this fluid becomes of a different confiftence and quality, till it is exchanged for milk after nativity. Haller. Phyfiol. V.
1. In the egg the white part, which is analogous to the liquor amnii
of quadrupeds, confifts of two diftinct parts ; one of which is more
vifcid, and probably more difficult of digeftion, and more nutritive
than the other ; and this latter is ufed in the laft week of incubation.
The yolk of the egg is a ftill ftronger or more nutritive fluid, which
is drawn up into the bowels of the chick juft at its exclufion from the
fhell, and ferves it for nourifhment for a day or two, till it is able to
digeft, and has learnt to chufe the harder feeds or grains, which are to
afford it fuftenance.   Nothing analogous to this yolk is found in the
fetus of lactiferous animals, as the milk is another nutritive fluid ready
prepared for the young progeny.

The yolk therefore is not neceffary to the fpawn of fifh, the eggs
of infects, or for the feeds of vegetables ; as their embryons have
probably their food prefented to them as foon as they are excluded
from their fhells, or have extended their roots. Whence it happens
that fome infects produce a living progeny in the fpring and fummer,
and eggs in the autumn ; and fome vegetables have living roots or
buds produced in the place of feeds, as the polygonum viviparum,
and magical onions. See Botanic Garden, p. ii. art. anthoxanthum.

There feems however to be a refervoir of nutriment prepared for
fome feeds befides their cotyledons or feed-leaves, which may be fuppofed in fome meafure analogous to the yolk of the egg. Such are the
faccharine juices of apples, grapes and other fruits, which fupply nu-

trition

trition to the feeds after they fall on the ground. And fuch is the milky juice in the centre of the coco-nut, and part of the kernel of it; the fame I fuppofe of all other monocotyledon feeds, as of the palms, graffes, and lilies.

II. 1. The procefs of generation is ftill involved in impenetrable obfcurity, conjectures may neverthelefs be formed concerning fome of its circumftances. Firft, the eggs of fifh and frogs are impregnated, after they leave the body of the female; becaufe they are depofited in a fluid, and are not therefore covered with a hard fhell. It is however remarkable, that neither frogs nor fifh will part with their fpawn without the prefence of the male; on which account female carp and gold-fifh in fmall ponds, where there are no males, frequently die from the diftention of their growing fpawn. 2. The eggs of fowls, which are laid without being impregnated, are feen to contain only the yolk and white, which are evidently the food or fuftenance for the future chick. 3. As the cicatricula of thefe eggs is given by the cock, and is evidently the rudiment of the new animal; we may conclude, that the embryon is produced by the male, and the proper food and nidus by the female. For if the female be fuppofed to form an equal part of the embryon, why fhould fhe form the whole of the apparatus for nutriment and for oxygenation? the male in many animals is larger, ftronger, and digefts more food than the female, and therefore fhould contribute as much or more towards the reproduction of the fpecies; but if he contributes only half the embryon, and none of the apparatus for fuftenance and oxygenation, the divifion is unequal; the ftrength of the male, and his confumption of food are too great for the effect, compared with that of the female, which is contrary to the ufual courfe of nature.

In objection to this theory of generation it may be faid, if the animalcula in femine, as feen by the microfcope, be all of them rudiments of homunculi, when but one of them can find a nidus, what a wafte nature has made of her productions? I do not affert

that

that thefe moving particles, vifible by the microfcope, are homunciones ; perhaps they may be the creatures of ftagnation or putridity, or perhaps no creatures at all ; but if they are fuppofed to be rudiments of homunculi, or embryons, fuch a profufion of them correfponds with the general efforts of nature to provide for the continuance of her fpecies of animals. Every individual tree produces innumerable feeds, and every individual fifh innumerable fpawn, in fuch inconceivable abundance as would in a fhort fpace of time crowd the earth and ocean with inhabitants ;' and thefe are much more perfect animals than the animalcula in femine can be fuppofed to be, and perifh in uncounted millions. This argument only fhews, that the productions of nature are governed by general laws ; and that by a wife fuperfluity of provifion fhe has enfured their continuance.

2. That the embryon is fecreted or produced by the male, and not by the conjunction of fluids from both male and female, appears from the analogy of vegetable feeds. In the large flowers, as the tulip, there is no fimilarity of apparatus between the anthers and the ftigma : the feed is produced according to the obfervations of Spallanzani long before the flowers open, and in confequence long before it can be impregnated, like the egg in the pullet. And after the prolific duft is fhed on the ftigma, the feed becomes coagulated in one point firft, like the cicatricula of the impregnated egg. See Botanic Garden, Part I. additional note 38. Now in thefe fimple products of nature, if the female contributed to produce the new embryon equally with the male, there would probably have been fome vifible fimilarity of parts for this purpofe, befides thofe neceffary for the nidus and fuftenance of the new progeny. Befides in many flowers the males are more numerous than the females, or than the feparate uterine cells in their germs, which would fhew, that the office of the male was at leaft as important as that of the female ; whereas if the female, befides producing the egg or feed, was to produce an equal part of the embryon, the office of reproduction would be unequally divided between them.

Add

Add to this, that in the moſt ſimple kind of vegetable reproduction, I mean the buds of trees, which are their viviparous offspring, the leaf is evidently the parent of the bud, which riſes in its boſom, according to the obſervation of Linnæus.   This leaf conſiſts of abſorbent veſſels, and pulmonary ones, to obtain its nutriment, and to impregnate it with oxygene.   This ſimple piece of living organization is alſo furniſhed with a power of reproduction ; and as the new offspring is thus ſupported adhering to its father, it needs no mother to ſupply it with a nidus, and nutriment, and oxygenation ; and hence no female leaf has exiſtence.

I conceive that the veſſels between the bud and the leaf communicate or inoſculate ; and that the bud is thus ſerved with vegetable blood, that is, with both nutriment and oxygenation, till the death of the parent-leaf in autumn.   And in this reſpect it differs from the fetus of viviparous animals.   Secondly, that then the bark-veſſels belonging to the dead-leaf, and in which I ſuppoſe a kind of manna to have been depoſited, become now the placental veſſels, if they may be ſo called, of the new bud.   From the vernal ſap thus produced of one ſugar-maple-tree in New-York and in Pennſylvania, five or ſix pounds of good ſugar may be made annually without deſtroying the tree.   Account of maple-ſugar by B. Ruſh.   London, Phillips. (See Botanic Garden, Part I. additional note on vegetable placentation.)

Theſe veſſels, when the warmth of the vernal ſun hatches the young bud, ſerve it with a ſaccharine nutriment, till it acquires leaves of its own, and ſhoots a new ſyſtem of abſorbents down the bark and root of the tree, juſt as the farinaceous or oily matter in ſeeds, and the ſaccharine matter in fruits, ſerve their embryons with nutriment, till they acquire leaves and roots.   This analogy is as forceable in ſo obſcure a ſubject, as it is curious, and may in large buds, as of the horſe-cheſnut, be almoſt ſeen by the naked eye ; if with a penknife the remaining rudiment of the laſt year's leaf, and of the new bud in its boſom, be cut away ſlice by ſlice.   The ſeven ribs of
the

the laft year's leaf will be feen to have arifen from the pith in feven diftinct points making a curve; and the new bud to have been produced in their centre, and to have pierced the alburnum and cortex, and grown without the affiftance of a mother. A fimilar procefs may be feen on diffecting a tulip-root in winter; the leaves, which inclofed the laft year's flower-ftalk, were not neceffary for the flower; but each of thefe was the father of a new bud, which may be now found at its bafe; and which, as it adheres to the parent, required no mother.

This paternal offspring of vegetables, I mean their buds and bulbs, is attended with a very curious circumftance; and that is, that they exactly refemble their parents, as is obfervable in grafting fruit-trees, and in propagating flower-roots; whereas the feminal offspring of plants, being fupplied with nutriment by the mother, is liable to perpetual variation. Thus alfo in the vegetable clafs dioicia, where the male flowers are produced on one tree, and the female ones on another; the buds of the male trees uniformly produce either male flowers, or other buds fimilar to themfelves; and the buds of the female trees produce either female flowers, or other buds fimilar to themfelves; whereas the feeds of thefe trees produce either male or female plants. From this analogy of the production of vegetable buds without a mother, I contend that the mother does not contribute to the formation of the living ens in animal generation, but is neceffary only for fupplying its nutriment and oxygenation.

There is another vegetable fact publifhed by M. Koelreuter, which he calls " a complete metamorphofis of one natural fpecies of plants into another," which fhews, that in feeds as well as in buds, the embryon proceeds from the male parent, though the form of the fubfequent mature plant is in part dependant on the female. M. Koelreuter impregnated a ftigma of the nicotiana ruftica with the farina of the nicotiana paniculata, and obtained prolific feeds from it. With the plants which fprung from thefe feeds, he repeated the experiment,

impregnating

impregnating them with the farina of the nicotiana paniculata. As the mule plants which he thus produced were prolific, he continued to impregnate them for many generations with the farina of the nicotiana paniculata, and they became more and more like the male parent, till he at length obtained fix plants in every refpect perfectly fimilar to the nicotiana paniculata ; and in no refpect refembling their female parent the nicotiana ruftica. *Blumenbach* on Generation.

3. It is probable that the infects, which are faid to require but one impregnation for fix generations, as the aphis (fee Amenit. Academ.) produce their progeny in the manner above defcribed, that is, without a mother, and not without a father ; and thus experience a lucina fine concubitu. Thofe who have attended to the habits of the polypus, which is found in the ftagnant water of our ditches in July, affirm, that the young ones branch out from the fide of the parent like the buds of trees, and after a time feparate themfelves from them. This is fo analogous to the manner in which the buds of trees appear to be produced, that thefe polypi may be confidered as all male animals, producing embryons, which require no mother to fupply them with a nidus, or with nutriment, and oxygenation.

This lateral or lineal generation of plants, not only obtains in the buds of trees, which continue to adhere to them, but is beautifully feen in the wires of knot-grafs, polygonum aviculare, and in thofe of ftrawberries, fragaria vefca. In thefe an elongated creeping bud is protruded, and, where it touches the ground, takes root, and produces a new plant derived from its father, from which it acquires both nutriment and oxygenation ; and in confequence needs no maternal apparatus for thefe purpofes. In viviparous flowers, as thofe of allium magicum, and polygonum viviparum, the anthers and the ftigmas become effete and perifh ; and the lateral or paternal offspring fucceeds inftead of feeds, which adhere till they are fufficiently mature, and then fall upon the ground, and take root like other bulbs.

The lateral production of plants by wires, while each new plant is
                                                                    thus

thus chained to its parent, and continues to put forth another and another, as the wire creeps onward on the ground, is exactly refembled by the tape-worm, or tænia, so often found in the bowels, stretching itself in a chain quite from the stomach to the rectum.   Linnæus afferts, " that it grows old at one extremity, while it continues to generate young ones at the other, proceeding ad infinitum, like a root of grafs.   The feparate joints are called gourd-worms, and propagate new joints like the parent without end, each joint being furnifhed with its proper mouth, and organs of digeftion."   Syftema naturæ. Vermes tenia.   In this animal there evidently appears a power of reproduction without any maternal apparatus for the purpofe of fupplying nutriment and oxygenation to the embryon, as it remains attached to its father till its maturity.   The volvox globator, which is a tranfparent animal, is faid by Linnæus to bear within it fons and grand-fons to the fifth generation.   Thefe are probably living fetufes, produced by the father, of different degrees of maturity, to be detruded at different periods of time, like the unimpregnated eggs of various fizes, which are found in poultry ; and as they are produced without any known copulation, contribute to evince, that the living embryon in other orders of animals is formed by the male-parent, and not by the mother, as one parent has the power to produce it.

This idea of the reproduction of animals from a fingle living filament of their fathers, appears to have been fhadowed or allegorized in the curious account in facred writ of the formation of Eve from a rib of Adam.

From all thefe analogies I conclude, that the embryon is produced folely by the male, and that the female fupplies it with a proper nidus, with fuftenance, and with oxygenation; and that the idea of the femen of the male conftituting only a ftimulus to the egg of the female, exciting it into life, (as held by fome philofophers) has no fupport from experiment or analogy.

III. 1. Many ingenious philofophers have found fo great difficulty

3 R

in

in conceiving the manner of the reproduction of animals, that they have fuppofed all the numerous progeny to have exifted in miniature in the animal originally created ; and that thefe infinitely minute forms are only evolved or diftended, as the embryon increafes in the womb. This idea, befides its being unfupported by any analogy we are acquainted with, afcribes a greater tenuity to organized matter, than we can readily admit ; as thefe included embryons are fuppofed each of them to confift of the various and complicate parts of animal bodies : they muft poffefs a much greater degree of minutenefs, than that which was afcribed to the devils that tempted St. Anthony; of whom 20,000 were faid to have been able to dance a faraband on the point of the fineft needle without incommoding each other.

2. Others have fuppofed, that all the parts of the embryon are formed in the male, previous to its being depofited in the egg or uterus ; and that it is then only to have its parts evolved or diftended as mentioned above ; but this is only to get rid of one difficulty by propofing another equally incomprehenfible : they found it difficult to conceive, how the embryon could be formed in the uterus or egg, and therefore wifhed it to be formed before it came thither. In anfwer to both thefe doctrines it may be obferved, 1ft, that fome animals, as the crab-fifh, can reproduce a whole limb, as a leg which has been broken off ; others, as worms and fnails, can reproduce a head, or a tail, when either of them has been cut away; and that hence in thefe animals at leaft a part can be formed anew, which cannot be fuppofed to have exifted previoufly in miniature.

Secondly, there are new parts or new veffels produced in many difeafes, as on the cornea of the eye in ophthalmy, in wens and cancers, which cannot be fuppofed to have had a prototype or original miniature in the embryon.

Thirdly, how could mule-animals be produced, which partake of the forms of both the parents, if the original embryon was a miniature exifting in the femen of the male parent ? if an embryon of the

male

male afs was only expanded, no refemblance to the mare could exift in the mule.

This miftaken idea of the extenfion of parts feems to have had its rife from the mature man refembling the general form of the fetus; and from thence it was believed, that the parts of the fetus were diftended into the man; whereas they have increafed 100 times in weight, as well as 100 times in fize; now no one will call the additional 99 parts a diftention of the original one part in refpect to weight. Thus the uterus during pregnancy is greatly enlarged in thicknefs and folidity as well as in capacity, and hence muft have acquired this additional fize by accretion of new parts, not by an extenfion of the old ones; the familiar act of blowing up the bladder of an animal recently flaughtered has led our imaginations to apply this idea of diftention to the increafe of fize from natural growth; which however muft be owing to the appofition of new parts; as it is evinced from the increafe of weight along with the increafe of dimenfion; and is even vifible to our eyes in the elongation of our hair from the colour of its ends; or when it has been dyed on the head; and in the growth of our nails from the fpecks fometimes obfervable on them; and in the increafe of the white crefcent at their roots, and in the growth of new flefh in wounds, which confifts of new nerves as well as of new blood-veffels.

3. Laftly, Mr. Buffon has with great ingenuity imagined the exiftence of certain organic particles, which are fuppofed to be partly alive, and partly mechanic fprings. The latter of thefe were difcovered by Mr. Needham in the milt or male organ of a fpecies of cuttle fifh, called calmar; the former, or living animalcula, are found in both male and female fecretions, in the infufions of feeds, as of pepper, in the jelly of roafted veal, and in all other animal and vegetable fubftances. Thefe organic particles he fuppofes to exift in the fpermatic fluids of both fexes, and that they are derived thither from every part of the body, and muft therefore refemble, as he fuppofes,

the

the parts from whence they are derived. Thefe organic particles he believes to be in conftant activity, till they become mixed in the womb, and then they inftantly join and produce an embryon or fetus fimilar to the two parents.

Many objections might be adduced to this fanciful theory, I fhall only mention two. Firft, that it is analogous to no known animal laws. And fecondly, that as thefe fluids, replete with organic particles derived both from the male and female organs, are fuppofed to be fimilar; there is no reafon why the mother fhould not produce a female embryon without the affiftance of the male, and realize the lucina fine concubitu.

IV. 1. I conceive the primordium, or rudiment of the embryon, as fecreted from the blood of the parent, to confift of a fimple living filament as a mufcular fibre; which I fuppofe to be an extremity of a nerve of loco-motion, as a fibre of the retina is an extremity of a nerve of fenfation; as for inftance one of the fibrils, which compofe the mouth of an abforbent veffel; I fuppofe this living filament, of whatever form it may be, whether fphere, cube, or cylinder, to be endued with the capability of being excited into action by certain kinds of ftimulus. By the ftimulus of the furrounding fluid, in which it is received from the male, it may bend into a ring; and thus form the beginning of a tube. Such moving filaments, and fuch rings, are defcribed by thofe, who have attended to microfcopic animalcula. This living ring may now embrace or abforb a nutritive particle of the fluid, in which it fwims; and by drawing it into its pores, or joining it by compreffion to its extremities, may increafe its own length or craffitude; and by degrees the living ring may become a living tube.

2. With this new organization, or accretion of parts, new kinds of irritability may commence; for fo long as there was but one living organ, it could only be fuppofed to poffefs irritability; fince fenfibility may be conceived to be an extenfion of the effect of irritability over the

the reft of the fyftem. Thefe new kinds of irritability and of fenfi-bility in confequence of new organization, appear from variety of facts in the more mature animal; thus the formation of the teftes, and confequent fecretion of the femen, occafion the paffion of luft; the lungs muft be previoufly formed before their exertions to obtain frefh air can exift; the throat or œfophagus muft be formed previous to the fenfation or appetites of hunger and thirft; one of which feems to refide at the upper end, and the other at the lower end of that canal.

Thus alfo the glans penis, when it is diftended with blood, ac-quires a new fenfibility, and a new appetency. The fame occurs to the nipples of the breafts of female animals, when they are diftended with blood, they acquire the new appetency of giving milk. So in-flamed tendons and membranes, and even bones, acquire new fenfa-tions; and the parts of mutilated animals, as of wounded fnails, and polypi, and crabs, are reproduced; and at the fame time acquire fen-fations adapted to their fituations. Thus when the head of a fnail is reproduced after decollation with a fharp rafor, thofe curious telefco-pic eyes are alfo reproduced, and acquire their fenfibility to light, as well as their adapted mufcles for retraction on the approach of injury.

With every new change, therefore, of organic form, or addition of organic parts, I fuppofe a new kind of irritability or of fenfibility to be produced; fuch varieties of irritability or of fenfibility exift in our adult ftate in the glands; every one of which is furnifhed with an irritability, or a tafte, or appetency, and a confequent mode of action peculiar to itfelf.

In this manner I conceive the veffels of the jaws to produce thofe of the teeth, thofe of the fingers to produce the nails, thofe of the fkin to produce the hair; in the fame manner as afterwards about the age of puberty the beard and other great changes in the form of the body, and difpofition of the mind, are produced in confequence of the

new

new fecretion of femen; for if the animal is deprived of this fecretion
thofe changes do not take place.   Thefe changes I conceive to be
formed not by elongation or diftention of primeval ftamina, but by
appofition of parts; as the mature crab-fifh, when deprived of a limb,
in a certain fpace of time has power to regenerate it; and the tadpole
puts forth its feet long after its exclufion from the fpawn; and the
caterpillar in changing into a butterfly acquires a new form, with
new powers, new fenfations, and new defires.

The natural hiftory of butterflies, and moths, and beetles, and
gnats, is full of curiofity; fome of them pafs many months, and
others even years, in their caterpillar or grub ftate; they then reft
many weeks without food, fufpended in the air, buried in the earth,
or fubmerfed in water; and change themfelves during this time into
an animal apparently of a different nature; the ftomachs of fome of
them, which before digefted vegetable leaves or roots, now only digeft honey; they have acquired wings for the purpofe of feeking this
new food, and a long probofcis to collect it from flowers, and I fuppofe a fenfe of fmell to detect the fecret places in flowers, where it is
formed.   The moths, which fly by night, have a much longer probofcis rolled up under their chins like a watch fpring; which they
extend to collect the honey from flowers in their fleeping ftate; when
they are clofed, and the nectaries in confequence more difficult to be
plundered.   The beetle kind are furnifhed with an external covering
of a hard material to their wings, that they may occafionally again
make holes in the earth, in which they paffed the former ftate of their
exiftence.

But what moft of all diftinguifhes thefe new animals is, that they
are new furnifhed with the powers of reproduction; and that they
now differ from each other in fex, which does not appear in their
caterpillar or grub ftate.   In fome of them the change from a caterpillar into a butterfly or moth feems to be accomplifhed for the fole
purpofe of their propagation; fince they immediately die after this is
finifhed,

finifhed, and take no food in the interim, as the filk-worm in this climate; though it is poffible, it might take honey as food, if it was prefented to it.   For in general it would feem, that food of a more ftimulating kind, the honey of vegetables inftead of their leaves, was neceffary for the purpofe of the feminal reproduction of thefe animals, exactly fimilar to what happens in vegetables; in thefe the juices of the earth are fufficient for their purpofe of reproduction by buds or bulbs; in which the new plant feems to be formed by irritative motions, like the growth of their other parts, as their leaves or roots; but for the purpofe of feminal or amatorial reproduction, where fenfation is required, a more ftimulating food becomes neceffary for the anther, and ftigma; and this food is honey; as explained in Sect. XIII. on Vegetable Animation.

The gnat and the tadpole refemble each other in their change from natant animals with gills into aerial animals with lungs; and in their change of the element in which they live; and probably of the food, with which they are fupported; and laftly, with their acquiring in their new ftate the difference of fex, and the organs of feminal or amatorial reproduction.   While the polypus, who is their companion in their former ftate of life, not being allowed to change his form and element, can only propagate like vegetable buds by the fame kind of irritative motions, which produces the growth of his own body, without the feminal or amatorial propagation, which requires fenfation; and which in gnats and tadpoles feems to require a change both of food and of refpiration.

From hence I conclude, that with the acquifition of new parts, new fenfations, and new defires, as well as new powers, are produced; and this by accretion to the old ones, and not by diftention of them.   And finally, that the moft effential parts of the fyftem, as the brain for the purpofe of diftributing the power of life, and the placenta for the purpofe of oxygenating the blood; and the additional abforbent veffels for the purpofe of acquiring aliment, are firft formed

by

by the irritations above mentioned, and by the pleafureable fenfations attending thofe irritations, and by the exertions in confequence of painful fenfations, fimilar to thofe of hunger and fuffocation. After thefe an apparatus of limbs for future ufes, or for the purpofe of moving the body in its prefent natant ftate, and of lungs for future refpiration, and of teftes for future reproduction, are formed by the irritations and fenfations, and confequent exertions of the parts previoufly exifting, and to which the new parts are to be attached.

3. In confirmation of thefe ideas it may be obferved, that all the parts of the body endeavour to grow, or to make additional parts to themfelves throughout our lives; but are reftrained by the parts immediately containing them; thus, if the fkin be taken away, the flefhy parts beneath foon fhoot out new granulations, called by the vulgar proud flefh. If the periofteum be removed, a fimilar growth commences from the bone. Now in the cafe of the imperfect embryon, the containing or confining parts are not yet fuppofed to be formed, and hence there is nothing to reftrain its growth.

4. By the parts of the embryon being thus produced by new appofitions, many phenomena both of animal and vegetable productions receive an eafier explanation; fuch as that many fetufes are deficient at the extremities, as in a finger or a toe, or in the end of the tongue, or in what is called a hare-lip with deficiency of the palate. For if there fhould be a deficiency in the quantity of the firft nutritive particles laid up in the egg for the reception of the firft living filament, the extreme parts, as being laft formed, muft fhew this deficiency by their being imperfect.

This idea of the growth of the embryon accords alfo with the production of fome monftrous births, which confift of a duplicature of the limbs, as chickens with four legs; which could not occur, if the fetus was formed by the diftention of an original ftamen, or miniature. For if there fhould be a fuperfluity of the firft nutritive particles laid up in the egg for the firft living filament; it is eafy to conceive,

3

ceive,

ceive, that a duplicature of fome parts may be formed. And that fuch fuperfluous nourifhment fometimes exifts, is evinced by the double yolks in fome eggs, which I fuppofe were thus formed previous to their impregnation by the exuberant nutriment of the hen.

This idea is confirmed by the analogy of the monfters in the vegetable world alfo; in which a duplicate or triplicate production of various parts of the flower is obfervable, as a triple nectary in fome columbines, and a triple petal in fome primrofes; and which are fuppofed to be produced by abundant nourifhment.

5. If the embryon be received into a fluid, whofe ftimulus is different in fome degree from the natural, as in the production of mule-animals, the new irritabilities or fenfibilities acquired by the increafing or growing organized parts may differ, and thence produce parts not fimilar to the father, but of a kind belonging in part to the mother; and thus, though the original ftamen or living ens was derived totally from the father, yet new irritabilities or fenfibilities being excited, a change of form correfponding with them will be produced. Nor could the production of mules exift, if the ftamen or miniature of all the parts of the embryon is previoufly formed in the male femen, and is only diftended by nourifhment in the female uterus. Whereas this difficulty ceafes, if the embryon be fuppofed to confift of a living filament, which acquires or makes new parts with new irritabilities, as it advances in its growth.

The form, folidity, and colour, of the particles of nutriment laid up for the reception of the firft living filament, as well as their peculiar kind of ftimulus, may contribute to produce a difference in the form, folidity, and colour of the fetus, fo as to refemble the mother, as it advances in life. This alfo may efpecially happen during the firft ftate of the exiftence of the embryon, before it has acquired organs, which can change thefe firft nutritive particles, as explained in No. 5. 2. of this Section. And as thefe nutritive particles are fup-

3 S   pofed

posed to be similar to those, which are formed for her own nutrition, it follows that the fetus should so far resemble the mother.

This explains, why hereditary diseases may be derived either from the male or female parent, as well as the peculiar form of either of their bodies. Some of these hereditary diseases are simply owing to a deficient activity of a part of the system, as of the absorbent vessels, which open into the cells or cavities of the body, and thus occasion dropsies. Others are at the same time owing to an increase of sensation, as in scrophula and consumption; in these the obstruction of the fluids is first caused by the inirritability of the vessels, and the inflammation and ulcers which succeed, are caused by the consequent increase of sensation in the obstructed part. Other hereditary diseases, as the epilepsy, and other convulsions, consist in too great voluntary exertions in consequence of disagreeable sensation in some particular diseased part. Now as the pains, which occasion these convulsions, are owing to defect of the action of the diseased part, as shewn in Sect. XXXIV. it is plain, that all these hereditary diseases may have their origin either from defective irritability derived from the father, or from deficiency of the stimulus of the nutriment derived from the mother. In either case the effect would be similar; as a scrophulous race is frequently produced among the poor from the deficient stimulus of bad diet, or of hunger; and among the rich, by a deficient irritability from their having been long accustomed to too great stimulus, as of vinous spirit.

6. From this account of reproduction it appears, that all animals have a similar origin, viz. from a single living filament; and that the difference of their forms and qualities has arisen only from the different irritabilities and sensibilities, or voluntarities, or associabilities, of this original living filament; and perhaps in some degree from the different forms of the particles of the fluids, by which it has been at first stimulated into activity. And that from hence, as Linnæus has

conjectured

conjectured in refpect to the vegetable world, it is not impoffible, but the great variety of fpecies of animals, which now tenant the earth, may have had their origin from the mixture of a few natural orders. And that thofe animal and vegetable mules, which could continue their fpecies, have done fo, and conftitute the numerous families of animals and vegetables which now exift; and that thofe mules, which were produced with imperfect organs of generation, perifhed without reproduction, according to the obfervation of Ariftotle; and are the animals, which we now call mules. See Botanic Garden, Part II. Note on Dianthus.

Such a promifcuous intercourfe of animals is faid to exift at this day in New South Wales by Captain Hunter. And that not only amongft the quadrupeds and birds of different kinds, but even amongft the fifh, and, as he believes, amongft the vegetables. He fpeaks of an animal between the opoffum and the kangaroo, from the fize of a fheep to that of a rat. Many fifh feemed to partake of the fhark; fome with a fkait's head and fhoulders, and the hind part of a fhark; others with a fhark's head and the body of a mullet; and fome with a fhark's head and the flat body of a fting-ray. Many birds partake of the parrot; fome have the head, neck, and bill of a parrot, with long ftraight feet and legs; others with legs and feet of a parrot, with head and neck of a fea-gull. Voyage to South Wales by Captain John Hunter, p. 68.

7. All animals therefore, I contend, have a fimilar caufe of their organization, originating from a fingle living filament, endued indeed with different kinds of irritabilities and fenfibilities, or of animal appetencies; which exift in every gland, and in every moving organ of the body, and are as effential to living organization as chemical affinities are to certain combinations of inanimate matter.

If I might be indulged to make a fimile in a philofophical work, I fhould fay, that the animal appétencies are not only perhaps lefs numerous originally than the chemical affinities; but that like thefe lat-

ter,

ter, they change with every new combination; thus vital air and azote, when combined, produce nitrous acid; which now acquires the property of diffolving filver; fo with every new additional part to the embryon, as of the throat or lungs, I fuppofe a new animal appetency to be produced.

In this early formation of the embryon from the irritabilities, fenfibilities, and affociabilities, and confequent appetencies, the faculty of volition can fcarcely be fuppofed to have had its birth. For about what can the fetus deliberate, when it has no choice of objects? But in the more advanced ftate of the fetus, it evidently poffeffes volition; as it frequently changes its attitude, though it feems to fleep the greateft part of its time; and afterwards the power of volition contributes to change or alter many parts of the body during its growth to manhood, by our early modes of exertion in the various departments of life. All thefe faculties then conftitute the vis fabricatrix, and the vis confervatrix, as well as the vis medicatrix of nature, fo much fpoken of, but fo little underftood by philofophers.

8. When we revolve in our minds, firft, the great changes, which we fee naturally produced in animals after their nativity, as in the production of the butterfly with painted wings from the crawling caterpillar; or of the refpiring frog from the fubnatant tadpole; from the feminine boy to the bearded man, and from the infant girl to the lactefcent woman; both which changes may be prevented by certain mutilations of the glands neceffary to reproduction.

Secondly, when we think over the great changes introduced into various animals by artificial or accidental cultivation, as in horfes, which we have exercifed for the different purpofes of ftrength or fwiftnefs, in carrying burthens or in running races; or in dogs, which have been cultivated for ftrength and courage, as the bull-dog; or for acutenefs of his fenfe of fmell, as the hound and fpaniel; or for the fwiftnefs of his foot, as the greyhound; or for his fwimming in the water, or for drawing fnow-fledges, as the rough-haired dogs of

the

the north; or laftly, as a play-dog for children, as the lap-dog; with the changes of the forms of the cattle, which have been domefticated from the greateft antiquity, as camels, and fheep; which have undergone fo total a transformation, that we are now ignorant from what fpecies of wild animals they had their origin. Add to thefe the great changes of fhape and colour, which we daily fee produced in fmaller animals from our domeftication of them, as rabbits, or pidgeons; or from the difference of climates and even of feafons; thus the fheep of warm climates are covered with hair inftead of wool; and the hares and partridges of the latitudes, which are long buried in fnow, become white during the winter months; add to thefe the various changes produced in the forms of mankind, by their early modes of exertion; or by the difeafes occafioned by their habits of life; both of which became hereditary, and that through many generations. Thofe who labour at the anvil, the oar, or the loom, as well as thofe who carry fedan-chairs, or who have been educated to dance upon the rope, are diftinguifhable by the fhape of their limbs; and the difeafes occafioned by intoxication deform the countenance with leprous eruptions, or the body with tumid vifcera, or the joints with knots and diftortions.

Thirdly, when we enumerate the great changes produced in the fpecies of animals before their nativity; thefe are fuch as refemble the form or colour of their parents, which have been altered by the cultivation or accidents above related, and are thus continued to their pofterity. Or they are changes produced by the mixture of fpecies as in mules; or changes produced probably by the exuberance of nourifhment fupplied to the fetus, as in monftrous births with additional limbs; many of thefe enormities of fhape are propagated, and continued as a variety at leaft, if not as a new fpecies of animal. I have feen a breed of cats with an additional claw on every foot; of poultry alfo with an additional claw, and with wings to their feet; and of others without rumps. Mr. Buffon mentions a breed of dogs without

tails,

tails, which are common at Rome and at Naples, which he fuppofes to have been produced by a cuftom long eftablifhed of cutting their tails clofe off.   There are many kinds of pidgeons, admired for their peculiarities, which are monfters thus produced and propagated.   And to thefe muft be added, the changes produced by the imagination of the male parent, as will be treated of more at large in No. VI. of this Section.

When we confider all thefe changes of animal form, and innumerable others, which may be collected from the books of natural hiftory; we cannot but be convinced, that the fetus or embryon is formed by appofition of new parts, and not by the diftention of a primordial neft of germs, included one within another, like the cups of a conjurer.

Fourthly, when we revolve in our minds the great fimilarity of ftructure, which obtains in all the warm-blooded animals, as well quadrupeds, birds, and amphibious animals, as in mankind; from the moufe and bat to the elephant and whale; one is led to conclude, that they have alike been produced from a fimilar living filament.   In fome this filament in its advance to maturity has acquired hands and fingers, with a fine fenfe of touch, as in mankind. In others it has acquired claws or talons, as in tygers and eagles.   In others, toes with an intervening web, or membrane, as in feals and geefe.   In others it has acquired cloven hoofs, as in cows and fwine; and whole hoofs in others, as in the horfe.   While in the bird kind this original living filament has put forth wings inftead of arms or legs, and feathers inftead of hair. In fome it has protruded horns on the forehead inftead of teeth in the fore part of the upper jaw; in others tufhes inftead of horns; and in others beaks inftead of either.   And all this exactly as is daily feen in the tranfmutations of the tadpole, which acquires legs and lungs, when he wants them; and lofes his tail, when it is no longer of fervice to him.

Fifthly, from their firft rudiment, or primordium, to the termination of their lives, all animals undergo perpetual transformations;
                                                                    which

which are in part produced by their own exertions in confequence of their defires and averfions, of their pleafures and their pains, or of irritations, or of affociations ; and many of thefe acquired forms or propenfities are tranfmitted to their pofterity.   See Sect. XXXI. 1.

As air and water are fupplied to animals in fufficient profufion, the three great objects of defire, which have changed the forms of many animals by their exertions to gratify them, are thofe of luft, hunger, and fecurity.   A great want of one part of the animal world has confifted in the defire of the exclufive poffeffion of the females ; and thefe have acquired weapons to combat each other for this purpofe, as the very thick, fhield-like, horny fkin on the fhoulder of the boar is a defence only againft animals of his own fpecies, who ftrike obliquely upwards, nor are his tufhes for other purpofes, except to defend himfelf, as he is not naturally a carnivorous animal.   So the horns of the ftag are fharp to offend his adverfary, but are branched for the purpofe of parrying or receiving the thrufts of horns fimilar to his own, and have therefore been formed for the purpofe of combating other ftags for the exclufive poffeffion of the females ; who are obferved, like the ladies in the times of chivalry, to attend the car of the victor.

The birds, which do not carry food to their young, and do not therefore marry, are armed with fpurs for the purpofe of fighting for the exclufive poffeffion of the females, as cocks and quails.   It is certain that thefe weapons are not provided for their defence againft other adverfaries, becaufe the females of thefe fpecies are without this armour.   The final caufe of this conteft amongft the males feems to be, that the ftrongeft and moft active animal fhould propagate the fpecies, which fhould thence become improved.

Another great want confifts in the means of procuring food, which has diverfified the forms of all fpecies of animals.   Thus the nofe of the fwine has become hard for the purpofe of turning up the foil in

fearch

search of insects and of roots. The trunk of the elephant is an elongation of the nose for the purpose of pulling down the branches of trees
for his food, and for taking up water without bending his knees.
Beasts of prey have acquired strong jaws or talons. Cattle have acquired a rough tongue and a rough palate to pull off the blades of grass,
as cows and sheep. Some birds have acquired harder beaks to crack
nuts, as the parrot. Others have acquired beaks adapted to break the
harder seeds, as sparrows. Others for the softer seeds of flowers, or
the buds of trees, as the finches. Other birds have acquired long
beaks to penetrate the moister soils in search of insects or roots, as
woodcocks; and others broad ones to filtrate the water of lakes, and
to retain aquatic insects. All which seem to have been gradually produced during many generations by the perpetual endeavour of the creatures to supply the want of food, and to have been delivered to their
posterity with constant improvement of them for the purposes required.

The third great want amongst animals is that of security, which
seems much to have diversified the forms of their bodies and the colour
of them; these consist in the means of escaping other animals more
powerful than themselves. Hence some animals have acquired wings
instead of legs, as the smaller birds, for the purpose of escape. Others
great length of fin, or of membrane, as the flying fish, and the bat.
Others great swiftness of foot, as the hare. Others have acquired
hard or armed shells, as the tortoise and the echinus marinus.

The contrivances for the purposes of security extend even to vegetables, as is seen in the wonderful and various means of their concealing or defending their honey from insects, and their seeds from
birds. On the other hand swiftness of wing has been acquired by
hawks and swallows to pursue their prey; and a proboscis of admirable
structure has been acquired by the bee, the moth, and the humming
bird, for the purpose of plundering the nectaries of flowers. All
which

which feem to have been formed by the original living filament, excited into action by the neceffities of the creatures, which poffefs them, and on which their exiftence depends.

From thus meditating on the great fimilarity of the ftructure of the warm-blooded animals, and at the fame time of the great changes they undergo both before and after their nativity; and by confidering in how minute a portion of time many of the changes of animals above defcribed have been produced; would it be too bold to imagine, that in the great length of time, fince the earth began to exift, perhaps millions of ages before the commencement of the hiftory of mankind, would it be too bold to imagine, that all warm-blooded animals have arifen from one living filament, which THE GREAT FIRST CAUSE endued with animality, with the power of acquiring new parts, attended with new propenfities, directed by irritations, fenfations, volitions, and affociations; and thus poffeffing the faculty of continuing to improve by its own inherent activity, and of delivering down thofe improvements by generation to its pofterity, world without end!

Sixthly, The cold-blooded animals, as the fifh-tribes, which are furnifhed with but one ventricle of the heart, and with gills inftead of lungs, and with fins inftead of feet or wings, bear a great fimilarity to each other; but they differ, neverthelefs, fo much in their general ftructure from the warm-blooded animals, that it may not feem probable at firft view, that the fame living filament could have given origin to this kingdom of animals, as to the former. Yet are there fome creatures, which unite or partake of both thefe orders of animation, as the whales and feals; and more particularly the frog, who changes from an aquatic animal furnifhed with gills to an aerial one furnifhed with lungs.

The numerous tribes of infects without wings, from the fpider to the fcorpion, from the flea to the lobfter; or with wings, from the gnat and the ant to the wafp and the dragon-fly, differ fo totally from

3 T   each

each other, and from the red-blooded claffes above defcribed, both in
the forms of their bodies, and their modes of life; befides the organ
of fenfe, which they feem to poffefs in their antennæ or horns, to
which it has been thought by fome naturalifts, that other creatures
have nothing fimilar; that it can fcarcely be fuppofed that this na-
tion of animals could have been produced by the fame kind of living
filament, as the red-blooded claffes above mentioned. And yet the
changes which many of them undergo in their early ftate to that of
their maturity, are as different, as one animal can be from another.
As thofe of the gnat, which paffes his early ftate in water, and then
ftretching out his new wings, and expanding his new lungs, rifes in
the air; as of the caterpillar, and bee-nymph, which feed on vegetable leaves or farina, and at length burfting from their felf-formed
graves, become beautiful winged inhabitants of the fkies, journey-
ing from flower to flower, and nourifhed by the ambrofial food of
honey.

There is ftill another clafs of animals, which are termed vermes
by Linnæus, which are without feet, or brain, and are hermaphro-
dites, as worms, leeches, fnails, fhell-fifh, coralline infects, and
fponges; which poffefs the fimpleft ftructure of all animals, and appear totally different from thofe already defcribed. The fimplicity
of their ftructure, however, can afford no argument againft their
having been produced from a living filament as above contended.

Laft of all the various tribes of vegetables are to be enumerated
amongft the inferior orders of animals. Of thefe the anthers and
ftigmas have already been fhewn to poffefs fome organs of fenfe, to be
nourifhed by honey, and to have the power of generation like infects,
and have thence been announced amongft the animal kingdom in
Sect. XIII. and to thefe muft be added the buds and bulbs which conftitute the viviparous offspring of vegetation. The former I fuppofe
to be beholden to a fingle living filament for their feminal or amatorial procreation; and the latter to the fame caufe for their lateral or
branching

branching generation, which they poffefs in common with the polypus, tænia, and volvox; and the fimplicity of which is an argument
in favour of the fimilarity of its caufe.

Linnæus fuppofes, in the Introduction to his Natural Orders, that
very few vegetables were at firft created, and that their numbers
were increafed by their intermarriages, and adds, fuadent hæc Crea-
toris leges a fimplicibus ad compofita.   Many other changes feem to
have arifen in them by their perpetual conteft for light and air above
ground, and for food or moifture beneath the foil.   As noted in Botanic Garden, Part II. Note on Cufcuta.   Other changes of vegetables from climate, or other caufes, are remarked in the Note on
Curcuma in the fame work.   From thefe one might be led to imagine, that each plant at firft confifted of a fingle bulb or flower to
each root, as the gentianella and daify; and that in the conteft for air
and light new buds grew on the old decaying flower ftem, fhooting
down their elongated roots to the ground, and that in procefs of ages
tall trees were thus formed, and an individual bulb became a fwarm
of vegetables.   Other plants, which in this conteft for light and air
were too flender to rife by their own ftrength, learned by degrees to
adhere to their neighbours, either by putting forth roots like the ivy,
or by tendrils like the vine, or by fpiral contortions like the honey-
fuckle; or by growing upon them like the mifleto, and taking nourifhment from their barks; or by only lodging or adhering on them,
and deriving nourifhment from the air, as tillandfia.

Shall we then fay that the vegetable living filament was originally
different from that of each tribe of animals above defcribed? And that
the productive living filament of each of thofe tribes was different originally from the other? Or, as the earth and ocean were probably
peopled with vegetable productions long before the exiftence of animals; and many families of thefe animals long before other families
of them, fhall we conjecture, that one and the fame kind of living
filaments is and has been the caufe of all organic life?

This

This idea of the gradual formation and improvement of the animal world accords with the obfervations of fome modern philofophers, who have fuppofed that the continent of America has been raifed out of the ocean at a later period of time than the other three quarters of the globe, which they deduce from the greater comparative heights of its mountains, and the confequent greater coldnefs of its refpective climates, and from the lefs fize and ftrength of its animals, as the tygers and allegators compared with thofe of Afia or Africa. And laftly, from the lefs progrefs in the improvements of the mind of its inhabitants in refpect to voluntary exertions.

This idea of the gradual formation and improvement of the animal world feems not to have been unknown to the ancient philofophers. Plato having probably obferved the reciprocal generation of inferior animals, as fnails and worms, was of opinion, that mankind with all other animals were originally hermaphrodites during the infancy of the world, and were in procefs of time feparated into male and female. The breafts and teats of all male quadrupeds, to which no ufe can be now affigned, adds perhaps fome fhadow of probability to this opinion. Linnæus excepts the horfe from the male quadrupeds, who have teats; which might have fhewn the earlier origin of his exiftence; but Mr. T. Hunter afferts, that he has difcovered the veftiges of them on his fheath, and has at the fame time enriched natural hiftory with a very curious fact concerning the male pidgeon; at the time of hatching the eggs both the male and female pidgeon undergo a great change in their crops; which thicken and become corrugated, and fecrete a kind of milky fluid, which coagulates, and with which alone they for a few days feed their young, and afterwards feed them with this coagulated fluid mixed with other food. How this refembles the breafts of female quadrupeds after the production of their young! and how extraordinary, that the male fhould at this time give milk as well as the female! See Botanic Garden, Part II. Note on Curcuma.

The

The late Mr. David Hume, in his pofthumous works, places the powers of generation much above thofe of our boafted reafon; and adds, that reafon can only make a machine, as a clock or a fhip, but the power of generation makes the maker of the machine; and probably from having obferved, that the greateft part of the earth has been formed out of organic recrements; as the immenfe beds of lime-ftone, chalk, marble, from the fhells of fifh; and the extenfive provinces of clay, fandftone, ironftone, coals, from decompofed vegetables; all which have been firft produced by generation, or by the fecretions of organic life; he concludes, that the world itfelf might have been generated, rather than created; that is, it might have been gradually produced from very fmall beginnings, increafing by the activity of its inherent principles, rather than by a fudden evolution of the whole by the Almighty fiat.—What a magnificent idea of the infinite power of THE GREAT ARCHITECT! THE CAUSE OF CAUSES! PARENT OF PARENTS! ENS ENTIUM!

For if we may compare infinities, it would feem to require a greater infinity of power to caufe the caufes of effects, than to caufe the effects themfelves. This idea is analogous to the improving excellence obfervable in every part of the creation; fuch as in the progreffive increafe of the folid or habitable parts of the earth from water; and in the progreffive increafe of the wifdom and happinefs of its inhabitants; and is confonant to the idea of our prefent fituation being a ftate of probation, which by our exertions we may improve, and are confequently refponfible for our actions.

V. 1. The efficient caufe of the various colours of the eggs of birds, and of the hair and feathers of animals, is a fubject fo curious, that I fhall beg to introduce it in this place. The colours of many animals feem adapted to their purpofes of concealing themfelves either to avoid danger, or to fpring upon their prey. Thus the fnake and wild cat, and leopard, are fo coloured as to refemble dark leaves and their lighter interftices; birds refemble the colour of the brown

ground,

ground, or the green hedges, which they frequent; and moths and butterflies are coloured like the flowers which they rob of their honey. Many inſtances are mentioned of this kind in Botanic Garden, p. 2. Note on Rubia.

Theſe colours have, however, in ſome inſtances another uſe, as the black diverging area from the eyes of the ſwan; which, as his eyes are placed leſs prominent than thoſe of other birds, for the convenience of putting down his head under water, .prevents the rays of light from being reflected into his eye, and thus dazzling his ſight, both in air and beneath the water; which muſt have happened, if that ſurface had been white like the reſt of his feathers.

There is a ſtill more wonderful thing concerning theſe colours adapted to the purpoſe of concealment; which is, that the eggs of birds are ſo coloured as to reſemble the colour of the adjacent objects and their interſtices. The eggs of hedge-birds are greeniſh with dark ſpots; thoſe of crows and magpies, which are ſeen from beneath through wicker neſts, are white with dark ſpots; and thoſe of larks and partridges are ruſſet or brown, like their neſts or ſituations.

A thing ſtill more aſtoniſhing is, that many animals in countries covered with ſnow become white in winter, and are ſaid to change their colour again in the warmer months, as bears, hares, and par-tridges. Our domeſticated animals loſe their natural colours, and break into great variety, as horſes, dogs, pidgeons. The final cauſe of theſe colours is eaſily underſtood, as they ſerve ſome purpoſes of the animal, but the efficient cauſe would ſeem almoſt beyond con-jecture.

Firſt, the choroid coat of the eye, on which the ſemitranſparent retina is expanded, is of different colour in different animals; in thoſe which feed on graſs it is green; from hence there would appear ſome connexion between the colour of the choroid coat and of that con-ſtantly painted on the retina by the green graſs. Now, when the ground becomes covered with ſnow, it would ſeem, that that action

of

of the retina, which is called whitenefs, being conftantly excited in the eye, may be gradually imitated by the extremities of the nerves of touch, or rete mucofum of the fkin. And. if it be fuppofed, that the action of the retina in producing the perception of any colour confifts in fo difpofing its own fibres or furface, as to reflect thofe colour- ed rays only, and tranfmit the others like foap-bubbles; then that part of the retina, which gives us the perception of fnow, muft at that time be white; and that which gives us the perception of grafs, muft be green.

Then if by the laws of imitation, as explained in Section XII. *33.* and XXXIX. 6. the extremities of the nerves of touch in the rete mucofum be induced into fimilar action, the fkin or feathers, or hair, may in like manner fo difpofe their extreme fibres, as to reflect white; for it is evident, that all thefe parts were originally obedient to irrita- tive motions during their growth, and probably continue to be fo; that thofe irritative motions are not liable in a healthy ftate to be fuc- ceeded by fenfation; which however is no uncommon thing in their difeafed ftate, or in their infant ftate, as in plica polonica, and in very young pen-feathers, which are ftill full of blood.

It was fhewn in Section XV. on the Production of Ideas, that the moving organ of fenfe in fome circumftances refembled the object which produced that motion. Hence it may be conceived, that the rete mucofum, which is the extremity of the nerves of touch, may by imitating the motions of the retina become coloured. And thus, like the fable of the camelion, all animals may poffefs a tendency to be coloured fomewhat like the colours they moft frequently infpect, and finally, that colours may be thus given to the egg-fhell by the imagination of the female parent; which fhell is previoufly a mucous membrane, indued with irritability, without which it could not cir- culate its fluids, and increafe in its bulk. Nor is this more wonder- ful than that a fingle idea of imagination fhould in an inftant colour the whole furface of the body of a bright fcarlet, as in the blufh of

fhame,

shame, though by a very different procefs. In this intricate fubject nothing but loofe analogical conjectures can be had, which may however lead to future difcoveries ; but certain it is that both the change of the colour of animals to white in the winters of fnowy countries, and the fpots on birds eggs, muft have fome efficient caufe ; fince the uniformity of their production shews it cannot arife from a fortuitous concurrence of circumftances ; and how is this efficient caufe to be detected, or explained, but from its analogy to other animal facts ?

2. The nutriment fupplied by the female parent in viviparous animals to their young progeny may be divided into three kinds, correfponding with the age of the new creature. 1. The nutriment contained in the ovum as previoufly prepared for the embryon in the ovary. 2. The liquor amnii prepared for the fetus in the uterus, and in which it fwims ; and laftly, the milk prepared in the pectoral glands for the new-born child. There is reafon to conclude that variety of changes may be produced in the new animal from all thefe fources of nutriment, but particularly from the firft of them.

The organs of digeftion and of fanguification in adults, and afterwards thofe of fecretion, prepare or feparate the particles proper for nourifhment from other combinations of matter, or recombine them into new kinds of matter, proper to excite into action the filaments, which abforb or attract them by animal appetency. In this procefs we muft attend not only to the action of the living filament which receives a nutritive particle to its bofom, but alfo to the kind of particle, in refpect to form, or fize, or colour, or hardnefs, which is thus previoufly prepared for it by digeftion, fanguification, and fecretion. Now as the firft filament of entity cannot be furnifhed with the preparative organs above mentioned, the nutritive particles, which are at firft to be received by it, are prepared by the mother ; and depofited in the ovum ready for its reception. Thefe nutritive particles muft be fuppofed to differ in fome refpects, when thus prepared by different animals. They may differ in fize, folidity, colour, and form ;

and

and yet may be sufficiently congenial to the living filament, to which they are applied, as to excite its activity by their stimulus, and its animal appetency to receive them, and to combine them with itself into organization.

By this first nutriment thus prepared for the embryon is not meant the liquor amnii, which is produced afterwards, nor the larger exterior parts of the white of the egg; but the fluid prepared, I suppose, in the ovary of viviparous animals, and that which immediately surrounds the cicatricula of an impregnated egg, and is visible to the eye in a boiled one.

Now these ultimate particles of animal matter prepared by the glands of the mother may be supposed to resemble the similar ultimate particles, which were prepared for her own nourishment; that is, to the ultimate particles of which her own organization consists. And that hence when these become combined with a new embryon, which in its early state is not furnished with stomach, or glands, to alter them; that new embryon will bear some resemblance to the mother.

This seems to be the origin of the compound forms of mules, which evidently partake of both parents, but principally of the male parent. In this production of chimeras the antients seem to have indulged their fancies, whence the sphinxes, griffins, dragons, centaurs, and minotaurs, which are vanished from modern credulity.

It would seem, that in these unnatural conjunctions, when the nutriment deposited by the female was so ill adapted to stimulate the living filament derived from the male into action, and to be received, or embraced by it, and combined with it into organization, as not to produce the organs necessary to life, as the brain, or heart, or stomach, that no mule was produced. Where all the parts necessary to life in these compound animals were formed sufficiently perfect, ex-

3 U

cept

cept the parts of generation, those animals were produced, which are now called mules.

The formation of the organs of sexual generation, in contradistinction to that by lateral buds, in vegetables, and in some animals, as the polypus, the tænia, and the volvox, seems the chef d'œuvre, the master-piece of nature; as appears from many flying insects, as in moths and butterflies, who seem to undergo a general change of their forms solely for the purpose of sexual reproduction, and in all other animals this organ is not complete till the maturity of the creature. Whence it happens that, in the copulation of animals of different species, the parts necessary to life are frequently completely formed; but those for the purpose of generation are defective, as requiring a nicer organization; or more exact coincidence of the particles of nutriment to the irritabilities or appetencies of the original-living filament. Whereas those mules, where all the parts could be perfectly formed, may have been produced in early periods of time, and may have added to the numbers of our various species of animals, as before observed.

As this production of mules is a constant effect from the conjunction of different species of animals, those between the horse and the female ass always resembling the horse more than the ass; and those, on the contrary, between the male ass and the mare, always resembling the ass more than the mare; it cannot be ascribed to the imagination of the male animal which cannot be supposed to operate so uniformly; but to the form of the first nutritive particles, and to their peculiar stimulus exciting the living filament to select and combine them with itself. There is a similar uniformity of effect in respect to the colour of the progeny produced between a white man, and a black woman, which, if I am well informed, is always of the mulatto kind, or a mixture of the two; which may perhaps be imputed to the peculiar form of the particles of nutriment supplied to the embryon by the mother at the early period of its existence, and their

peculiar

peculiar ftimulus; as this effect, like that of the mule progeny above treated of, is uniform and confiftent, and cannot therefore be afcribed to the imagination of either of the parents.

When the embryon has produced a placenta, and furnifhed itfelf with veffels for felection of nutritious particles, and for oxygenation of them, no great change in its form or colour is likely to be produced by the particles of fuftenance it now takes from the fluid, in which it is immerfed; becaufe it has now acquired organs to alter or new combine them.   Hence it continues to grow, whether this fluid, in which it fwims, be formed by the uterus or by any other cavity of the body, as in extra-uterine geftation; and which would feem to be produced by the ftimulus of the fetus on the fides of the cavity, where it is found, as mentioned before.   And thirdly, there is ftill lefs reafon to expect any unnatural change to happen to the child after its birth from the difference of the milk it now takes; becaufe it has acquired a ftomach, and lungs, and glands, of fufficient power to decompofe and recombine the milk; and thus to prepare from it the various kinds of nutritious particles, which the appetencies of the various fibrils or nerves may require.

From all this reafoning I would conclude, that though the imagination of the female may be fuppofed to affect the embryon by producing a difference in its early nutriment; yet that no fuch power can effect it after it has obtained a placenta, and other organs; which may felect or change the food, which is prefented to it either in the liquor amnii, or in the milk.   Now as the eggs in pullets, like the feeds in vegetables, are produced gradually, long before they are impregnated, it does not appear how any fudden effect of imagination of the mother at the time of impregnation can produce any confiderable change in the nutriment already thus laid up for the expected or defired embryon.   And that hence any changes of the embryon, except thofe uniform ones in the production of mules and mulattoes, more probably depend on the imagination of the male parent.   At the fame

3 U 2   time

time it feems manifeft, that thofe monftrous births, which confift in fome deficiencies only, or fome redundancies of parts, originate from the deficiency or redundance of the firft nutriment prepared in the ovary, or in the part of the egg immediately furrounding the cicatricula, as defcribed above ; and which continues fome time to excite the firft living filament into action, after the fimple animal is completed ; or ceafes to excite it, before the complete form is accomplifhed. The former of thefe circumftances is evinced by the eggs with double yolks, which frequently happen to our domefticated poultry, and which, I believe, are fo formed before impregnation, but which would be well worth attending to, both before and after impregnation ; as it is probable, fomething valuable on this fubject might be learnt from them. The latter circumftance, or that of deficiency of original nutriment, may be deduced from reverfe analogy.

There are, however, other kinds of monftrous births, which neither depend on deficiency of parts, nor fupernumerary ones ; nor are owing to the conjunction of animals of different fpecies ; but which appear to be new conformations, or new difpofitions of parts in refpect to each other, and which, like the variation of colours and forms of our domefticated animals, and probably the fexual parts of all animals, may depend on the imagination of the male parent, which we now come to confider.

VI. 1. The nice actions of the extremities of our various glands, are exhibited in their various productions, which are believed to be made by the gland, and not previoufly to exift as fuch in the blood. Thus the glands, which conftitute the liver, make bile ; thofe of the ftomach make gaftric acid ; thofe beneath the jaw, faliva ; thofe of the ears, ear-wax ; and the like. Every kind of gland muft poffefs a peculiar irritability, and probably a fenfibility, at the early ftate of its exiftence ; and muft be furnifhed with a nerve of fenfe, or of motion, to perceive, and to felect, and to combine the particles, which compofe the fluid it fecretes. And this nerve of fenfe which perceives

the

the different articles which compofe the blood, muft at leaft be conceived to be as fine and fubtile an organ, as the optic or auditory nerve, which perceive light or found.   See Sect. XIV. 9.

But in nothing is this nice action of the extremities of the bloodveffels fo wonderful, as in the production of contagious matter.   A fmall drop of variolous contagion diffufed in the blood, or perhaps only by being inferted beneath the cuticle, after a time, (as about a quarter of a lunation,) excites the extreme veffels of the fkin into certain motions, which produce a fimilar contagious material, filling with it a thoufand puftules.   So that by irritation, or by fenfation in confequence of irritation, or by affociation of motions, a material is formed by the extremities of certain cutaneous veffels, exactly fimilar to the ftimulating material, which caufed the irritation, or confequent fenfation, or affociation.

Many glands of the body have their motions, and in confequence their fecreted fluids, affected by pleafurable or painful ideas, fince they are in many inftances influenced by fenfitive affociations, as well as by the irritations of the particles of the paffing blood.   Thus the idea of meat, excited in the minds of hungry dogs, by their fenfe of vifion, or of fmell, increafes the difcharge of faliva, both in quantity and vifcidity; as is feen in its hanging down in threads from their mouths, as they ftand round a dinner-table.   The fenfations of pleafure, or of pain, of peculiar kinds, excite in the fame manner a great difcharge of tears; which appear alfo to be more faline at the time of their fecretion, from their inflaming the eyes and eye-lids.   The palenefs from fear, and the blufh of fhame, and of joy, are other inftances of the effects of painful, or pleafurable fenfations, on the extremities of the arterial fyftem.

It is probable, that the pleafurable fenfation excited in the ftomach by food, as well as its irritation, contributes to excite into action the gaftric glands, and to produce a greater fecretion of their fluids.   The fame probably occurs in the fecretion of bile; that is, that the pleafur-

able

able fenfation excited in the ftomach, affects this fecretion by fenfitive affociation, as well as by irritative affociation.

And laftly it would feem, that all the glands in the body have their fecreted fluids affected, in quantity and quality, by the pleafurable or painful fenfations, which produce or accompany thofe fecretions. And that the pleafurable fentations arifing from thefe fecretions may conftitute the unnamed pleafure of exiftence, which is contrary to what is meant by tædium vitæ, or ennui; and by which we fome-times feel ourfelves happy, without being able to afcribe it to any mental caufe, as after an agreeable meal, or in the beginning of in-toxication.

Now it would appear, that no fecretion or excretion of fluid is at-tended with fo much agreeable fenfation, as that of the femen; and it would thence follow, that the glands, which perform this fecretion, are more likely to be much affected by their catenations with pleafur-able fenfations. This circumftance is certain, that much more of this fluid is produced in a given time, when the object of its exclufion is agreeable to the mind.

2. A forceable argument, which fhews the neceffity of pleafurable fenfation to copulation, is, that the act cannot be performed without it; it is eafily interrupted by the pain of fear or bafhfulnefs; and no efforts of volition or of irritation can effect this procefs, except fuch as induce pleafurable ideas or fenfations. See Sect. XXXIII. 1. 1.

A curious analogical circumftance attending hermaphrodite infects, as fnails and worms, ftill further illuftrates this theory; if the fnail or worm could have impregnated itfelf, there might have been a faving of a large male apparatus; but as this is not fo ordered by nature, but each fnail and worm reciprocally receives and gives impregnation, it appears, that a pleafurable excitation feems alfo to have been required.

This wonderful circumftance of many infects being hermaphro-dites, and at the fame time not having power to impregnate them-felves, is attended to by Dr. Lifter, in his Exercitationes Anatom. de

Limacibus,

Limacibus, p. 145; who, amongſt many other final cauſes, which he adduces to account for it, adds, ut tam triſtibus et frigidis animalibus majori cum voluptate perficiatur venus.

There is, however, another final cauſe, to which this circumſtance may be imputed: it was obſerved above, that vegetable buds and bulbs, which are produced without a mother, are always exact reſemblances of their parent; as appears in grafting fruit-trees, and in the flower-buds of the dioiceous plants, which are always of the ſame ſex on the ſame tree; hence thoſe hermaphrodite inſects, if they could have produced young without a mother, would not have been capable of that change or improvement, which is ſeen in all other animals, and in thoſe vegetables, which are procreated by the male embryon received and nouriſhed by the female. And it is hence probable, that if vegetables could only have been produced by buds and bulbs, and not by ſexual generation, that there would not at this time have exiſted one thouſandth part of their preſent number of ſpecies; which have probably been originally mule-productions; nor could any kind of improvement or change have happened to them, except by the difference of ſoil or climate.

3. I conclude, that the imagination of the male at the time of copulation, or at the time of the ſecretion of the ſemen, may ſo affect this ſecretion by irritative or ſenſitive aſſociation, as deſcribed in No. 5. 1. of this ſection, as to cauſe the production of ſimilarity of form and of features, with the diſtinction of ſex; as the motions of the chiſſel of the turner imitate or correſpond with thoſe of the ideas of the artiſt. It is not here to be underſtood, that the firſt living fibre, which is to form an animal, is produced with any ſimilarity of form to the future animal; but with propenſities, or appetencies, which ſhall produce by accretion of parts the ſimilarity of form, feature, or ſex, correſponding to the imagination of the father.

Our ideas are movements of the nerves of ſenſe, as of the optic nerve in recollecting viſible ideas, ſuppoſe of a triangular piece of

3   ivory.

ivory. The fine moving fibres of the retina act in a manner to which I give the name of white; and this action is confined to a defined part of it; to which figure I give the name of triangle. And it is a pre-ceding pleasurable sensation exifting in my mind, which occasions me to produce this particular motion of the retina, when no triangle is present. Now it is probable, that the acting fibres of the ultimate terminations of the secreting apertures of the vessels of the testes, are as fine as those of the retina; and that they are liable to be thrown into that peculiar action, which marks the sex of the secreted em-bryon, by sympathy with the pleasurable motions of the nerves of vision, or of touch; that is, with certain ideas of imagination. From hence it would appear, that the world has long been mistaken in ascribing great power to the imagination of the female, whereas from this account of it, the real power of imagination, in the act of gene-ration, belongs solely to the male. See Sect. XII. 3. 3.

It may be objected to this theory, that a man may be supposed to have in his mind, the idea of the form and features of the female, rather than his own, and therefore there should be a greater number of female births. On the contrary, the general idea of our own form occurs to every one almost perpetually, and is termed consciousness of our existence, and thus may effect, that the number of males surpasses that of females. See Sect. XV. 3. 4. and XVIII. 13. And what further confirms this idea is, that the male children most frequently resemble the father in form, or feature, as well as in sex; and the fe-male most frequently resemble the mother, in feature, and form, as well as in sex.

It may again be objected, if a female child sometimes resembles the father, and a male child the mother, the ideas of the father, at the time of procreation, must suddenly change from himself to the mo-ther, at the very instant, when the embryon is secreted or formed. This difficulty ceases when we consider, that it is as easy to form an idea of feminine features with male organs of reproduction, or of male features with female ones, as the contrary; as we conceive the idea of a sphinx or mermaid, as easily and as distinctly as of a woman. Add to this, that at the time of procreation the idea of the male organs, and of the female features, are often both excited at the same time, by contact, or by vision.

I ask, in my turn, is the sex of the embryon produced by accident? Certainly whatever is produced has a cause; but when this cause is too minute for our comprehension, the effect is said in common language to happen by chance, as in throwing a certain number on dice. Now what cause can occasionally produce the male or female character of the embryon, but the peculiar actions of those glands, which form the embryon? And what can influence or govern these actions of the gland, but its associations or catenations with other sensitive motions? Nor is this more extraordinary, than that the catenations of irritative motions with the apparent vibrations of objects at sea should produce sickness of the stomach; or that a nauseous story should occasion vomiting.

4. An argument, which evinces the effect of imagination on the first rudiment of the embryon, may be deduced from the production of some peculiar monsters. Such, for instance, as those which have two heads joined to one body, and those which have two bodies joined to one head; of which frequent examples occur amongst our domesticated quadrupeds, and poultry. It is absurd to suppose, that such forms could exist in primordial germs, as explained in No. IV. 4. of this section. Nor is it possible, that such deformities could be produced by the growth of two embryons, or living filaments; which should afterwards adhere together; as the head and tail part of different polypi are said to do (Blumenbach on generation, Cadel, London); since in that case one embryon, or living filament, must have begun to form one part first, and the other another part first. But such monstrous conformations become less difficult to comprehend, when they are considered as an effect of the imagination, as before

explained,

explained, on the living filament at the time of its secretion; and that such duplicature of limbs were produced by accretion of new parts, in consequence of propensities, or animal appetencies thus acquired from the male parent.

For instance, I can conceive, if a turkey-cock should behold a rabbit, or a frog, at the time of procreation, that it might happen, that a forcible or even a pleasurable idea of the form of a quadruped might so occupy his imagination, as to cause a tendency in the nascent filament to resemble such a form, by the apposition of a duplicature of limbs. Experiments on the production of mules and monsters would be worthy the attention of a Spallanzani, and might throw much light upon this subject, which at present must be explained by conjectural analogies.

The wonderful effect of imagination, both in the male and female parent, is shewn in the production of a kind of milk in the crops both of the male and female pigeons after the birth of their young, as observed by Mr. Hunter, and mentioned before. To this should be added, that there are some instances of men having had milk secreted in their breasts, and who have given suck to children, as recorded by Mr. Buffon. This effect of imagination, of both the male and female parent, seems to have been attended to in very early times; Jacob is said not only to have placed rods of trees, in part stripped of their bark, so as to appear spotted, but also to have placed spotted lambs before the flocks, at the time of their copulation. Genesis, chap. xxx. verse 40.

5. In respect to the imagination of the mother, it is difficult to comprehend, how this can produce any alteration in the fetus, except by affecting the nutriment laid up for its first reception, as described in No. V. 2. of this section, or by affecting the nourishment or oxygenation with which she supplies it afterwards. Perpetual anxiety may probably affect the secretion of the liquor amnii into the uterus, as it enfeebles the whole system; and sudden fear is a frequent

cause

caufe of mifcarriage; for fear, contrary to joy, decreafes for a time the action of the extremities of the arterial fyftem; hence fudden palenefs fucceeds, and a fhrinking or contraction of the veffels of the fkin, and other membranes. By this circumftance, I imagine, the terminations of the placental veffels are detached from their adhefions, or infertions, into the membrane of the uterus; and the death of the child fucceeds, and confequent mifcarriage.

Of this I recollect a remarkable inftance, which could be afcribed to no other caufe, and which I fhall therefore relate in few words. A healthy young woman, about twenty years of age, had been about five months pregnant, and going down into her cellar to draw fome beer, was frighted by a fervant-boy ftarting up from behind the barrel, where he had concealed himfelf with defign to alarm the maidfervant, for whom he miftook his miftrefs. She came with difficulty up ftairs, began to flood immediately, and mifcarried in a few hours. She has fince borne feveral children, nor ever had any tendency to mifcarry of any of them.

In refpect to the power of the imagination of the male over the form, colour, and fex of the progeny, the following inftances have fallen under my obfervation, and may perhaps be found not very unfrequent, if they were more attended to. I am acquainted with a gentleman, who has one child with dark hair and eyes, though his lady and himfelf have light hair and eyes; and their other four children are like their parents. On obferving this diffimilarity of one child to the others he affured me, that he believed it was his own imagination, that produced the difference; and related to me the following ftory. He faid, that when his lady lay in of her third child, he became attached to a daughter of one of his inferior tenants, and offered her a bribe for her favours in vain; and afterwards a greater bribe, and was equally unfuccefsful; that the form of this girl dwelt much in his mind for fome weeks, and that the next child, which was the dark-ey'd young

3 X 2   lady

lady above mentioned, was exceedingly like, in both features and colour, to the young woman who refufed his addreffes.

To this inftance I muft add, that I have known two families, in which, on account of an intailed eftate in expectation, a male heir was moft eagerly defired by the father; and on the contrary, girls were produced to the feventh in one, and to the ninth in another; and then they had each of them a fon. I conclude, that the great defire of a male heir by the father produced rather a difagreeable than an agreeable fenfation; and that his ideas dwelt more on the fear of generating a female, than on the pleafurable fenfations or ideas of his own male form or organs at the time of copulation, or of the fecretion of the femen; and that hence the idea of the female character was more prefent to his mind than that of the male one; till at length in defpair of generating a male thefe ideas ceafed, and thofe of the male character prefided at the genial hour.

7. Hence I conclude, that the act of generation cannot exift without being accompanied with ideas, and that a man muft have at that time either a general idea of his own male form, or of the form of his male organs; or an idea of the female form, or of her organs; and that this marks the fex, and the peculiar refemblances of the child to either parent. From whence it would appear, that the phalli, which were hung round the necks of the Roman ladies, or worn in their hair, might have effect in producing a greater proportion of male children; and that the calipædia, or art of begetting beautiful children, and of procreating either males or females, may be taught by affecting the imagination of the male-parent; that is, by the fine extremities of the feminal glands imitating the actions of the organs of fenfe either of fight or touch. But the manner of accomplifhing this cannot be unfolded with fufficient delicacy for the public eye; but may be worth the attention of thofe, who are ferioufly interefted in the procreation of a male or female child.

Recapitulation.

Recapitulation.

VII. 1. A certain quantity of nutritive particles are produced by the female parent before impregnation, which require no further digestion, secretion, or oxygenation. Such are seen in the unimpregnated eggs of birds, and in the unimpregnated seed-veffels of vegetables.

2. A living filament is produced by the male, which being inferted amidft thefe firft nutritive particles, is ftimulated into action by them; and in confequence of this action, fome of the nutritive particles are embraced, and added to the original living filament; in the fame manner as common nutrition is performed in the adult animal.

3. Then this new organization, or additional part, becomes ftimulated by the nutritive particles in its vicinity, and fenfation is now fuperadded to irritation; and other particles are in confequence embraced, and added to the living filament; as is feen in the new granulations of flefh in ulcers.

By the power of affociation, or by irritation, the parts already produced continue their motions, and new ones are added by fenfation, as above mentioned; and laftly by volition, which laft fenforial power is proved to exift in the fetus in its maturer age, becaufe it has evidently periods of activity and of fleeping; which laft is another word for a temporary fufpenfion of volition.

The original living filament may be conceived to poffefs a power of repulfing the particles applied to certain parts of it, as well as of embracing others, which ftimulate other parts of it; as thefe powers exift in different parts of the mature animal; thus the mouth of every gland embraces the particles of fluid, which fuits its appetency; and its excretory duct repulfes thofe particles, which are difagreeable to it.

4. Thus

4. Thus the outline or miniature of the new animal is produced gradually, but in no great length of time; becaufe the original nutritive particles require no previous preparation by digeftion, fecretion, and oxygenation: but require fimply the felection and appofition, which is performed by the living filament. Mr. Blumenbach fays, that he poffeffes a human fetus of only five weeks old, which is the fize of a common bee, and has all the features of the face, every finger, and every toe, complete; and in which the organs of generation are diftinctly feen. P. 76. In another fetus, whofe head was not larger than a pea, the whole of the bafis of the fkull with all its depreffions, apertures, and proceffes, were marked in the moft fharp and diftinct manner, though without any offification. Ib.

5. In fome cafes by the nutriment originally depofited by the mother the filament acquires parts not exactly fimilar to thofe of the father, as in the production of mules and mulattoes. In other cafes, the deficiency of this original nutriment caufes deficiencies of the extreme parts of the fetus, which are laft formed, as the fingers, toes, lips. In other cafes, a duplicature of limbs are caufed by the fuperabundance of this original nutritive fluid, as in the double yolks of eggs, and the chickens from them with four legs and four wings. But the production of other monfters, as thofe with two heads, or with parts placed in wrong fituations, feems to arife from the imagination of the father being in fome manner imitated by the extreme veffels of the feminal glands; as the colours of the fpots on eggs, and the change of the colour of the hair and feathers of animals by domeftication, may be caufed in the fame manner by the imagination of the mother.

6. The living filament is a part of the father, and has therefore certain propenfities, or appetencies, which belong to him; which may have been gradually acquired during a million of generations, even from the infancy of the habitable earth; and which now poffeffes fuch properties, as would render, by the appofition of nutritious particles,

particles, the new fetus exactly fimilar to the father; as occurs in the buds and bulbs of vegetables, and in the polypus, and
tænia or tape-worm.   But as the firft nutriment is fupplied by the
mother, and therefore refembles fuch nutritive particles, as have
been ufed for her own nutriment or growth, the progeny takes in
part the likenefs of the mother.

Other fimilarity of the excitability, or of the form of the male
parent, fuch as the broad or narrow fhoulders, or fuch as conftitute
certain hereditary difeafes, as fcrophula, epilepfy, infanity, have
their origin produced in one or perhaps two generations; as in the
progeny of thofe who drink much vinous fpirits; and thofe hereditary propenfities ceafe again, as I have obferved, if one or two fober
generations fucceed; otherwife the family becomes extinct.

This living filament from the father is alfo liable to have its propenfities, or appetencies, altered at the time of its production by the
imagination of the male parent; the extremities of the feminal glands
imitating the motions of the organs of fenfe; and thus the fex of the
embryon is produced; which may be thus made a male or a female
by affecting the imagination of the father at the time of impregnation.
See Sect. XXXIX. 6. 3. and 7.

7. After the fetus is thus completely formed together with its umbilical veffels and placenta, it is now fupplied with a different kind of
food, as appears by the difference of confiftency of the different
parts of the white of the egg, and of the liquor amnii, for it has
now acquired organs for digeftion or fecretion, and for oxygenation,
though they are as yet feeble; which can in fome degree change, as
well as felect, the nutritive particles, which are now prefented to it.
But may yet be affected by the deficiency of the quantity of nutrition fupplied by the mother, or by the degree of oxygenation fupplied
to its placenta by the maternal blood.

The augmentation of the complete fetus by additional particles of
nutriment is not accomplifhed by diftention only, but by appofition

to

to every part both external and internal; each of which acquires by animal appetencies the new addition of the particles which it wants. And hence the enlarged parts are kept fimilar to their prototypes, and may be faid to be extended; but their extenfion muft be conceived only as a neceffary confequence of the enlargement of all their parts by appofition of new particles.

Hence the new appofition of parts is not produced by capillary attraction, becaufe the whole is extended; whereas capillary attraction would rather tend to bring the fides of flexible tubes together, and not to diftend them. Nor is it produced by chemical affinities, for then a folution of continuity would fucceed, as when fugar is diffolved in water; but it is produced by an animal procefs, which is the confequence of irritation, or fenfation; and which may be termed animal appetency.

This is further evinced from experiments, which have been inftituted to fhew, that a living mufcle of an animal body requires greater force to break it, than a fimilar mufcle of a dead body. Which 'evinces, that befides the attraction of cohefion, which all matter poffeffes, and befides the chemical attractions of affinities, which hold many bodies together, there is an animal adhefion, which adds vigour to thefe common laws of the inanimate world.

8. At the nativity of the child it depofits the placenta or gills, and by expanding its lungs acquires more plentiful oxygenation from the currents of air, which it muft now continue perpetually to refpire to the end of its life; as it now quits the liquid element, in which it was produced, and like the tadpole, when it changes into a frog, becomes an aerial animal.

9. As the habitable parts of the earth have been, and continue to be, perpetually increafing by the production of fea-fhells and coral-lines, and by the recrements of other animals, and vegetables; fo from the beginning of the exiftence of this terraqueous globe, the

animals,

animals, which inhabit it, have conftantly improved, and are ftill in a ftate of progreffive improvement.

This idea of the gradual generation of all things feems to have been as familiar to the ancient philofophers as to the modern ones; and to have given rife to the beautiful hieroglyphic figure of the πρωτον ωον, or firft great egg, produced by NIGHT, that is, whofe origin is involved in obfcurity, and animated by εϱος, that is, by DI-VINE LOVE; from whence proceeded all things which exift.

Conclufion.

VIII. 1. CAUSE AND EFFECT may be confidered as the pro-greffion, or fucceffive motions, of the parts of the great fyftem of Nature. The ftate of things at this moment is the effect of the ftate of things, which exifted in the preceding moment; and the caufe of the ftate of things, which fhall exift in the next moment.

Thefe caufes and effects may be more eafily comprehended, if mo-tion be confidered as a change of the figure of a group of bodies, as propofed in Sect. XIV. 2. 2. inafmuch as our ideas of vifible or tan-gible objects are more diftinct, than our abftracted ideas of their mo-tions. Now the change of the configuration of the fyftem of nature at this moment muft be an effect of the preceding configuration, for a change of configuration cannot exift without a previous configuration; and the proximate caufe of every effect muft immediately precede that effect. For example, a moving ivory ball could not proceed onwards, unlefs it had previoufly began to proceed; or unlefs an impulfe had been previoufly given it; which previous motion or impulfe-conftitutes a part of the laft fituation of things.

As the effects produced in this moment of time become caufes in the next, we may confider the progreffive motions of objects as a

chain

chain of caufes only; whofe firft link proceeded from the great Creator, and which have exifted from the beginning of the created univerfe, and are perpetually proceeding.

2. Thefe caufes may be conveniently divided into two kinds, efficient and inert caufes, according with the two kinds of entity fuppofed to exift in the natural world, which may be termed matter and
fpirit, as propofed in Sect. I. and further treated of in Sect. XIV.
The efficient caufes of motion, or new configuration, confift either
of the principle of general gravitation, which actuates the fun and
planets; or of the principle of particular gravitation, as in electricity,
magnetifm, heat; or of the principle of chemical affinity, as in combuftion, fermentation, combination; or of the principle of organic
life, as in the contraction of vegetable and animal fibres. The inert
caufes of motion, or new configuration, confift of the parts of matter, which are introduced within the fpheres of activity of the principles above defcribed. Thus, when an apple falls on the ground,
the principle of gravitation is the efficient caufe, and the matter of
the apple the inert caufe. If a bar of iron be approximated to a magnet, it may be termed the inert caufe of the motion, which brings
thefe two bodies into contact; while the magnetic principle may be
termed the efficient caufe. In the fame manner the fibres, which
conftitute the retina may be called the inert caufe of the motions of
that organ in vifion, while the fenforial power may be termed the efficient caufe.

3. Another more common diftribution of the perpetual chain of
caufes and effects, which conftitute the motions, or changing configurations, of the natural world, is into active and paffive. Thus,
if a ball in motion impinges againft another ball at reft, and communicates its motion to it, the former ball is faid to act, and the latter
to be acted upon. In this fenfe of the words a magnet is faid to attract iron; and the prick of a fpur to ftimulate a horfe into exertion;
fo that in this view of the works of nature all things may be faid either

ther fimply to exift, or to exift as caufes, or to exift as effects; that is, to exift either in an active or paffive ftate.

This diftribution of objects, and their motions, or changes of pofition, has been found fo convenient for the purpofes of common life, that on this foundation refts the whole conftruction or theory of language. The names of the things themfelves are termed by grammarians Nouns, and their modes of exiftence are termed Verbs. The nouns are divided into fubftantives, which denote the principal things fpoken of; and into adjectives, which denote fome circumftances, or lefs kinds of things, belonging to the former. The verbs are divided into three kinds, fuch as denote the exiftence of things fimply, as, to be; or their exiftence in an active ftate, as, to eat; or their exiftence in a paffive ftate, as, to be eaten. Whence it appears, that all languages confift only of nouns and verbs, with their abbreviations for the greater expedition of communicating our thoughts; as explained in the ingenious work of Mr. Horne Tooke, who has unfolded by a fingle flafh of light the whole theory of language, which had fo long lain buried beneath the learned lumber of the fchools. Diverfions of Purley. Johnfon. London.

4. A third divifion of caufes has been into proximate and remote; thefe have been much fpoken of by the writers on medical fubjects, but without fufficient precifion. If to proximate and remote caufes we add proximate and remote effects, we fhall include four links of the perpetual chain of caufation; which will be more convenient for the difcuffion of many philofophical fubjects.

Thus if a particle of chyle be applied to the mouth of a lacteal veffel, it may be termed the remote caufe of the motions of the fibres, which compofe the mouth of that lacteal veffel; the fenforial power is the proximate caufe; the contraction of the fibres of the mouth of the veffel is the proximate effect; and their embracing the particle of chyle is the remote effect; and thefe four links of caufation conftitute abforption.

Thus when we attend to the rifing fun, firft the yellow rays of

light

light ftimulate the fenforial power refiding in the extremities of the optic nerve, this is the remote caufe. 2. The fenforial power is excited into a ftate of activity, this is the proximate caufe. 3. The fibrous extremities of the optic nerve are contracted, this is the proximate effect. 4. A pleafureable or painful fenfation is produced in confequence of the contraction of thefe fibres of the optic nerve, this is the remote effect ; and thefe four links of the chain of caufation conftitute the fenfitive idea, or what is commonly termed the fenfation of the rifing fun.

5. Other caufes have been announced by medical writers under the names of caufa procatarctica, and caufa proegumina, and caufa fine quâ non.   All which are links more or lefs diftant of the chain of remote caufes.

To thefe muft be added the final caufe, fo called by many authors, which means the motive, for the accomplifhment of which the preceding chain of caufes was put into action.   The idea of a final caufe, therefore, includes that of a rational mind, which employs means to effect its purpofes ; thus the defire of preferving himfelf from the pain of cold, which he has frequently experienced, induces the favage to conftruct his hut; the fixing ftakes into the ground for walls, branches of trees for rafters, and turf for a cover, are a feries of fucceffive voluntary exertions ; which are fo many means to produce a certain effect.   This effect of preferving himfelf from cold, is termed the final caufe ; the conftruction of the hut is the remote effect ; the action of the mufcular fibres of the man, is the proximate effect ; the volition, or activity of defire to preferve himfelf from cold, is the proximate caufe ; and the pain of cold, which excited that defire, is the remote caufe.

6. This perpetual chain of caufes and effects, whofe firft link is rivetted to the throne of GOD, divides itfelf into innumerable diverging branches, which, like the nerves arifing from the brain, permeate the moft minute and moft remote extremities of the fyftem, diffufing

motion

motion and fenfation to the whole.   As every caufe is fuperior in power to the effect, which it has produced, fo our idea of the power of the Almighty Creator becomes more elevated and fublime, as we trace the operations of nature from caufe to caufe, climbing up the links of thefe chains of being, till we afcend to the Great Source of all things.

Hence the modern difcoveries in chemiftry and in geology, by having traced the caufes of the combinations of bodies to remoter origins, as well as thofe in aftronomy, which dignify the prefent age, contribute to enlarge and amplify our ideas of the power of the Great Firft Caufe.   And had thofe ancient philofophers, who contended that the world was formed from atoms, afcribed their combinations to certain immutable properties received from the hand of the Creator, fuch as general gravitation, chemical affinity, or animal appetency, inftead of afcribing them to a blind chance ; the doctrine of atoms, as conftituting or compofing the material world by the variety of their combinations, fo far from leading the mind to atheifm, would ftrengthen the demonftration of the exiftence of a Deity, as the firft caufe of all things; becaufe the analogy refulting from our perpetual experience of caufe and effect would have thus been exemplified through univerfal nature.

The heavens declare the glory of GOD, *and the firmament fheweth his handywork! One day telleth another, and one night certifieth another; they have neither fpeech nor language, yet their voice is gone forth into all lands, and their words into the ends of the world.   Manifold are thy works,* O LORD! *in wifdom haft thou made them all.*   Pfal. xix. civ.

SECT.

SECT. XL.

On the OCULAR SPECTRA of Light and Colours, by Dr. R. W. Darwin, of Shrewſbury. Reprinted, by Permiſſion, from the Philoſophical Tranſactions, Vol. LXXVI. p. 313.

Spectra of four kinds. 1. *Activity of the retina in viſion.* 2. *Spectra from defect of ſenſibility.* 3. *Spectra from exceſs of ſenſibility.* 4. *Of direct ocular ſpectra.* 5. *Greater ſtimulus excites the retina into ſpaſmodic action.* 6. *Of reverſe ocular ſpectra.* 7. *Greater ſtimulus excites the retina into various ſucceſſive ſpaſmodic actions.* 8. *Into fixed ſpaſmodic action.* 9. *Into temporary paralyſis.* 10. *Miſcellaneous remarks*; 1. *Direct and reverſe ſpectra at the ſame time. A ſpectral halo. Rule to predetermine the colours of ſpectra.* 2. *Variation of ſpectra from extraneous light.* 3. *Variation of ſpectra in number, figure, and remiſſion.* 4. *Circulation of the blood in the eye is viſible.* 5. *A new way of magnifying objects. Concluſion.*

WHEN any one has long and attentively looked at a bright object, as at the ſetting ſun, on cloſing his eyes, or removing them, an image, which reſembles in form the object he was attending to, continues ſome time to be viſible: this appearance in the eye we ſhall call the ocular ſpectrum of that object.

Theſe ocular ſpectra are of four kinds: 1ſt, Such as are owing to a leſs ſenſibility of a defined part of the retina; or *ſpectra from defect of ſenſibility.* 2d, Such as are owing to a greater ſenſibility of a defined part of the retina; or *ſpectra from exceſs of ſenſibility.* 3d, Such as reſemble their object in its colour as well as form; which may be

termed

termed *direct ocular spectra.*   4th, Such as are of a colour contrary to that of their object; which may be termed *reverse ocular spectra.*

The laws of light have been most successfully explained by the great Newton, and the perception of visible objects has been ably investigated by the ingenious Dr. Berkeley and M. Malebranche; but these minute phænomena of vision have yet been thought reducible to no theory, though many philosophers have employed a considerable degree of attention upon them: among these are Dr. Jurin, at the end of Dr. Smith's Optics; M. Æpinus, in the Nov. Com. Petropol. V. 10.; M. Beguelin, in the Berlin Memoires, V. II. 1771; M. d'Arcy, in the Histoire de l'Acad. des Scienc. 1765; M. de la Hire; and, lastly, the celebrated M. de Buffon, in the Memoires de l'Acad. des Scien. who has termed them accidental colours, as if subjected to no established laws, Ac. Par. 1743. M. p. 215.

I must here apprize the reader, that it is very difficult for different people to give the same names to various shades of colours; whence, in the following pages, something must be allowed, if on repeating the experiments the colours here mentioned should not accurately correspond with his own names of them.

I. *Activity of the Retina in Vision.*

FROM the subsequent experiments it appears, that the retina is in an active not in a passive state during the existence of these ocular spectra; and it is thence to be concluded, that all vision is owing to the activity of this organ.

1. Place a piece of red silk, about an inch in diameter, as in plate 1, at Sect. III. 1, on a sheet of white paper, in a strong light; look steadily upon it from about the distance of half a yard for a minute; then closing your eyelids cover them with your hands, and a green

spectrum

ſpectrum will be ſeen in your eyes, reſembling in form the piece of red ſilk: after ſome time, this ſpectrum will diſappear and ſhortly re-appear; and this alternately three or four times, if the experiment is well made, till at length it vaniſhes entirely.

2. Place on a ſheet of white paper a circular piece of blue ſilk, about four inches in diameter, in the ſunſhine; cover the center of this with a circular piece of yellow ſilk, about three inches in dia-meter; and the center of the yellow ſilk with a circle of pink ſilk, about two inches in diameter; and the center of the pink ſilk with a circle of green ſilk, about one inch in diameter; and the centre of this with a circle of indigo, about half an inch in diameter; make a ſmall ſpeck with ink in the very center of the whole, as in plate 3, at Sect. III. 3. 6.; look ſteadily for a minute on this central ſpot, and then cloſing your eyes, and applying your hand at about an inch diſ-tance before them, ſo as to prevent too much or too little light from paſſing through the eyelids, you will ſee the moſt beautiful circles of colours that imagination can conceive, which are moſt reſembled by the colours occaſioned by pouring a drop or two of oil on a ſtill lake in a bright day; but theſe circular iriſes of colours are not only different from the colours of the ſilks above mentioned, but are at the ſame time perpetually changing as long as they exiſt.

3. When any one in the dark preſſes either corner of his eye with his finger, and turns his eye away from his finger, he will ſee a circle of colours like thoſe in a peacock's tail: and a ſudden flaſh of light is excited in the eye by a ſtroke on it. (Newton's Opt. Q. 16.)

4. When any one turns round rapidly on one foot, till he becomes dizzy, and falls upon the ground, the ſpectra of the ambient objects continue to preſent themſelves in rotation, or appear to librate, and he ſeems to behold them for ſome time ſtill in motion.

From all theſe experiments it appears, that the ſpectra in the eye are not owing to the mechanical impulſe of light impreſſed on the retina, nor to its chemical combination with that organ, nor to the

abſorption

abforption and emiffion of light, as is obferved in many bodies; for in all thefe cafes the fpectra muft either remain uniformly, or gradually diminifh; and neither their alternate prefence and evanefcence as in the firft experiment, nor the perpetual changes of their colours as in the fecond, nor the flafh of light or colours in the preffed eye as in the third, nor the rotation or libration of the fpectra as in the fourth, could exift.

It is not abfurd to conceive, that the retina may be ftimulated into motion, as well as the red and white mufcles which form our limbs and veffels; fince it confifts of fibres, like thofe, intermixed with its medullary fubftance. To evince this ftructure, the retina of an ox's eye was fufpended in a glafs of warm water, and forcibly torn in a few places; the edges of thefe parts appeared jagged and hairy, and did not contract, and become fmooth like fimple mucus, when it is diftended till it breaks; which fhews that it confifts of fibres; and this its fibrous conftruction became ftill more diftinct to the fight, by adding fome cauftic alkali to the water, as the adhering mucus was firft eroded, and the hair-like fibres remained floating in the veffel. Nor does the degree of tranfparency of the retina invalidate the evidence of its fibrous ftructure, fince Leeuwenhoek has fhewn that the cryftalline humour itfelf confifts of fibres. (Arcana Naturæ, V. 1. p. 70.)

Hence it appears, that as the mufcles have larger fibres intermixed with a fmaller quantity of nervous medulla, the organ of vifion has a greater quantity of nervous medulla intermixed with fmaller fibres; and it is probable that the locomotive mufcles, as well as the vafcular ones, of microfcopic animals have much greater tenuity than thefe of the retina.

And befides the fimilar laws, which will be fhewn in this paper to govern alike the actions of the retina and of the mufcles, there are many other analogies which exift between them. They are both originally excited into action by irritations, both act nearly in the fame

quantity

quantity of time, are alike ſtrengthened or fatigued by exertion, are alike painful if excited into action when they are in an inflamed ſtate, are alike liable to paralyſis, and to the torpor of old age.

II. Of spectra from defect of sensibility.

The retina is not ſo eaſily excited into action by leſs irritation after having been lately ſubjected to greater.

1. When any one paſſes from the bright daylight into a darkened room, the iriſes of his eyes expand themſelves to their utmoſt extent in a few ſeconds of time; but it is very long before the optic nerve, after having been ſtimulated by the greater light of the day, becomes ſenſible of the leſs degree of it in the room; and, if the room is not too obſcure, the iriſes will again contract themſelves in ſome degree, as the ſenſibility of the retina returns.

2. Place about half an inch ſquare of white paper on a black hat, and looking ſteadily on the center of it for a minute, remove your eyes to a ſheet of white paper; and after a ſecond or two a dark ſquare will be ſeen on the white paper, which will continue ſome time. A ſimilar dark ſquare will be ſeen in the cloſed eye, if light be admitted through the eyelids.

So after looking at any luminous object of a ſmall ſize, as at the ſun, for a ſhort time, ſo as not much to fatigue the eyes, this part of the retina becomes leſs ſenſible to ſmaller quantities of light; hence, when the eyes are turned on other leſs luminous parts of the ſky, a dark ſpot is ſeen reſembling the ſhape of the ſun, or other luminous object which we laſt beheld. This is the ſource of one kind of the dark-coloured *muſcæ volitantes*. If this dark ſpot lies above the center of the eye, we turn our eyes that way, expecting to bring it into

3

the

the center of the eye, that we may view it more diftinctly; and in this cafe the dark fpectrum feems to move upwards. If the dark fpectrum is found beneath the centre of the eye, we purfue it from the fame motive, and it feems to move downwards. This has given rife to various conjectures of fomething floating in the aqueous humours of the eyes; but whoever, in attending to thefe fpots, keeps his eyes unmoved by looking fteadily at the corner of a cloud, at the fame time that he obferves the dark fpectra, will be thoroughly convinced, that they have no motion but what is given to them by the movement of our eyes in purfuit of them. Sometimes the form of the fpectrum, when it has been received from a circular luminous body, will become oblong; and fometimes it will be divided into two circular fpectra, which is not owing to our changing the angle made by the two optic axifes, according to the diftance of the clouds or other bodies to which the fpectrum is fuppofed to be contiguous, but to other caufes mentioned in No. X. 3. of this fection. The apparent fize of it will alfo be variable according to its fuppofed diftance.

. As thefe fpectra are more eafily obfervable when our eyes are a little weakened by fatigue, it has frequently happened, that people of delicate conftitutions have been much alarmed at them, fearing a beginning decay of their fight, and have thence fallen into the hands of ignorant oculifts; but I believe they never are a prelude to any other difeafe of the eye, and that it is from habit alone, and our want of attention to them, that we do not fee them on all objects every hour of our lives. But as the nerves of very weak people lofe their fenfibility, in the fame manner as their mufcles lofe their activity, by a fmall time of exertion, it frequently happens, that fick people in the extreme debility of fevers are perpetually employed in picking fomething from the bed-clothes, occafioned by their miftaking the appearance of thefe *mufcæ volitantes* in their eyes. Benvenuto Celini, an Italian artift, a man of ftrong abilities, relates, that having paffed the whole night on a diftant mountain with fome companions and a conjurer,

and

and performed many ceremonies to raife the devil, on their return in the morning to Rome, and looking up when the fun began to rife, they faw numerous devils run on the tops of the houfes, as they paffed along; fo much were the fpectra of their weakened eyes magnified by fear, and made fubfervient to the purpofes of fraud or fuperftition. (Life of Ben. Celini.)

3. Place a fquare inch of white paper on a large piece of ftraw-coloured filk; look fteadily fome time on the white paper, and then move the center of your eyes on the filk, and a fpectrum of the form of the paper will appear on the filk, of a deeper yellow than the other part of it: for the central part of the retina, having been fome time expofed to the ftimulus of a greater quantity of white light, is become lefs fenfible to a fmaller quantity of it, and therefore fees only the yellow rays in that part of the ftraw-coloured filk.

Facts fimilar to thefe are obfervable in other parts of our fyftem: thus, if one hand be made warm, and the other expofed to the cold, and then both of them immerfed in fubtepid water, the water is perceived warm to one hand, and cold to the other; and we are not able to hear weak founds for fome time after we have been expofed to loud ones; and we feel a chillinefs on coming into an atmofphere of temperate warmth, after having been fome time confined in a very warm room: and hence the ftomach, and other organs of digeftion, of thofe who have been habituated to the greater ftimulus of fpirituous liquor, are not excited into their due action by the lefs ftimulus of common food alone; of which the immediate confequence is indigeftion and hypochondriacifm.

III. Of spectra from excess of sensibility.

The retina is more easily excited into action by greater irritation after having been lately subjected to less.

1. If the eyes are closed, and covered perfectly with a hat, for a minute or two, in a bright day; on removing the hat a red or crimson light is seen through the eyelids. In this experiment the retina, after being some time kept in the dark, becomes so sensible to a small quantity of light, as to perceive distinctly the greater quantity of red rays than of others which pass through the eyelids. A similar coloured light is seen to pass through the edges of the fingers, when the open hand is opposed to the flame of a candle.

2. If you look for some minutes steadily on a window in the beginning of the evening twilight, or in a dark day, and then move your eyes a little, so that those parts of the retina, on which the dark frame-work of the window was delineated, may now fall on the glass part of it, many luminous lines, representing the frame-work, will appear to lie across the glass panes : for those parts of the retina, which were before least stimulated by the dark frame-work, are now more sensible to light than the other parts of the retina which were exposed to the more luminous parts of the window.

3. Make with ink on white paper a very black spot, about half an inch in diameter, with a tail about an inch in length, so as to represent a tadpole, as in plate 2, at Sect. III. 8. 3.; look steadily for a minute on this spot, and, on moving the eye a little, the figure of the tadpole will be seen on the white part of the paper, which figure of the tadpole will appear whiter or more luminous than the other parts of the white paper; for the part of the retina on which the tadpole was delineated, is now more sensible to light than the other parts

of

of it, which were expofed to the white paper. This experiment is mentioned by Dr. Irwin, but is not by him afcribed to the true caufe, namely, the greater fenfibility of that part of the retina which has been expofed to the black fpot, than of the other parts which had received the white field of paper, which is put beyond a doubt by the next experiment.

4. On clofing the eyes after viewing the black fpot on the white paper, as in the foregoing experiment, a red fpot is feen of the form of the black fpot: for that part of the retina, on which the black fpot was delineated, being now more fenfible to light than the other parts of it, which were expofed to the white paper, is capable of perceiving the red rays which penetrate the eyelids. If this experiment be made by the light of a tallow candle, the fpot will be yellow inftead of red; for tallow candles abound much with yellow light, which paffes in greater quantity and force through the eyelids than blue light; hence the difficulty of diftinguifhing blue and green by this kind of candle light. The colour of the fpectrum may poffibly vary in the daylight, according to the different colour of the meridian or the morning or evening light.

M. Beguelin, in the Berlin Memoires, V. II. 1771, obferves, that, when he held a book fo that the fun fhone upon his half-clofed eyelids, the black letters, which he had long infpected, became red, which muft have been thus occafioned. Thofe parts of the retina which had received for fome time the black letters, were fo much more fenfible than thofe parts which had been oppofed to the white paper, that to the former the red light, which paffed through the eyelids, was perceptible. There is a fimilar ftory told, I think, in M. de Voltaire's Hiftorical Works, of a Duke of Tufcany, who was playing at dice with the general of a foreign army, and, believing he faw bloody fpots upon the dice, portended dreadful events, and retired in confufion. The obferver, after looking for a minute on the black

fpots

spots of a die, and carelefsly clofing his eyes, on a bright day, would fee the image of a die with red spots upon it, as above explained.

5. On emerging from a dark cavern, where we have long continued, the light of a bright day becomes intolerable to the eye for a confiderable time, owing to the excefs of fenfibility exifting in the eye, after having been long expofed to little or no ftimulus. This occafions us immediately to contract the iris to its fmalleft aperture, which becomes again gradually dilated, as the retina becomes accuftomed to the greater ftimulus of the daylight.

The twinkling of a bright ftar, or of a diftant candle in the night, is perhaps owing to the fame caufe. While we continue to look upon thefe luminous objects, their central parts gradually appear paler, owing to the decreafing fenfibility of the part of the retina expofed to their light; whilft, at the fame time, by the unfteadinefs of the eye, the edges of them are perpetually falling on parts of the retina that were juft before expofed to the darknefs of the night, and therefore tenfold more fenfible to light than the part on which the ftar or candle had been for fome time delineated. This pains the eye in a fimilar manner as when we come fuddenly from a dark 'room into bright daylight, and gives the appearance of bright fcintillations. Hence the ftars twinkle moft when the night is darkeft, and do not twinkle through telefcopes, as obferved by Muffchenbroeck; and it will afterwards be feen why this twinkling is fometimes of different colours when the object is very bright, as Mr. Melvill obferved in looking at Sirius. For the opinions of others on this fubject, fee Dr. Prieftley's valuable Hiftory of Light and Colours, p. 494.

Many facts obfervable in the animal fyftem are fimilar to thefe; as the hot glow occafioned by the ufual warmth of the air, or our clothes, on coming out of a cold bath; the pain of the fingers on approaching the fire after having handled fnow; and the inflamed heels from walking in fnow. Hence thofe who have been expofed to much cold have died on being brought to a fire, or their limbs have become

fo

fo much inflamed as to mortify. Hence much food or wine given fuddenly to thofe who have almoſt periſhed by hunger has deſtroyed them; for all the organs of the famiſhed body are now become fo much more irritable to the ſtimulus of food and wine, which they have long been deprived of, that inflammation is excited, which terminates in gangrene or fever.

IV. OF DIRECT OCULAR SPECTRA.

A quantity of ſtimulus ſomewhat greater than natural excites the retina into ſpaſmodic action, which ceaſes in a few ſeconds.

A CERTAIN duration and energy of the ſtimulus of light and colours excites the perfect action of the retina in viſion; for very quick motions are imperceptible to us, as well as very ſlow ones, as the whirling of a top, or the ſhadow on a ſun-dial. So perfect darkneſs does not affect the eye at all; and exceſs of light produces pain, not viſion.

1. When a fire-coal is whirled round in the dark, a lucid circle remains a conſiderable time in the eye; and that with fo much vivacity of light, that it is miſtaken for a continuance of the irritation of the object. In the fame manner, when a fiery meteor ſhoots acroſs the night, it appears to leave a long lucid train behind it, part of which, and perhaps fometimes the whole, is owing to the continuance of the action of the retina after having been thus vividly excited. This is beautifully illuſtrated by the following experiment: fix a paper fail, three or four inches in diameter, and made like that of a fmoke jack, in a tube of paſteboard; on looking through the tube at a diſtant proſpect, fome disjointed parts of it will be feen through the narrow intervals between the fails; but as the fly begins to revolve, theſe intervals appear

appear larger; and when it revolves quicker, the whole profpect is feen quite as diftinct as if nothing intervened, though lefs luminous.

2. Look through a dark tube, about half a yard long, at the area of a yellow circle of half an inch diameter, lying upon a blue area of double that diameter, for half a minute; and on clofing your eyes the colours of the fpectrum will appear fimilar to the two areas, as in fig. 3.; but if the eye is kept too long upon them, the colours of the fpectrum will be the reverfe of thofe upon the paper, that is, the internal circle will become blue, and the external area yellow; hence fome attention is required in making this experiment.

3. Place the bright flame of a fpermaceti candle before a black object in the night; look fteadily at it for a fhort time, till it is obferved to become fomewhat paler; and on clofing the eyes, and covering them carefully, but not fo as to comprefs them, the image of the blazing candle will continue diftinctly to be vifible.

4. Look fteadily, for a fhort time, at a window in a dark day, as in Exp. 2. Sect. III. and then clofing your eyes, and covering them with your hands, an exact delineation of the window remains for fome time vifible in the eye. This experiment requires a little practice to make it fucceed well; fince, if the eyes are fatigued by looking too long on the window, or the day be too bright, the luminous parts of the window will appear dark in the fpectrum, and the dark parts of the frame-work will appear luminous, as in Exp. 2. Sect. III. And it is even difficult for many, who firft try this experiment, to perceive the fpectrum at all; for any hurry of mind, or even too great attention to the fpectrum itfelf, will difappoint them, till they have had a little experience in attending to fuch fmall fenfations.

The fpectra defcribed in this fection, termed direct ocular fpectra, are produced without much fatigue of the eye; the irritation of the luminous object being foon withdrawn, or its quantity of light being not fo great as to produce any degree of uneafinefs in the organ of vifion; which diftinguifhes them from the next clafs of ocular fpectra,

4 A   which

which are the confequence of fatigue.　Thefe direct fpectra are beft obferved in fuch circumftances that no light, but what comes from the object, can fall upon the eye; as in looking through a tube, of half a yard long, and an inch wide, at a yellow paper on the fide of a room, the direct fpectrum was eafily produced on clofing the eye without taking it from the tube: but if the lateral light is admitted through the eyelids, or by throwing the fpectrum on white paper, it becomes a reverfe fpectrum, as will be explained below.

The other fenfes alfo retain for a time the impreffions that have been made upon them, or the actions they have been excited into. So if a hard body is preffed upon the palm of the hand, as is practifed in tricks of legerdemain, it is not eafy to diftinguifh for a few feconds whether it remains or is removed; and taftes continue long to exift vividly in the mouth, as the fmoke of tobacco, or the tafte of gentian, after the fapid material is withdrawn.

V. *A quantity of ftimulus fomewhat greater than the laft mentioned excites the retina into fpafmodic action, which ceafes and recurs alternately.*

1. On looking for a time on the fetting fun, fo as not greatly to fatigue the fight, a yellow fpectrum is feen when the eyes are clofed and covered, which continues for a time, and then difappears and recurs repeatedly before it entirely vanifhes.　This yellow fpectrum of the fun when the eyelids are opened becomes blue; and if it is made to fall on the green grafs, or on other coloured objects, it varies its own colour by an intermixture of theirs, as will be explained in another place.

2. Place a lighted fpermaceti candle in the night about one foot from your eye, and look fteadily on the centre of the flame, till your

eye

eye becomes much more fatigued than in Sect. IV. Exp. 3.; and on clofing your eyes a reddifh fpectrum will be perceived, which will ceafe and return alternately.

The action of vomiting in like manner ceafes, and is renewed by intervals, although the emetic drug is thrown up with the firft effort: fo after-pains continue fome time after parturition; and the alternate pulfations of the heart of a viper are renewed for fome time after it is cleared from its blood.

VI. Of reverse ocular spectra.

The retina after having been excited into action by a ftimulus fomewhat greater than the laft mentioned falls into oppofite fpafmodic action.

The actions of every part of animal bodies may be advantageoufly compared with each other. This ftrict analogy contributes much to the inveftigation of truth; while thofe loofer analogies, which compare the phenomena of animal life with thofe of chemiftry or mechanics, only ferve to miflead our inquiries.

When any of our larger mufcles have been in long or in violent action, and their antagonifts have been at the fame time extended, as foon as the action of the former ceafes, the limb is ftretched the contrary way for our eafe, and a pandiculation or yawning takes place.

By the following obfervations it appears, that a fimilar circumftance obtains in the organ of vifion; after it has been fatigued by one kind of action, it fpontaneoufly falls into the oppofite kind.

1. Place a piece of coloured filk, about an inch in diameter, on a fheet of white paper, about half a yard from your eyes; look fteadily upon it for a minute; then remove your eyes upon another part of the white paper, and a fpectrum will be feen of the form of the filk thus infpected, but of a colour oppofite to it. A fpectrum nearly

fimilar

fimilar will appear if the eyes are clofed, and the eyelids fhaded by approaching the hand near them, fo as to permit fome, but to prevent
too much light falling on them.

> Red filk produced a green fpectrum.
>
> Green produced a red one.
>
> Orange produced blue.
>
> Blue produced orange.
>
> Yellow produced violet.
>
> Violet produced yellow.

That in thefe experiments the colours of the fpectra are the reverfe
of the colours which occafioned them, may be feen by examining the
third figure in Sir Ifaac Newton's Optics, L. II. p. 1, where thofe
thin laminæ of air, which reflected yellow, tranfmitted violet; thofe
which reflected red, tranfmitted a blue-green; and fo of the reft,
agreeing with the experiments above related.

2. Thefe reverfe fpectra are fimilar to a colour, formed by a combination of all the primary colours except that with which the eye
has been fatigued in making the experiment: thus the reverfe
fpectrum of red muft be fuch a green as would be produced by a combination of all the other prifmatic colours. To evince this fact the
following fatisfactory experiment was made. The prifmatic colours
were laid on a circular pafteboard wheel, about four inches in diameter, in the proportions defcribed in Dr. Prieftley's Hiftory of Light
and Colours, pl. 12. fig. 83. except that the red compartment was
entirely left out, and the others proportionably extended fo as to complete the circle. Then, as the orange is a mixture of red and yellow,
and as the violet is a mixture of red and indigo, it became neceffary
to put yellow on the wheel inftead of orange, and indigo inftead of
violet, that the experiment might more exactly quadrate with the
theory it was defigned to eftablifh or confute; becaufe in gaining a
green fpectrum from a red object, the eye is fuppofed to have become
infenfible to red light. This wheel, by means of an axis, was made

to

to whirl like a top; and on its being put in motion, a green colour was produced, correfponding with great exactnefs to the reverfe fpectrum of red.

3. In contemplating any one of thefe reverfe fpectra in the clofed and covered eye, it difappears and re-appears feveral times fucceffively, till at length it entirely vanifhes, like the direct fpectra in Sect. V.; but with this additional circumftance, that when the fpectrum becomes faint or evanefcent, it is inftantly revived by removing the hand from before the eyelids, fo as to admit more light: becaufe then not only the fatigued part of the retina is inclined fpontaneoufly to fall into motions of a contrary direction, but being ftill fenfible to all other rays of light, except that with which it was lately fatigued, is by thefe rays at the fame time ftimulated into thofe motions which form the reverfe fpectrum.

From thefe experiments there is reafon to conclude, that the fatigued part of the retina throws itfelf into a contrary mode of action, like ofcitation or pandiculation, as foon as the ftimulus which has fatigued it is withdrawn; and that it ftill remains fenfible, that is, liable to be excited into action by any other colours at the fame time, except the colour with which it has been fatigued.

VII. *The retina after having been excited into action by a ftimulus fome-what greater than the laft mentioned falls into various fucceffive fpafmodic actions.*

1. ON looking at the meridian fun as long as the eyes can well bear its brightnefs, the difk firft becomes pale, with a luminous crefcent, which feems to librate from one edge of it to the other, owing to the unfteadinefs of the eye; then the whole phafis of the fun becomes blue, furrounded with a white halo; and on clofing the eyes, and covering them with the hands, a yellow fpectrum is feen, which in a little time changes into a blue one.

M. de

M. de la Hire obferved, after looking at the bright fun, that the impreffion in his eye firft affumed a yellow appearance, and then green, and then blue; and wifhes to afcribe thefe appearances to fome affection of the nerves. (Porterfield on the Eye, Vol. I. p. 343.)

2. After looking fteadily on about an inch fquare of pink filk, placed on white paper, in a bright funfhine, at the diftance of a foot from my eyes, and clofing and covering my eyelids, the fpectrum of the filk was at firft a dark green, and the fpectrum of the white paper became of a pink. The fpectra then both difappeared; and then the internal fpectrum was blue; and then, after a fecond difappearance, became yellow, and laftly pink, whilft the fpectrum of the field varied into red and green.

Thefe fucceffions of different coloured fpectra were not exactly the fame in the different experiments, though obferved, as near as could be, with the fame quantity of light, and other fimilar circumftances; owing, I fuppofe, to trying too many experiments at a time; fo that the eye was not quite free from the fpectra of the colours which were previoufly attended to.

The alternate exertions of the retina in the preceding fection refembled the ofcitation or pandiculation of the mufcles, as they were performed in directions contrary to each other, and were the confequence of fatigue rather than of pain. And in this they differ from the fucceffive diffimilar exertions of the retina, mentioned in this fection, which refemble in miniature the more violent agitations of the limbs in convulfive difeafes, as epilepfy, chorea S. Viti, and opifthotonos; all which difeafes are perhaps, at firft, the confequence of pain, and have their periods afterwards eftablifhed by habit.

VIII. *The*

VIII. *The retina, after having been excited into action by a stimulus somewhat greater than the last mentioned, falls into a fixed spasmodic
action, which continues for some days.*

1. After having looked long at the meridian sun, in making some
of the preceding experiments, till the disks faded into a pale blue, I
frequently observed a bright blue spectrum of the sun on other objects·
all the next and the succeeding day, which constantly occurred when I
attended to it, and frequently when I did not previously attend to it.
When I closed and covered my eyes, this appeared of a dull yellow;
and at other times mixed with the colours of other objects on which
it was thrown. It may be imagined, that this part of the retina was
become insensible to white light, and thence a bluish spectrum became
visible on all luminous objects; but as a yellowish spectrum was also
seen in the closed and covered eye, there can remain no doubt of this
being the spectrum of the sun. A similar appearance was observed by
M. Æpinus, which he acknowledges he could give no account of.
(Nov. Com. Petrop. V. 10. p. 2. and 6.)

The locked jaw, and some cataleptic spasms, are resembled by this
phenomenon; and from hence we may learn the danger to the eye by
inspecting very luminous objects too long a time.

IX. *A quantity of stimulus greater than the preceding induces a temporary
paralysis of the organ of vision.*

1. Place a circular piece of bright red silk, about half an inch in
diameter, on the middle of a sheet of white paper; lay them on the
floor in a bright sunshine, and fixing your eyes steadily on the center
of the red circle, for three or four minutes, at the distance of four or

six

fix feet from the object, the red filk will gradually become paler, and finally ceafe to appear red at all.

2. Similar to thefe are many other animal facts; as purges, opiates, and even poifons, and contagious matter, ceafe to ftimulate our fyftem, after we have been habituated to their ufe.   So fome people fleep undifturbed by a clock, or even by a forge hammer in their neighbourhood: and not only continued irritations, but violent exertions of any kind, are fucceeded by temporary paralyfis.   The arm drops down after violent action, and continues for a time ufelefs; and it is probable, that thofe who have perifhed fuddenly in fwimming, or in fcating on the ice, have owed their deaths to the paralyfis, or extreme fatigue, which fucceeds every violent and continued exertion.

X. Miscellaneous Remarks.

There were fome circumftances occurred in making thefe experiments, which were liable to alter the refults of them, and which I fhall here mention for the affiftance of others, who may wifh to repeat them.

1. *Of direct and inverfe fpectra exifting at the fame time; of reciprocal direct fpectra; of a combination of direct and inverfe fpectra; of a fpectral halo; rules to pre-determine the colours of fpectra.*

a. When an area, about fix inches fquare, of bright pink Indian paper, had been viewed on an area, about a foot fquare, of white writing paper, the internal fpectrum in the clofed eye was green, being the reverfe fpectrum of the pink paper; and the external fpectrum was pink, being the direct fpectrum of the pink paper.   The
fame

fame circumftance happened when the internal area was white, and external one pink; that is, the internal fpectrum was pink, and the external one green. All the fame appearances occurred when the pink paper was laid on a black hat.

b. When fix inches fquare of deep violet polifhed paper was viewed on a foot fquare of white writing paper, the internal fpectrum was yellow, being the reverfe fpectrum of the violet paper, and the external one was violet, being the direct fpectrum of the violet paper.

c. When fix inches fquare of pink paper was viewed on a foot fquare of blue paper, the internal fpectrum was blue, and the external fpectrum was pink; that is, the internal one was the direct fpectrum of the external object, and the external one was the direct fpectrum of the internal object, inftead of their being each the reverfe fpectrum of the objects they belonged to.

d. When fix inches fquare of blue paper were viewed on a foot fquare of yellow paper, the interior fpectrum became a brilliant yellow, and the exterior one a brilliant blue. The vivacity of the fpectra was owing to their being excited both by the ftimulus of the interior and exterior objects; fo that the interior yellow fpectrum was both the reverfe fpectrum of the blue paper, and the direct one of the yellow paper; and the exterior blue fpectrum was both the reverfe fpectrum of the yellow paper, and the direct one of the blue paper.

e. When the internal area was only a fquare half-inch of red paper, laid on a fquare foot of dark violet paper, the internal fpectrum was green, with a reddifh-blue halo. When the red internal paper was two inches fquare, the internal fpectrum was a deeper green, and the external one redder. When the internal paper was fix inches fquare, the fpectrum of it became blue, and the fpectrum of the external paper was red.

4 B

f. When

f. When a square half-inch of blue paper was laid on a six-inch square of yellow paper, the spectrum of the central paper in the closed eye was yellow, incircled with a blue halo. On looking long on the meridian sun, the disk fades into a pale blue surrounded with a whitish halo.

These circumstances, though they very much perplexed the experiments till they were investigated, admit of a satisfactory explanation; for while the rays from the bright internal object in exp. *a.* fall with their full force on the center of the retina, and, by fatiguing that part of it, induce the reverse spectrum, many scattered rays, from the same internal pink paper, fall on the more external parts of the retina, but not in such quantity as to occasion much fatigue, and hence induce the direct spectrum of the pink colour in those parts of the eye. The same reverse and direct spectra occur from the violet paper in exp. *b.*: and in exp. *c.* the scattered rays from the central pink paper produce a direct spectrum of this colour on the external parts of the eye, while the scattered rays from the external blue paper produce a direct spectrum of that colour on the central part of the eye, instead of these parts of the retina falling reciprocally into their reverse spectra. In exp. *d.* the colours being the reverse of each other, the scattered rays from the exterior object falling on the central parts of the eye, and there exciting their direct spectrum, at the same time that the retina was excited into a reverse spectrum by the central object, and this direct and reverse spectrum being of similar colour, the superior brilliancy of this spectrum was produced. In exp. *e.* the effect of various quantities of stimulus on the retina, from the different respective sizes of the internal and external areas, induced a spectrum of the internal area in the center of the eye, combined of the reverse spectrum of that internal area and the direct one of the external area, in various shades of colour, from a pale green to a deep blue, with similar changes in the spectrum of the external area. For

the

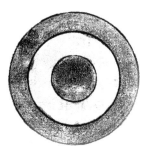

the fame reafons, when an internal bright object was fmall, as in exp. *f.* inftead of the whole of the fpectrum of the external object being reverfe to the colour of the internal object, only a kind of halo, or radiation of colour, fimilar to that of the internal object, was fpread a little way on the external fpectrum.   For this internal blue area being fo fmall, the fcattered rays from it extended but a little way on the image of the external area of yellow paper, and could therefore produce only a blue halo round the yellow fpectrum in the center.

If any one fhould fufpect that the fcattered rays from the exterior coloured object do not intermix with the rays from the interior coloured object, and thus affect the central part of the eye, let him look through an opake tube, about two feet in length, and an inch in diameter, at a coloured wall of a room with one eye, and with the other eye naked; and he will find, that by fhutting out the lateral light, the area of the wall feen through a tube appears as if illuminated by the funfhine, compared with the other parts of it; from whence arifes the advantage of looking through a dark tube at diftant paintings.

Hence we may fafely deduce the following rules to determine before-hand the colours of all fpectra.   1. The direct fpectrum without any lateral light is an evanefcent reprefentation of its object in the unfatigued eye.   2. With fome lateral light it becomes of a colour combined of the direct fpectrum of the central object, and of the circumjacent objects, in proportion to their refpective quantity and brilliancy.   3. The reverfe fpectrum without lateral light is a reprefentation in the fatigued eye of the form of its objects, with fuch a colour as would be produced by all the primary colours, except that of the object.   4. With lateral light the colour is compounded of the reverfe fpectrum of the central object, and the direct fpectrum of the circumjacent objects, in proportion to their refpective quantity and brilliancy.

2. *Variation*

2. *Variation and vivacity of the spectra occasioned by extraneous light.*

The reverse spectrum, as has been before explained, is similar to a colour, formed by a combination of all the primary colours, except that with which the eye has been fatigued in making the experiment : so the reverse spectrum of red is such a green as would be produced by a combination of all the other prismatic colours. Now it must be observed, that this reverse spectrum of red is therefore the direct spectrum of a combination of all the other prismatic colours, except the red ; whence, on removing the eye from a piece of red silk to a sheet of white paper, the green spectrum, which is perceived, may either be called the reverse spectrum of the red silk, or the direct spectrum of all the rays from the white paper, except the red ; for in truth it is both. Hence we see the reason why it is not easy to gain a direct spectrum of any coloured object in the day-time, where there is much lateral light, except of very bright objects, as of the setting sun, or by looking through an opake tube ; because the lateral external light falling also on the central part of the retina, contributes to induce the reverse spectrum, which is at the same time the direct spectrum of that lateral light, deducting only the colour of the central object which we have been viewing. And for the same reason, it is difficult to gain the reverse spectrum, where there is no lateral light to contribute to its formation. Thus, in looking through an opake tube on a yellow wall, and closing my eye, without admitting any lateral light, the spectra were all at first yellow ; but at length changed into blue. And on looking in the same manner on red paper, I did at length get a green spectrum ; but they were all at first red ones : and the same after looking at a candle in the night.

The reverse spectrum was formed with greater facility when the

eye

eye was thrown from the object on a sheet of white paper, or when light was admitted through the closed eyelids; because not only the fatigued part of the retina was inclined spontaneously to fall into motions of a contrary direction; but being still sensible to all other rays of light except that with which it was lately fatigued, was by these rays stimulated at the same time into those motions which form the reverse spectrum. Hence, when the reverse spectrum of any colour became faint, it was wonderfully revived by admitting more light through the eyelids, by removing the hand from before them: and hence, on covering the closed eyelids, the spectrum would often cease for a time, till the retina became sensible to the stimulus of the smaller quantity of light, and then it recurred. Nor was the spectrum only changed in vivacity, or in degree, by this admission of light through the eyelids; but it frequently happened, after having viewed bright objects, that the spectrum in the closed and covered eye was changed into a third spectrum, when light was admitted through the eyelids: which third spectrum was composed of such colours as could pass through the eyelids, except those of the object. Thus, when an area of half an inch diameter of pink paper was viewed on a sheet of white paper in the sunshine, the spectrum with closed and covered eyes was green; but on removing the hands from before the closed eyelids, the spectrum became yellow, and returned instantly again to green, as often as the hands were applied to cover the eyelids, or removed from them: for the retina being now insensible to red light, the yellow rays passing through the eyelids in greater quantity than the other colours, induced a yellow spectrum; whereas if the spectrum was thrown on white paper, with the eyes open, it became only a lighter green.

Though a certain quantity of light facilitates the formation of the reverse spectrum, a greater quantity prevents its formation, as the more powerful stimulus excites even the fatigued parts of the eye into action; otherwise we should see the spectrum of the last viewed object as often as we turn our eyes. Hence the reverse spectra are best seen

by

by gradually approaching the hand near the clofed eyelids to a certain diftance only, which muft be varied with the brightnefs of the day, or the energy of the fpectrum. Add to this, that all dark fpectra, as black, blue, or green, if light be admitted through the eyelids, after they have been fome time covered, give reddifh fpectra, for the reafons given in Sect. III. Exp. 1.

From thefe circumftances of the extraneous light coinciding with the fpontaneous efforts of the fatigued retina to produce a reverfe fpectrum, as was obferved before, it is not eafy to gain a direct fpectrum, except of objects brighter than the ambient light; fuch as a candle in the night, the fetting fun, or viewing a bright object through an opake tube; and then the reverfe fpectrum is inftantaneoufly produced by the admiffion of fome external light; and is as inftantly converted again to the direct fpectrum by the exclufion of it. Thus, on looking at the fetting fun, on clofing the eyes, and covering them, a yellow fpectrum is feen, which is the direct fpectrum of the fetting fun; but on opening the eyes on the fky, the yellow fpectrum is immediately changed into a blue one, which is the reverfe fpectrum of the yellow fun, or the direct fpectrum of the blue fky, or a combination of both. And this is again transformed into a yellow one on clofing the eyes, and fo reciprocally, as quick as the motions of the opening and clofing eyelids. Hence, when Mr. Melvill obferved the fcintillations of the ftar Sirius to be fometimes coloured, thefe were probably the direct fpectrum of the blue fky on the parts of the retina fatigued by the white light of the ftar. (Effays Phyfical and Literary, p. 81. V. 2.)

When a direct fpectrum is thrown on colours darker than itfelf, it mixes with them; as the yellow fpectrum of the fetting fun, thrown on the green grafs, becomes a greener yellow. But when a direct fpectrum is thrown on colours brighter than itfelf, it becomes inftantly changed into the reverfe fpectrum, which mixes with thofe brighter colours. So the yellow fpectrum of the fetting fun thrown on the

luminous

luminous fky becomes blue, and changes with the colour or bright-nefs of the clouds on which it appears. But the reverfe fpectrum mixes with every kind of colour on which it is thrown, whether brighter than itfelf or not: thus the reverfe fpectrum, obtained by viewing a piece of yellow filk, when thrown on white paper, was a lucid blue green; when thrown on black Turkey leather, becomes a deep violet. And the fpectrum of blue filk, thrown on white paper, was a light yellow; on black filk was an obfcure orange; and the blue fpectrum, obtained from orange-coloured filk, thrown on yellow, became a green.

In thefe cafes the retina is thrown into activity or fenfation by the ftimulus of external colours, at the fame time that it continues the activity or fenfation which forms the fpectra; in the fame manner as the prifmatic colours, painted on a whirling top, are feen to mix to-gether. When thefe colours of external objects are brighter than the direct fpectrum which is thrown upon them, they change it into the reverfe fpectrum, like the admiffion of external light on a direct fpec-trum, as explained above. When they are darker than the direct fpec-trum, they mix with it, their weaker ftimulus being infufficient to induce the reverfe fpectrum.

3. *Variation of fpectra in refpect to number, and figure, and remiffion.*

When we look long and attentively at any object, the eye cannot always be kept entirely motionlefs; hence, on infpecting a circular area of red filk placed on white paper, a lucid crefcent or edge is feen to librate on one fide or other of the red circle: for the exterior parts of the retina fometimes falling on the edge of the central filk, and fometimes on the white paper, are lefs fatigued with red light than the central part of the retina, which is conftantly expofed to it; and

3.   therefore,

therefore, when they fall on the edge of the red filk, they perceive it
more vividly.   Afterwards, when the eye becomes fatigued, a green
fpectrum in the form of a crefcent is feen to librate on one fide or
other of the central circle, as by the unfteadinefs of the eye a part of
the fatigued retina falls on the white paper ; and as by the increafing
fatigue of the eye the central part of the filk appears paler, the edge on
which the unfatigued part of the retina occafionally falls will appear
of a deeper red than the original filk, becaufe it is compared with the
pale internal part of it.   M. de Buffon in making this experiment obferved, that the red edge of the filk was not only deeper coloured than
the original filk ; but, on his retreating a little from it, it became oblong, and at length divided into two, which muft have been owing to
his obferving it either before or behind the point of interfection of the
two optic axifes.   Thus, if a pen is held up before a diftant candle, when
we look intenfely at the pen two candles are feen behind it ; when we
look intenfely at the candle two pens are feen.   If the fight be unfteady at the time of beholding the fun, even though one eye only be
ufed, many images of the fun will appear, or luminous lines, when
the eye is clofed.   And as fome parts of thefe will be more vivid than
others, and fome parts of them will be produced nearer the center of
the eye than others, thefe will difappear fooner than the others ; and
hence the number and fhape of thefe fpectra of the fun will continually
vary, as long as they exift.   The caufe of fome being more vivid than
others, is the unfteadinefs of the eye of the beholder, fo that fome
parts of the retina have been longer expofed to the funbeams.   That
fome parts of a complicated fpectrum fade and return before other parts
of it, the following experiment evinces.   Draw three concentric circles ; the external one an inch and a half in diameter, the middle one
an inch, and the internal one half an inch ; colour the external and
internal areas blue, and the remaining one yellow, as in Fig. 4. ; after
having looked about a minute on the center of thefe circles, in a bright
light, the fpectrum of the external area appears firft in the clofed eye,
                                                                    then

then the middle area, and laftly the central one; and then the central one difappears, and the others in inverted order. If concentric circles of more colours are added, it produces the beautiful ever changing fpectrum in Sect. I. Exp. 2.

From hence it would feem, that the center of the eye produces quicker remiffions of fpectra, owing perhaps to its greater fenfibility; that is, to its more energetic exertions. Thefe remiffions of fpectra bear fome analogy to the tremors of the hands, and palpitations of the heart, of weak people: and perhaps a criterion of the ftrength of any mufcle or nerve may be taken from the time it can be continued in exertion.

4. *Variation of fpectra in refpect to brilliancy; the vifibility of the circulation of the blood in the eye.*

1. The meridian or evening light makes a difference in the colours of fome fpectra; for as the fun defcends, the red rays, which are lefs refrangible by the convex atmofphere, abound in great quantity. Whence the fpectrum of the light parts of a window at this time, or early in the morning, is red; and becomes blue either a little later or earlier; and white in the meridian day; and is alfo variable from the colour of the clouds or fky which are oppofed to the window.

2. All thefe experiments are liable to be confounded, if they are made too foon after each other, as the remaining fpectrum will mix with the new ones. This is a very troublefome circumftance to painters, who are obliged to look long upon the fame colour; and in particular to thofe whofe eyes, from natural debility, cannot long continue the fame kind of exertion. For the fame reafon, in making thefe experiments, the refult becomes much varied if the eyes, after viewing any object, are removed on other objects for but an inftant of time, before we clofe them to view the fpectrum; for the light from

4 C

the

the object, of which we had only a transient view, in the very time of closing our eyes acts as a stimulus on the fatigued retina; and for a time prevents the desired spectrum from appearing, or mixes its own spectrum with it. Whence, after the eyelids are closed, either a dark field, or some unexpected colours, are beheld for a few seconds, before the desired spectrum becomes distinctly visible.

3. The length of time taken up in viewing an object, of which we are to observe the spectrum, makes a great difference in the appearance of the spectrum, not only in its vivacity, but in its colour; as the direct spectrum of the central object, or of the circumjacent ones, and also the reverse spectra of both, with their various combinations, as well as the time of their duration in the eye, and of their remissions or alternations, depend upon the degree of fatigue the retina is subjected to. The Chevalier d'Arcy constructed a machine by which a coal of fire was whirled round in the dark, and found, that when a luminous body made a revolution in eight thirds of time, it presented to the eye a complete circle of fire; from whence he concludes, that the impression continues on the organ about the seventh part of a second. (Mém. de l'Acad. des Sc. 1765.) This, however, is only to be considered as the shortest time of the duration of these direct spectra; since in the fatigued eye both the direct and reverse spectra, with their intermissions, appear to take up many seconds of time, and seem very variable in proportion to the circumstances of fatigue or energy.

4. It sometimes happens, if the eyeballs have been rubbed hard with the fingers, that lucid sparks are seen in quick motion amidst the spectrum we are attending to. This is similar to the flashes of fire from a stroke on the eye in fighting, and is resembled by the warmth and glow, which appears upon the skin after friction, and is probably owing to an acceleration of the arterial blood into the vessels emptied by the previous pressure. By being accustomed to observe such small sensations in the eye, it is easy to see the circulation of the blood in this organ. I have attended to this frequently, when I have observed

my

my eyes more than commonly fenfible to other fpectra. The circulation may be feen either in both eyes at a time, or only in one of
them; for as a certain quantity of light is neceffary to produce this
curious phenomenon, if one hand be brought nearer the clofed eyelids
than the other, the circulation in that eye will for a time difappear.
For the eafier viewing the circulation, it is fometimes neceffary to
rub the eyes with a certain degree of force after they are clofed, and
to hold the breath rather longer than is agreeable, which, by accu-
mulating more blood in the eye, facilitates the experiment; but in
general it may be feen diftinctly after having examined other fpectra
with your back to the light, till the eyes become weary; then having
covered your clofed eyelids for half a minute, till the fpectrum is
faded away which you were examining, turn your face to the light,
and removing your hands from the eyelids, by and by again fhade them
a little, and the circulation becomes curioufly diftinct. The ftreams
of blood are however generally feen to unite, which fhews it to be the
venous circulation, owing, I fuppofe, to the greater opacity of the
colour of the blood in thefe veffels; for this venous circulation is alfo
much more eafily feen by the microfcope in the tail of a tadpole.

5. *Variation of fpectra in refpect to diftinctnefs and fize; with a new way
of magnifying objects.*

1. It was before obferved, that when the two colours viewed together were oppofite to each other, as yellow and blue, red and green,
&c. according to the table of reflections and tranfmiffions of light in
Sir Ifaac Newton's Optics, B. II. Fig. 3. the fpectra of thofe colours
were of all others the moft brilliant, and beft defined; becaufe they
were combined of the reverfe fpectrum of one colour, and of the direct
fpectrum of the other. Hence, in books printed with fmall types, or
in the minute graduation of thermometers, or of clock-faces, which

are

are to be feen at a diftance, if the letters or figures are coloured with orange, and the ground with indigo; or the letters with red, and the ground with green; or any other lucid colour is ufed for the letters, the fpectrum of which is fimilar to the colour of the ground; fuch letters will be feen much more diftinctly, and with lefs confufion, than in black or white: for as the fpectrum of the letter is the fame colour with the ground on which they are feen, the unfteadinefs of the eye in long attending to them will not produce coloured lines by the edges of the letters, which is the principal caufe of their confufion. The beauty of colours lying in vicinity to each other, whofe fpectra are thus reciprocally fimilar to each colour, is owing to this greater eafe that the eye experiences in beholding them diftinctly; and it is probable, in the organ of hearing, a fimilar circumftance may conftitute the pleafure of melody. Sir Ifaac Newton obferves, that gold and indigo were agreeable when viewed together; and thinks there may be fome analogy between the fenfations of light and found. (Optics, Qu. 14.)

In viewing the fpectra of bright objects, as of an area of red filk of half an inch diameter on white paper, it is eafy to magnify it to tenfold its fize: for if, when the fpectrum is formed, you ftill keep your eye fixed on the filk area, and remove it a few inches further from you, a green circle is feen round the red filk: for the angle now fubtended by the filk is lefs than it was when the fpectrum was formed, but that of the fpectrum continues the fame, and our imagination places them at the fame diftance. Thus when you view a fpectrum on a fheet of white paper, if you approach the paper to the eye, you may diminifh it to a point; and if the paper is made to recede from the eye, the fpectrum will appear magnified in proportion to the diftance.

I was furprifed, and agreeably amufed, with the following experiment. I covered a paper about four inches fquare with yellow, and with a pen filled with a blue colour wrote upon the middle of it the

word

BANKS.

word BANKS in capitals, as in Fig. 5, and fitting with my back to
the fun, fixed my eyes for a minute exactly on the center of the letter N in the middle of the word; after clofing my eyes, and fhading
them fomewhat with my hand, the word was diftinctly feen in the
fpectrum in yellow letters on a blue field; and then, on opening my
eyes on a yellowifh wall at twenty feet diftance, the magnified name
of BANKS appeared written on the wall in golden characters.

Conclufion.

It was obferved by the learned M. Sauvages (Nofol. Method.
Cl. VIII. Ord. 1.) that the pulfations of the optic artery might be perceived by looking attentively on a white wall well illuminated. A
kind of net-work, darker than the other parts of the wall, appears and
vanifhes alternately with every pulfation. This change of the colour
of the wall he well afcribes to the compreffion of the retina by the diaftole of the artery. The various colours produced in the eye by the
preffure of the finger, or by a ftroke on it, as mentioned by Sir Ifaac
Newton, feem likewife to originate from the unequal preffure on various parts of the retina. Now as Sir Ifaac Newton has fhewn, that
all the different colours are reflected or tranfmitted by the laminæ of
foap bubbles, or of air, according to their different thicknefs or thinnefs, is it not probable, that the effect of the activity of the retina may
be to alter its thicknefs or thinnefs, fo as better to adapt it to reflect or
tranfmit the colours which ftimulate it into action? May not mufcular fibres exift in the retina for this purpofe, which may be lefs minute than the locomotive mufcles of microfcopic animals? May not
thefe mufcular actions of the retina conftitute the fenfation of light
and colours; and the voluntary repetitions of them, when the object

is

is withdrawn, conftitute our memory of them? And laftly, may not the laws of the fenfations of light, here inveftigated, be applicable to all our other fenfes, and much contribute to elucidate many phenomena of animal bodies both in their healthy and difeafed ftate; and thus render this inveftigation well worthy the attention of the phyfician, the metaphyfician, and the natural philofopher?

November 1, 1785.

ADDITIONS.

ADDITIONS.

At Page 120, *after Line* 19, *pleafe to add.*

FROM the experiments above mentioned of Galvani, Volta, Fowler, and others, it appears, that a plate of zinc and a plate of filver have greater effect than lead and filver. If one edge of a plate of filver about the fize of half a crown-piece be placed upon the tongue, and one edge of a plate of zinc about the fame fize beneath the tongue, and if their oppofite edges are then brought into contact before the point of the tongue, a tafte is perceived at the moment of their coming into contact; fecondly, if one of the above plates be put between the upper lip and the gum of the fore-teeth, and the other be placed under the tongue, and, their exterior edges be then brought into contact in a darkifh room, a flafh of light is perceived in the eyes.

Thefe effects I imagine only fhew the fenfibility of our nerves of fenfe to very fmall quantities of the electric fluid, as it paffes through them; for I fuppofe thefe fenfations are occafioned by flight electric fhocks produced in the following manner. By the experiments publifhed by Mr. Bennet, with his ingenious doubler of electricity, which is the greateft difcovery made in that fcience fince the coated jar, and the eduction of lightning from the fkies, it appears, that zinc was

3

always

always found minus, and filver was always found plus, when both of them were in their feparate ftate. Hence, when they are placed in the manner above defcribed, as foon as their exterior edges come nearly into contact, fo near as to have an extremely thin plate of air between them, that plate of air becomes charged in the fame manner as a plate of coated glafs; and is at the fame inftant difcharged through the nerves of tafte or of fight, and gives the fenfations, as above defcribed, of light or of faporocity; and only fhews the great fenfibility of thefe organs of fenfe to the ftimulus of the electric fluid in fuddenly paffing through them.

At Page 160, *after Line* 29, *pleafe to add.*

Thefe animals feem to poffefs fomething like an additional fenfe by means of their whifkers; which have perhaps fome analogy to the antennæ of moths and butterflies. The whifkers of cats confift not only of the long hairs on their upper lips, but they have alfo four or five long hairs ftanding up from each eyebrow, and alfo two or three on each cheek; all which, when the animal erects them, make with their points fo many parts of the periphery of a circle, of an extent at leaft equal to the circumference of any part of their own bodies. With this inftrument, I conceive, by a little experience, they can at once determine, whether any aperture amongft hedges or fhrubs, in which animals of this genus live in their wild ftate, is large enough to admit their bodies; which to them is a matter of the greateft confequence, whether purfuing or purfued. They have likewife a power of erecting and bringing forward the whifkers on their lips; which probably is for the purpofe of feeling, whether a dark hole be further permeable.

The

The antennæ, or horns, of butterflies and moths, who have awkward wings, the minute feathers of which are very liable to injury, ferve, I fuppofe, a fimilar purpofe of meafuring, as they fly or creep amongft the leaves of plants and trees, whither their wings can pafs without touching them.

In Sect. XXXIX.   *pleafe to add.*

Dr. Thunberg obferves, in his Journey to the Cape of Good Hope, that there are fome families, which have defcended from blacks in the female line for three generations. The firft generation proceeding from an European, who married a tawny flave, remains tawny, but approaches to a white complexion; but the children of the third generation, mixed with Europeans, become quite white, and are often remarkably beautiful. V. i. p. 112.

Additional Obfervations on VERTIGO, *which ought to have been inferted in Sect.* XX. 6. *after the Words* " optic nerve," *at the End of the fecond Paragraph.*

After revolving with your eyes open till you become vertiginous, as foon as you ceafe to revolve, not only the circum-ambient objects appear to circulate round you in a direction contrary to that, in which you have been turning, but you are liable to roll your eyes forwards and backwards; as is well obferved, and ingenioufly demonftrated by Dr. Wells in a late publication on vifion. The fame occurs, if you revolve with your eyes clofed, and open them immediately at the time of your ceafing to turn; and even during the whole time of revolving, as may be felt by your hand preffed lightly on your clofed

4 D   eyelids.

eyelids. To thefe movements of the eyes, of which he fuppofes the obferver to be inconfcious, Dr. Wells afcribes the apparent circumgyration of objects on ceafing to revolve.

The caufe of thus turning our eyes forwards, and then back again, after our body is at reft, depends, I imagine, on the fame circumftance, which induces us to follow the indiftinct fpectra, which are formed on one fide of the center of the retina, when we obferve them apparently on clouds, as defcribed in Sect. XL. 2. 2.; and then not being able to gain a more diftinct vifion of them, we turn our eyes back, and again and again purfue the flying fhade:

But this rolling of the eyes, after revolving till we become vertiginous, cannot caufe the apparent circumgyration of objects, in a direction contrary to that in which we have been revolving, for the following reafons. 1. Becaufe in purfuing a fpectrum in the fky, or on the ground, as above mentioned, we perceive no retrograde motions of objects. 2. Becaufe the apparent retrograde motions of objects, when we have revolved till we are vertiginous, continues much longer than the rolling of the eyes above defcribed.

3. When we have revolved from right to left, the apparent motion of objects, when we ftop, is from left to right; and when we have revolved from left to right, the apparent circulation of objects is from right to left; yet in both thefe cafes the eyes of the revolver are feen equally to roll forwards and backwards.

4. Becaufe this rolling of the eyes backwards and forwards takes place during our revolving, as may be perceived by the hand lightly preffed on the clofed eyelids, and therefore exifts before the effect afcribed to it.

And fifthly, I now come to relate an experiment, in which the rolling of the eyes does not take place at all after revolving, and yet the vertigo is more diftreffing than in the fituations above mentioned. If any one looks fteadily at a fpot in the ceiling over his head, or indeed at his own finger held up high over his head, and in that fitu-

ation

ation turns round till he becomes giddy ; and then ftops, and looks horizontally ; he now finds, that the apparent rotation of objects is from above downwards, or from below upwards ; that is, that the apparent circulation of objects is now vertical inftead of horizontal, making part of a circle round the axis of his eye ; and this without any rolling of his eyeballs. The reafon of there being no rolling of the eyeballs perceived after this experiment, is, becaufe the images of objects are formed in rotation round the axis of the eye, and not from one fide to the other of the axis of it ; fo that, as the eyeball has not power to turn in its focket round its own axis, it cannot follow the apparent motions of thefe evanefcent fpectra, either before or after the body is at reft. From all which arguments it is manifeft, that thefe apparent retrograde gyrations of objects are not caufed by the rolling of the eyeballs ; firft, becaufe no apparent retrogreffion of ob-jects is obferved in other rollings of the eyes : fecondly, becaufe the apparent retrogreffion of objects continues many feconds after the rolling of the eyeballs ceafes. Thirdly, becaufe the apparent retro-greffion of objects is fometimes one way, and fometimes another, yet the rolling of the eyeballs is the fame. Fourthly, becaufe the rolling of the eyeballs exifts before the apparent retrograde motions of objects is obferved ; that is, before the revolving perfon ftops. And fifthly, becaufe the apparent retrograde gyration of objects is produced, when there is no rolling of the eyeballs at all.

Doctor Wells imagines, that no fpectra can be gained in the eye, if a perfon revolves with his eyelids clofed, and thinks this a fufficient argument againft the opinion, that the apparent progreffion of the fpectra of light or colours in the eye can caufe the apparent retro-greffion of objects in the vertigo above defcribed ; but it is certain, when any perfon revolves in a light room with his eyes clofed, that he neverthelefs perceives differences of light both in quantity and co-lour through his eyelids, as he turns round ; and readily gains fpectra of thofe differences. And thefe fpectra are not very different except

in

in vivacity from thofe, which he acquires, when he revolves with unclofed eyes, fince if he then revolves very rapidly the colours and forms of furrounding objects are as it were mixed together in his eye; as when the prifmatic colours are painted on a wheel, they appear white as they revolve. The truth of this is evinced by the ftaggering or vertigo of men perfectly blind, when they turn round; which is not attended with apparent circulation of objects, but is a vertiginous diforder of the fenfe of touch. Blind men balance themfelves by their fenfe of touch; which, being lefs adapted for perceiving fmall deviations from their perpendicular, occafions them to carry themfelves more erect in walking. This method of balancing themfelves by the direction of their preffure againft the floor, becomes difordered by the unufual mode of action in turning round, and they begin to lofe their perpendicularity, that is, they become vertiginous; but without any apparent circular motions of vifible objects.

It will appear from the following experiments, that the apparent progreffion of the ocular fpectra of light or colours is the caufe of the apparent retrogreffion of objects, after a perfon has revolved, till he is vertiginous.

Firft, when a perfon turns round in a light room with his eyes open, but clofes them before he ftops, he will feem to be carried forwards in the direction he was turning for a fhort time after he ftops. But if he opens his eyes again, the objects before him inftantly appear to move in a retrograde direction, and he lofes the fenfation of being carried forwards. The fame occurs if a perfon revolves in a light room with his eyes clofed; when he ftops, he feems to be for a time carried forwards, if his eyes are ftill clofed; but the inftant he opens them, the furrounding objects appear to move in retrograde gyration. From hence it may be concluded, that it is the fenfation or imagination of our continuing to go forwards in the direction in which we were turning, that caufes the apparent retrograde circulation of objects.

3

Secondly,

Secondly, though there is an audible vertigo, as is known by the battèment, or undulations of found in the ears, which many vertiginous people experience; and though there is alfo a tangible vertigo, as when a blind perfon turns round, as mentioned above; yet as this circumgyration of objects is an hallucination or deception of the fenfe of fight, we are to look for the caufe of our appearing to move forward, when we ftop with our eyes clofed after gyration, to fome affection of this fenfe. Now, thirdly, if the fpectra formed in the eye during our rotation, continue to change, when we ftand ftill, like the fpectra defcribed in Sect. III. 3. 6. fuch changes muft fuggeft to us the idea or fenfation of our ftill continuing to turn round; as is the cafe, when we revolve in a light room, and clofe our eyes before we ftop. And laftly, on opening our eyes in the fituation above defcribed, the objects we chance to view amid thefe changing fpectra in the eye, muft feem to move in a contrary direction; as the moon fometimes appears to move retrograde, when fwift-gliding clouds are paffing forwards fo much nearer the eye of the beholder.

To make obfervations on faint ocular fpectra requires fome degree of habit, and compofure of mind, and even patience; fome of thofe defcribed in Sect. XL. were found difficult to fee, by many, who tried them; now it happens, that the mind, during the confufion of vertigo, when all the other irritative tribes of motion, as well as thofe of vifion, are in fome degree difturbed, together with the fear of falling, is in a very unfit ftate for the contemplation of fuch weak fenfations, as are occafioned by faint ocular fpectra. Yet after frequently revolving, both with my eyes clofed, and with them open, and attending to the fpectra remaining in them, by fhading the light from my eyelids more or lefs with my hand, I at length ceafed to have the idea of going forward, after I ftopped with my eyes clofed; and faw changing fpectra in my eyes, which feemed to move, as it were, over the field of vifion; till at length, by repeated trials on funny days, I perfuaded myfelf, on opening my eyes, after revolving fome

time,

time, on a fhelf of gilded books in my library, that I could perceive the fpectra in my eyes move forwards over one or two of the books, like the vapours in the air of a fummer's day; and could fo far undeceive myfelf, as to perceive the books to ftand ftill. After more trials I fometimes brought myfelf to believe, that I faw changing fpectra of lights and fhades moving in my eyes, after turning round for fome time, but did not imagine either the fpectra or the objects to be in a ftate of gyration. I fpeak, however, with diffidence of thefe facts, as I could not always make the experiments fucceed, when there was not a ftrong light in my room, or when my eyes were not in the moft proper ftate for fuch obfervations.

The ingenious and learned M. Savage has mentioned other theories to account for the apparent circumgyration of objects in vertiginous people. As the retrograde motions of the particles of blood in the optic arteries, by fpafm, or by fear, as is feen in the tails of tadpoles, and membranes between the fingers of frogs. Another caufe he thinks may be from the librations to one fide, and to the other, of the cryftalline lens in the eye, by means of involuntary actions of the mufcles, which conftitute the ciliary procefs. Both thefe theories lie under the fame objection as that of Dr. Wells before mentioned; namely, that the apparent motions of objects, after the obferver has revolved for fome time, fhould appear to vibrate this way and that; and not to circulate uniformly in a direction contrary to that, in which the obferver had revolved.

M. Savage has, laftly, mentioned the theory of colours left in the eye, which he has termed impreffions on the retina. He fays, "Experience teaches us, that impreffions made on the retina by a vifible object remain fome feconds after the object is removed; as appears from the circle of fire which we fee, when a fire-ftick is whirled round in the dark; therefore when we are carried round our own axis in a circle, we undergo a temporary vertigo, when we ftop; becaufe the impreffions of the circumjacent objects remain for a time afterwards

wards on the retina." Nofolog. Method. Claf. VIII. 1. 1. We have before obferved, that the changes of thefe colours remaining in the eye, evinces them to be motions of the fine terminations of the retina, and not impreffions on it; as impreffions on a paffive fubftance muft either remain, or ceafe intirely.

Dùm, Liber! aftra petis volitans trepidantibus alis,
Irruis immemori, parvula gutta, mari.
Me quoque, me currente rotâ revolubilis ætas
Volverit in tenebras,—i, Liber, ipfe fequor.

END OF THE FIRST VOLUME.

INDEX

TO THE

SECTIONS OF PART FIRST.

Affociation

4 E 2

Emotions

4 F

Vegetables

DIRECTIONS TO THE BINDER.

1. Pleafe to place the Plate confifting of one red fpot, at Sect. III. 1.

2. ——— Confifting of one black fpot, at Sect. III. 3. 3.

3. ——— Confifting of five concentric coloured circles, at Sect. III. 3. 6.

4. ——— Confifting of one yellow circle furrounded by one blue one, at Sect. XL. 4. 2.

5. ——— Confifting of one yellow circle and two blue ones, at Sect. XL. 10. 3.

6. ——— Confifting of the word BANKS in blue on a yellow ground, at Sect. XL. 10. 5.

E R R A T A.

Page 178. line 24. *for* autennæ, *read* antennæ.
   183. — 1. *for* have, *read* has.
   141. in line fixth of the Latin verfes, *for* incutitur, *read* impellit in.

ADVERTISEMENT.

THE publication of the Second Volume of this Work, containing a diſtribution of the diſeaſes, both of mind and body, into four natural claſſes, with their ſubſequent orders, genera, and ſpecies, their immediate cauſes, and their methods of cure, together with a new arrangement of the articles of the Materia Medica, their qualities, and modes of operation, is poſtponed till next year, on account of the more neceſſary avocations of the writer; and that by reviſing it during the ſummer months he may make it more worthy the acceptance of the Public.